CPSS Power Electronics Series

This series comprises advanced textbooks, research monographs, professional books, and reference works covering different aspects of power electronics, such as Variable Frequency Power Supply, DC Power Supply, Magnetic Technology, New Energy Power Conversion, Electromagnetic Compatibility as well as Wireless Power Transfer Technology and Equipment. The series features leading Chinese scholars and researchers and publishes authored books as well as edited compilations. It aims to provide critical reviews of important subjects in the field, publish new discoveries and significant progress that has been made in development of applications and the advancement of principles, theories and designs, and report cutting-edge research and relevant technologies. The CPSS Power Electronics series has an editorial board with members from the China Power Supply Society and a consulting editor from Springer.

Readership: Research scientists in universities, research institutions and the industry, graduate students, and senior undergraduates.

Indexed by EI Compendex

Bo Zhang · Xujian Shu · Lihao Wu

Parity-Time Symmetric Wireless Power Transfer

Springer

Bo Zhang
School of Electric Power Engineering
South China University of Technology
Guangzhou, Guangdong, China

Xujian Shu
College of Electrical Engineering
Fuzhou University
Fuzhou, China

Lihao Wu
School of Electric Power Engineering
South China University of Technology
Guangzhou, Guangdong, China

ISSN 2520-8853 ISSN 2520-8861 (electronic)
CPSS Power Electronics Series
ISBN 978-981-96-3148-3 ISBN 978-981-96-3149-0 (eBook)
https://doi.org/10.1007/978-981-96-3149-0

This Springer imprint is published by the registered company Springer Nature Singapore Pte Ltd.
The registered company address is: 152 Beach Road, #21-01/04 Gateway East, Singapore 189721, Singapore

Preface

Wireless power transfer (WPT) technology has been a research hotspot at home and abroad in recent years, it does not need wires or other physical contact, directly converts electrical energy into electromagnetic waves, light waves, sound waves, and other energy forms, and transmits energy from the power supply to the load through space, which has significant advantages such as safety, convenience and easy intelligence, and the industrialization prospect is very promising. However, wireless charging for electrical devices against the variation of coupling conditions poses a great challenge to the robustness of WPT systems.

A robust magnetic-field coupled wireless power transfer (MC-WPT) system using a nonlinear parity-time (PT)-symmetric circuit is proven to realize the stable output power and efficiency. Inspired by this system and technique, we proposed a highly efficient realization of the nonlinear PT-symmetric circuit, and applied it to the drone-in-flight wireless charging platform, achieving the first application of PT-symmetric MC-WPT system. In addition, we have realized the constant controlled power, constant voltage, constant current, etc. of the PT-symmetric MC-WPT system, and extended it to the application of multi-load and multi-relay-coil. Meanwhile, we have further presented a variety of techniques for improving the performances of PT-symmetric WPT systems. More importantly, we introduced PT symmetry into the electric-field coupled wireless power transfer (EC-WPT) system and electric-magnetic coupled hybrid wireless power transfer system (HWPT) for the first time, further expanding the application prospects of the PT-symmetric WPT system.

This monograph consists of eleven chapters. Chapter 1 mainly expounds the development, research status and existing problems of wireless power transfer technology, and then elaborates the introduction of parity-time symmetry into wireless power transfer system. Chapter 2 introduces the basic principles of PT symmetry and the representation of PT symmetry in circuits, and states the coupled RLC circuits with PT symmetry and the realization method of negative resistor. Chapter 3 mainly focuses on PT-symmetric MC-WPT system, introduces the basic structure of the system, establishes the system model, and analyzes the system characteristics. Chapter 4 introduces the PT-symmetric MC-WPT system with constant output power against distance variation and misalignment, including system structure, characteristic analysis, control strategy, system design and verification. Chapter 5 introduces the PT-symmetric MC-WPT system with constant controllable power against distance variation and misalignment, including system modeling and characteristic analysis, control and implementation, system design and verification. Chapter 6 introduces the PT-symmetric MC-WPT system with constant voltage (CV) and constant current (CC), including system structure and characteristic analysis, control and implementation, system design and verification. Chapter 7 focuses on multi-load applications, analyzes the system structure and characteristics of the multi-load PT-symmetric MC-WPT systems, and gives the system design and implementation. Chapter 8 is dedicated to the PT-symmetric MC-WPT system with relay coils, analyzes the PT-symmetric condition, operating frequency, output power and transfer efficiency of the multi-relay-coil system. Chapter 9 introduces several techniques for improving performances of PT-symmetric WPT system, such as improvement of transfer distance, reduction of switching frequency and switching losses, three-phase PT-symmetric WPT and quasi-PT-symmetric WPT. Chapter 10 extends the PT-symmetric system to the PT-symmetric EC-WPT system, analyzes the structure, modeling and characteristics of the PT-symmetric EC-WPT system, and gives the system design and verification. Chapter 11 extends the PT-symmetric system to the PT-symmetric HWPT system, analyzes the system structure, modeling and characteristics of the PT-symmetric HWPT system, and gives the system design and verification.

This monograph is the result of our research on parity-time-symmetric wireless power transfer, we hope that it is helpful for more graduate students and researchers involved in the area of electrical engineering to study parity-time-symmetric wireless power transfer systems.

Guangzhou, China Bo Zhang
Fuzhou, China Xujian Shu
Guangzhou, China Lihao Wu

Acknowledgments

For the completion of this monograph, the authors would like to acknowledge the support of the Key Program of the National Natural Science Foundation of China under Grant 52130705 and the National Natural Science Foundation of China under Grant 52207193 to the research. We would also like to express our gratitude to Prof. Dongyuan Qiu, Prof. Yanfeng Chen, and Prof. Yanwei Jiang for carefully reading the earlier version of this monograph and offering many helpful comments. In addition, we would like to sincerely thank all the editors of the CPSS Power Electronics Series of Springer. Without their invitation, encouragement, recommendations, and comments, this monograph would not have been possible.

Guangzhou, China
November 2024

Bo Zhang
Xujian Shu
Lihao Wu

Contents

About the Authors

Bo Zhang (Fellow, IEEE) was born in Shanghai, China, in 1962. He received the B.S. degree in electrical engineering from Zhejiang University, Hangzhou, China, in 1982, the M.S. degree in power electronics from Southwest Jiaotong University, Chengdu, China, in 1988, and the Ph.D. degree in power electronics from the Nanjing University of Aeronautics and Astronautics, Nanjing, China, in 1994.

He is currently a Distinguished Professor with the School of Electric Power, South China University of Technology, Guangzhou, China. He has authored or co-authored seven books in IEEE-Wiley and Springer and over 600 technical papers, and he holds over 230 patents. His current research interests include nonlinear analysis, modeling and control of power electronic converters, and wireless power transfer applications.

Xujian Shu was born in Anhui, China. She received the B.S. degree in electrical engineering and automation from China University of Mining and Technology, Xuzhou, China, in 2015, and received the Ph.D. degree in power electronics from South China University of Technology, Guangzhou, China, in 2021.

She is currently a lecturer of Fuzhou University, Fuzhou, China. Her research interests include wireless power transfer applications, and fractional-order circuits.

Lihao Wu was born in Fujian, China, in 1994. He received the B.S. degree in electrical engineering from Fuzhou University, Fuzhou, China, in 2017 and the Ph.D. degree in power electronics at School of Electric Power, South China University of Technology, Guangzhou, China, in 2022.

His research interests include wireless power transfer applications and power electronics converters.

Chapter 1
Introduction to PT-Symmetric Wireless Power Transfer

Abstract Since the Second Industrial Revolution in the mid-nineteenth century, mankind has ushered in the electric age. From manual operation to automation, from wired telephones to wireless communications, etc., human beings are inseparable from electricity and related technologies in the process of pursuing progress and freedom, and the use of electrical energy has greatly promoted the development of human society. With the widespread use of various electrical equipment in human production and life, electric energy has become the main form of energy in the energy system and is closely related to the daily life of human beings.

Since the Second Industrial Revolution in the mid-nineteenth century, mankind has ushered in the electric age. From manual operation to automation, from wired telephones to wireless communications, etc., human beings are inseparable from electricity and related technologies in the process of pursuing progress and freedom, and the use of electrical energy has greatly promoted the development of human society. With the widespread use of various electrical equipment in human production and life, electric energy has become the main form of energy in the energy system and is closely related to the daily life of human beings.

Wireless power transfer (WPT) technology is a kind of electric energy transmission technology, which converts electric energy into electromagnetic waves, light waves, sound waves, and other forms, and then directly transmits electric energy to the load through space, getting rid of the shackles of wires to power electrical equipment. Compared with wired conduction mode, WPT technology has the advantages of cost saving, convenience, flexibility, safety, and reliability, and it is widely used in a variety of wireless power supply occasions, including the internet of things battery-free sensors or devices [1, 2], wearable devices, mobile phones and other portable consumer electronics [3–5], electric vehicles, automation equipment such as robots [6], as well as electrical equipment in implantable biomedical, underwater, mines, and aerospace fields [7, 8]. The terminals of WPT systems cover electrical loads ranging from microwatts to megawatts.

According to the wireless charging market research report released by Global Market Insight in January 2022, the global wireless charging market reached

B. Zhang et al., *Parity-Time Symmetric Wireless Power Transfer*, CPSS Power Electronics Series, https://doi.org/10.1007/978-981-96-3149-0_1

US$12.68 billion in 2020 and is expected to increase to US$185.9 billion by the end of 2030, with a CAGR of about 30.0% from 2021 to 2030 [9]. As a forward-looking technology, the MIT Technology Review named "wireless power transfer" and "aerial power extraction" as one of the top 10 breakthrough technologies in 2008 and 2016, respectively [10, 11]. It can be seen that WPT technology has huge market space and research significance, and promoting the development and commercialization of WPT technology will surely bring significant social and economic benefits. Meanwhile, WPT has become an important area of research and has received great attention in both academia and industry.

1.1 Development and Status of WPT Technology

1.1.1 Development History

WPT technology can be traced back to the 1890s. The development of electromagnetic theory first provided a theoretical basis for WPT technology. The proposal of Ampere's loop law in 1826 and Faraday's law of electromagnetic induction in 1831 laid the foundation for electromagnetic theory. Then in 1864, James C. Maxwell summarized Maxwell's equation and predicted the existence of electromagnetic waves, and John H. Poynting further proposed Poynting's theorem on the conservation of energy in electromagnetic fields to quantify the energy flow of electromagnetic fields in 1884. By the 1880s, the theory of electromagnetism had been basically perfected, paving the way for the development of WPT. In 1888, Heinrich R. Hertz successfully experimentally confirmed the existence of electromagnetic waves, and this research results aroused great interest in Nicola Tesla. After that, N. Tesla began to withdraw from the commercial promotion of AC transmission systems, resumed research work, and conducted a series of experiments. In 1891, N. Tesla gave the first experimental demonstration of WPT at a meeting of the United States Institute of Electrical Engineers at Colombia College, where N. Tesla successfully made the Geisler tube glow without any wires by standing between two metal plates connected to the Tesla coil output with two long Geisler tubes in his hand, as shown in Fig. 1.1 [11].

At the end of the nineteenth century, N. Tesla attempted to build a WPT tower, the famous Wardenclyffe Tower as shown in Fig. 1.2, to use the resonance between the Earth and the Earth's ionosphere to enable wireless signal and power transmission across the Atlantic. Regrettably, due to Guglielmo Marconi's more outstanding achievements in the development of radiocommunication systems using electromagnetic waves [12], and the lack of follow-up funds, N. Tesla's global WPT project was aborted. Although N. Tesla's dream of global wireless power transmission did not come true, it is undeniable that N. Tesla has brought great inspiration to future generations, and is a pioneer in promoting the development of WPT. In 1894, Hu Ting and LeBlanc proposed a device that uses an AC power source with a frequency of

Fig. 1.1 N. Tesla first demonstrated WPT at Columbia College in 1891. a Site diagram. b Experimental device

about 3kHz to inductively power the traction tram [13], which is called an electric track transformer system, as shown in Fig. 1.3, which is similar in structure to the wireless power supply system of modern rail transit. However, with the development of internal combustion engines, trams have lost their popularity, so the research on wireless power supply for trams has not received more attention. However, the ideas and experimental devices of these early wireless transmission pioneers, such as electromagnetic induction and resonance circuits, inspired later researchers.

Fig. 1.2 Wardenclyffe tower

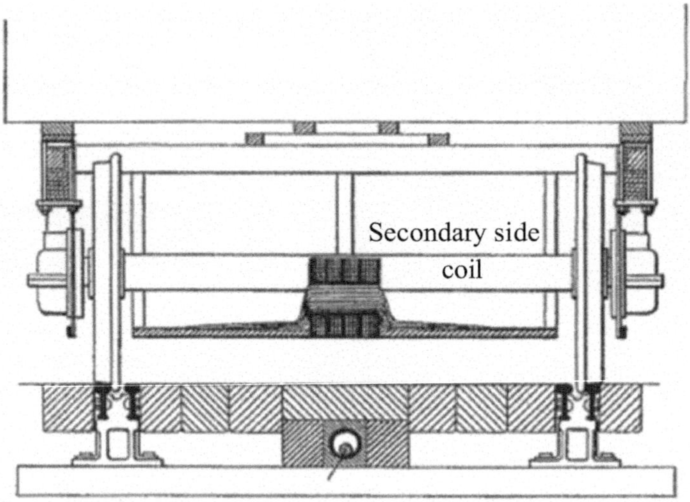

Fig. 1.3 Wireless power supply device for trams

At the beginning of the twentieth century, with the rapid development of radio communication technology, people paid more attention to the wireless transmission of signals rather than energy, so for a long time, the research of WPT technology was basically dormant. In the 1930s, with the invention of high-power magnetrons and klystrons, the technology of generating high-power microwaves has made great progress. In 1964, William C. Brown pioneered microwave wireless power transfer (MWPT) technology, which uses 2.45 GHz microwaves to power a model helicopter that can hover in the air for up to ten hours [14]. Since then, W. C. Brown has been working on the development of new rectifier antennas to improve the overall efficiency of the system, and by 1975 it had achieved 495 W of power in the 2.4–2.5 GHz ISM band at 54% DC-DC efficiency [15]. Although microwave scholars in Japan, Canada, the United States, and other countries have done a large number of MWPT experiments after the 1960s, MWPT is still difficult to enter commercial applications, because large-size antennas are necessary to obtain high-efficiency long-distance transmission [16]. This deficiency is acceptable for energy transmission in space. In 1971, Peter Glaser presented the concept of solar satellite power plants based on W. Glaser. C. Brown's research, as shown in Fig. 1.4, where the satellite collects space solar energy, converts it into microwave energy, and launches it to the earth's surface for human use [17]. In 1978, United States NASA planned to transmit 750 MW of solar energy collected by geostationary satellites via 2.45 GHz microwaves, with a transmitting antenna diameter of 1km and a receiver antenna diameter of 10km, and subsequently conducted several kilowatt-scale simulation experiments. This plan was eventually scrapped due to low efficiency, high energy density on the receiving side, potential hazards, and low cost-effectiveness [18]. At the end of the 1990s, people began to explore a safer way to collect space solar energy [19]. In 2003, United States

NASA's Marshall Space Flight Center developed a laser-powered drone with a total efficiency of 6.8% from the input of the laser emitter to the output of the solar cell, as shown in Fig. 1.5. Compared with microwave transmission, laser wireless power transfer (LWPT) has almost no electromagnetic interference problem [20].

With the development of long-distance WPT technologies such as microwave and laser, short-distance WPT has begun to attract attention. It is worth noting that transformers, which are closely related to WPT, began to play an important role in electrical energy conversion in the 1880s [21]. Inspired by the principle of transformers, M. Hutin and M. Le-Blanc applied for a patent for a transformer system for electric trains in 1894, which was essentially an inductive wireless power supply system [13]. Since the electrification of power equipment was not widespread at the time, follow-up research on this wireless power supply system was not followed.

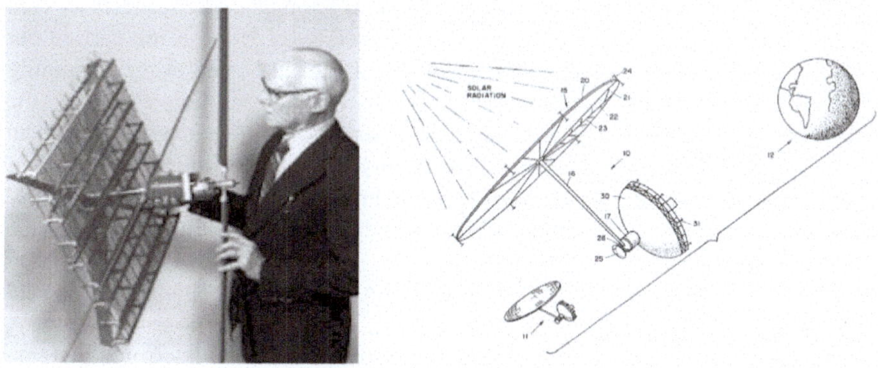

(a) (b)

Fig. 1.4 Microwave wireless power transfer. **a** Helicopter propeller connected to a rectenna. **b** Solar satellite power station

(a) (b)

Fig. 1.5 Laser wireless power transfer. **a** Energy transmitter. **b** Energy receiver

Until the 1990s, scholars began to systematically study the high-power applications of WPT, among which John Boys and Grant Covic of the University of Auckland were the most prominent, analyzing and studying the WPT system of electric vehicles from multiple aspects such as modeling, power conversion topology, and control, and doing a large number of experimental demonstrations [22], the schematic of the WPT system of electric vehicles is shown in Fig. 1.6. The system consists of the track winding (a large conducting loop) and the pick winding, and the pickup is able to move relative to the track. On the other hand, at the same time, low-power applications of WPT systems also began to rise. In 1958, the first human pacemaker was successfully implanted [23], prompting the discussion and research on WPT for implantable medical devices in the 1960s [24, 25]. In order to overcome the problem of electromagnetic interference and to miniaturize the receiving device as much as possible, Cochran et al. attempted to replace electromagnetic waves with ultrasound in 1985 to transmit electrical energy to stimulate osteogenesis in vivo [26]. In the 1990s, Ron S.Y. Hui of the City University of Hong Kong carried out a lot of research work on planar PCB ironless transformers, and at the beginning of the twenty-first century, and the research results were successfully applied to the wireless charging platform for portable consumer electronic devices, as shown in Fig. 1.7. In general, by the beginning of the twenty-first century, the development of semiconductor devices and power electronics technology, and the demand for power supply for electrification equipment have virtually provided a suitable soil for the rise of WPT technology.

In 2007, André Kurs et al. at the Massachusetts Institute of Technology proposed the concept of strong magnetic resonance [27], and used the concept of strong magnetic resonance to light up a 60W light bulb suspended 2m away from the transmitting coil through a four-coil wireless power transmission structure, as shown in Fig. 1.8, where the parasitic capacitance of the coil is used for resonant tuning, and the operating frequency is 9.9MHz. Although the overall efficiency is only 40%, this study is the first time to clearly elucidate the physical mechanism of efficient wireless power transfer, which has caused a sensation in academia and industry, and is of great significance as a milestone.

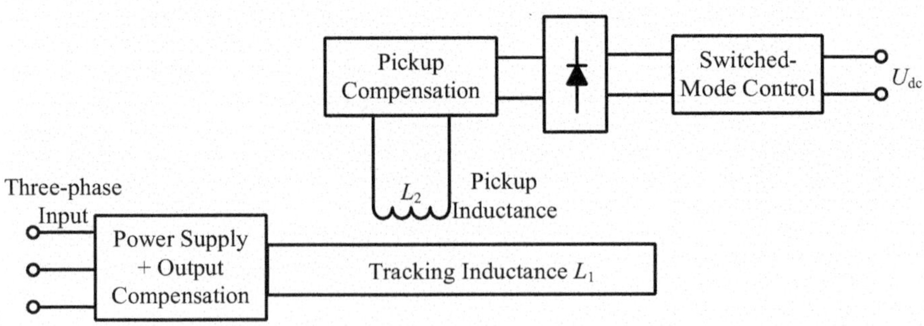

Fig. 1.6 Schematic of inductive WPT system

Fig. 1.7 Portable consumer electronic device charging platform

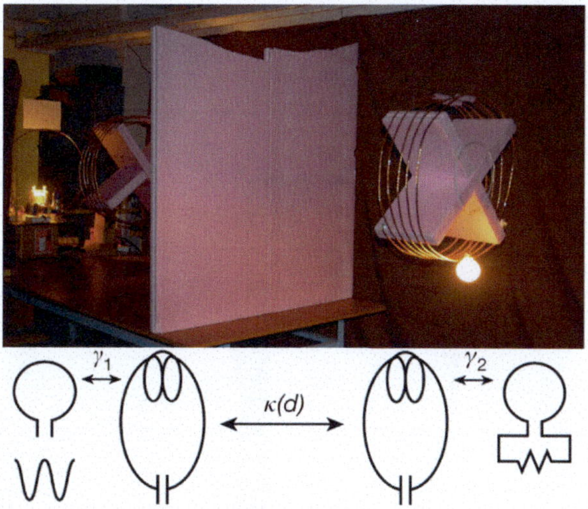

Fig. 1.8 Four-coil resonant WPT experiment of MIT

This research result of MIT has led to a rapid increase in WPT research activities. Chun T. Rim and other researchers at the Korea Advanced Institute of Science and Technology (KAIST) in Korea have carried out research on dynamic WPT technology and Synthesized Magnetic Field Focusing (SMF) technology. A new dipole coil resonant WPT technology was proposed [28, 29]. Meanwhile, Chunting Chris Mi and others from San Diego State University conducted a comprehensive study on compensation networks and capacitively coupled WPT technologies [30–32]. Prof. Bo Zhang's team from South China University of Technology presented the circuit model of the magnetic resonance wireless power transmission system, analyzed the maximum efficiency and frequency splitting phenomenon, and proposed a frequency tracking control method [30–33]. Professor Sun Yue's team at Chongqing University has conducted research on stability control and load detection of WPT systems and has done many applications [33, 34]. In addition, the team of Professor Zhu Chunbo of Harbin Institute of Technology [35], the team of Professor Yang Qingxin of Tianjin Polytechnic University [36], the team of Professor Huang Xueliang of Southeast University [37], and the team of Professor Chen Qianhong of Nanjing University of Aeronautics and Astronautics [38] have also carried out a large number of studies and experiments, which have promoted the development of WPT technology research in China.

In 2017, Shanhui Fan et al. from Stanford University introduced the principle of parity-time (PT) symmetry in quantum mechanics into the WPT system, and constructed a PT-symmetric WPT system using nonlinear saturated gain elements, as shown in Fig. 1.9, which automatically maintains stable transmission efficiency when the transmission distance changes [39], and the output power is only 10mW. However, it has further stimulated researchers to explore the mechanism and application of WPT technology.

In 2019, the team of Professor Zhang Bo of South China University of Technology proposed a new WPT technology based on fractional-order capacitors, which uses fractional-order capacitors to construct an autonomous WPT system, and can automatically keep the transmission efficiency constant when the transfer distance changes within a certain range, and can still automatically adjust to maintain stable and efficient power output in the case of detuning of the receiving coil [40].

In recent years, WPT research has shown the characteristics of multidisciplinary integration, which is manifested in the following aspects:

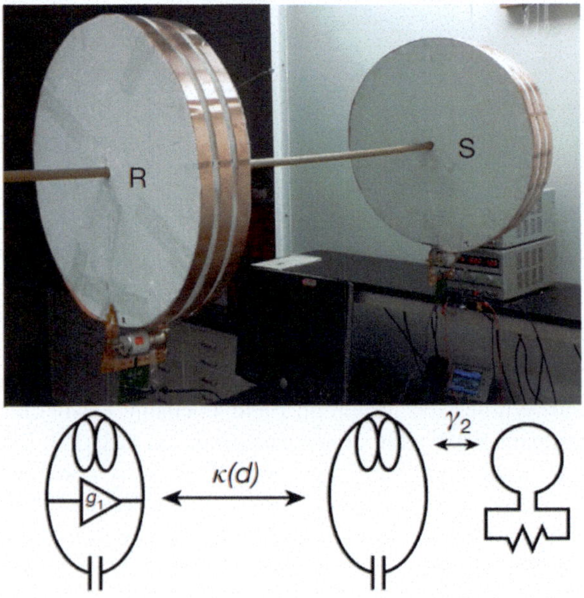

Fig. 1.9 PT-symmetric WPT experiment

1. Physics provides support for the basic theoretical research of transmission mechanisms and new materials, including the team of Professor Marin Soljačić of the Massachusetts Institute of Technology [27, 41], the team of Professor Shanhui Fan of Stanford University [39], and the team of Professor Andrea Alù of the City University of New York [42, 43], etc.
2. Microwave radio frequency science uses electromagnetic theory and integrated circuit technology to devote itself to the research of free space transmission and device miniaturization that take into account electromagnetic safety [44, 45].
3. The research in the field of power electronics has gradually developed into the main body of WPT technology application, promoting the industrialization of medium and high-power WPT systems, and the more active universities and research institutions in the world include the University of Auckland [46–49], the Korea Advanced Institute of Science and Technology [50–52], the University of Tokyo [53–55], San Diego State University [56–58], Imperial College London [59–61], Oak Ridge National Laboratory of United States [62–65] and the Disney Institute [66], among others. In 2017, Apple released a new generation of mobile phones with built-in wireless charging, further promoting the large-scale commercial application of wireless charging technology for consumer electronics. In recent years, with the maturity of wireless charging standards [67], the industrial transformation of scientific and technological achievements has also begun to accelerate, and it is foreseeable that wireless charging will be ubiquitous in the future.

1.1.2 Current Research Status of Application

WPT technology has been applied to wireless charging of portable electronic devices since 2000. The Wireless Power Consortium (WPC), established in 2008, has released several versions of the Qi standard for wireless charging of portable electronic products since 2010. The standard is based on the principle of magnetic coupling induction, supports charging power from 1 to 15W, and is currently available in more than 3700 licensed products, including wirelessly charged smartphones from companies such as Apple, Samsung, and Huawei, as shown in Fig. 1.10a. Another global wireless charging organization is the Airfuel Alliance. Figure 1.10b, c illustrate the Dell wireless charging laptop and the WattUp long-range wireless charger based on the Airfuel standard. In addition to these standard technologies, Wi-Charge showcased a wireless charger for portable devices that uses infrared light at the 2018 Consumer Electronics Show (CES) to enable wireless charging over a wide range of areas by placing it on the ceiling, as shown in Fig. 1.10d. Wenergy Technology adopts the self-developed Wi-Po low-frequency magnetic resonance technology, which is compatible with the Qi standard and increases the transfer distance to 3.5 cm. In 2019, Spansive introduced the Spansive Source charger, which utilizes a multi-coil magnetic field focusing algorithm to wirelessly charge up to six phones simultaneously with only 5 watts of power. At present, portable electronic products have become the most commercialized application field of WPT technology.

In the field of household appliances, WPT technology has initially begun to be commercialized, and Fig. 1.11 shows some tailless appliances. Fulton Innovation has been committed to applying eCoupled WPT technology to home appliances since the 1990s, and by 2008, it had realized the wireless power supply of medium and high-power kitchen appliances. In 2009, Sony developed a 60 W TV resonant WPT system, which can increase the transfer distance by 0.8 m by adding a relay unit. In 2010, Haier company showcased a 100W tailless TV developed in cooperation with WiTricity at the 43rd CES, with a maximum transmission distance of 1 m. Subsequently, in 2012, it launched a wireless rice cooker and blender, with a transfer distance of 4–5 cm, a maximum power of 1.2 kW, and an efficiency of more than 85%, and was successfully launched in 2013. Since 2014, home appliance giant Philips has launched a series of magnetically coupled wireless kitchen appliances

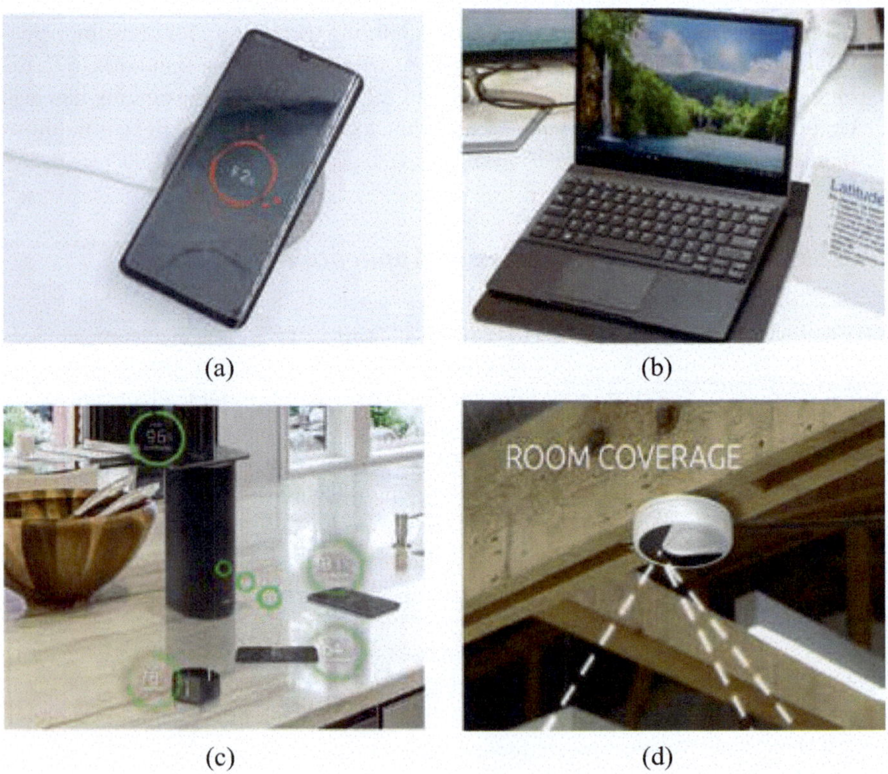

Fig. 1.10 WPT applications for portable electronic devices. **a** Huawei 27 W wireless fast charging. **b** Dell laptop wireless charging. **c** Wattup RF wireless charging. **d** Wi-charge infrared wireless charging

that support up to 2.5 kW. China's Chushan Electronics, Zhonghui, Chuangzhi, and other companies have also developed WPT systems for medium and high-power household appliances, with a power of up to 2 kW and an efficiency of more than 80%. In terms of standards, WPC is currently developing a medium-power (30–200 W) wireless charging standard for household vacuum cleaners, lawn mowers, and portable power tools. In 2019, the WPC announced that it would name the wireless power supply specification standard for kitchen appliances that is being developed as the Ki standard, which supports up to 2.2 kW. In recent years, some scientific research institutions have also carried out relevant applied research. Researchers from Harbin Institute of Technology designed a tailless appliance coupling mechanism and used LCL compensation to implement a WPT system compatible with induction heating. Researchers from Chongqing University have studied the frequency stability control and design methods of constant frequency and constant voltage parameters of WPT systems for home appliances. Researchers at the Delft University of Technology (TU Delft) have proposed a topology design to reduce the EMI of the inductive WPT system for home appliances.

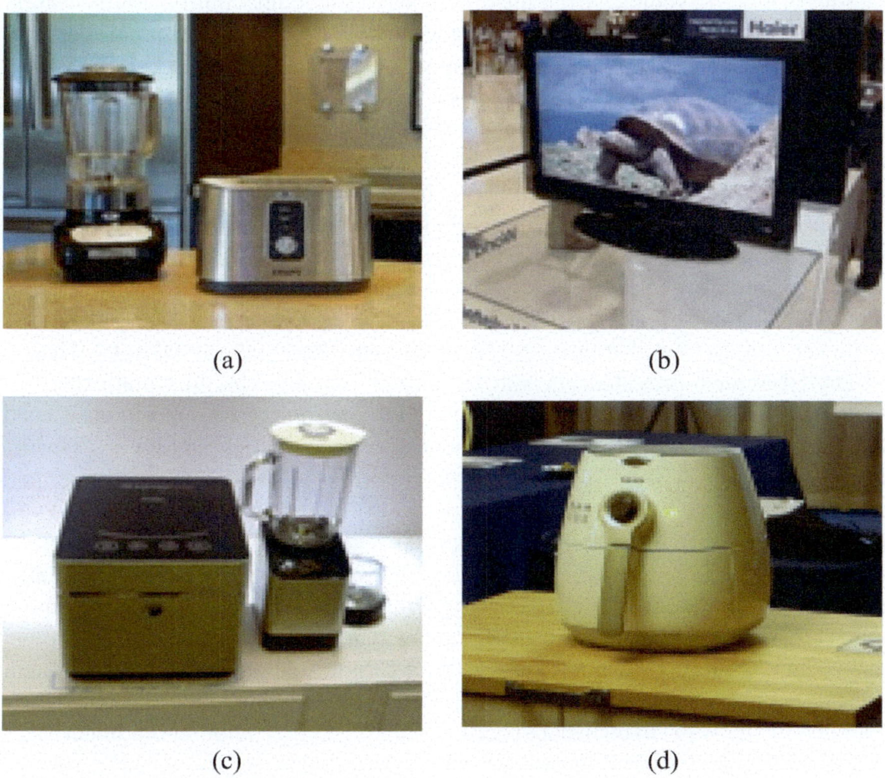

Fig. 1.11 WPT plications for home appliances. **a** Fulton innovation tailless kitchen appliances. **b** Haier tailless TVs. **c** Haier tailless kitchen appliances. **d** Philips tailless kitchen appliances

Electric vehicles are an inevitable choice to deal with the future energy crisis, and many researchers, technology companies, and automakers are trying to explore the commercial application of WPT technology for electric vehicles, some of which are illustrated in Fig. 1.12. In 2018, BMW launched the BMW 530e iPerformance electric car, which is the world's first car with its own wireless charging function, and the charging power is 3.6 kW [68]. Evatran's Plugless wireless charging system supports charging power of 3.6 and 7.2 kW, with an efficiency of about 83%, and can be used as an additional wireless charging device for electric vehicles such as Tesla Model S and BMW I3. WiTricity has launched the Drive 11 wireless charging system, which has a transfer distance of 10–25 cm, can provide 3.6–11 kW charging power, and achieve 90–93% efficiency from the grid side to the battery side [68]. Qualcomm's Halo electric vehicle wireless charging system has a charging power of 3.3–20 kW and an efficiency of more than 90%. Qualcomm Halo has also developed a dynamic wireless charging system that allows a maximum driving speed of 20 km/h. In 2018, the State Grid built a WPT system for road-mobile electric vehicles, with a transfer power of 20 kW, an efficiency of 80%, and a maximum driving speed of 60 km/h. In addition, many scientific research institutions have carried out research on the application of wireless charging of electric vehicles from the directions of modeling control, compensation network, coupling mechanism, electromagnetic shielding and bidirectional charging. United States Oak Ridge National Laboratory successfully developed a 20kW and 120kW wireless charging system for electric vehicles in 2016 and 2018, respectively, with an efficiency of up to 97% [69]. Since then, relevant international standards such as SAE J2954, IEC 61,980, and ISO 19363 have standardized the wireless charging system for electric vehicles, promoting practical application and commercialization, and combined with the development trend of autonomous driving, WPT technology for electric vehicles will be widely used.

WPT technology has also found many applications in the fields of industrial robots, the Internet of Things (IoT), medical devices, underwater equipment, and the power supply of rail transit. At present, companies such as Daifuku, Heads, and DAIHEN have provided kilowatt-level AGV wireless charging platforms, but the distribution of dynamic charging units and electromagnetic interference need to be further studied. With the development of 5G technology, ubiquitous sensors and smart devices will be merged into the Internet of Things in the future, and wireless information and power synchronous transmission technology (SWIPT) will play an important role in the power supply and information interaction of sensor networks [70], and researchers at the National University of Singapore have revealed the fundamental trade-offs when designing wireless multiple-input, multiple-output (MIMO) systems to maximize the efficiency of synchronous information and power transmission. Researchers at the University of Connecticut installed a WPT system in a smart wave energy converter (Smart-WEC) and delivered the electrical energy generated by the WEC to distributed ocean systems (such as autonomous underwater vehicle AUVs, underwater sensor networks, ocean sensors, and tsunami buoys), effectively solving the power supply problem for distributed ocean applications [71]. Bombardier has developed a 250 kW wirelessly powered tram that complies with EMF guidelines and can reach speeds of up to 80 km/h. Researchers at the Korea

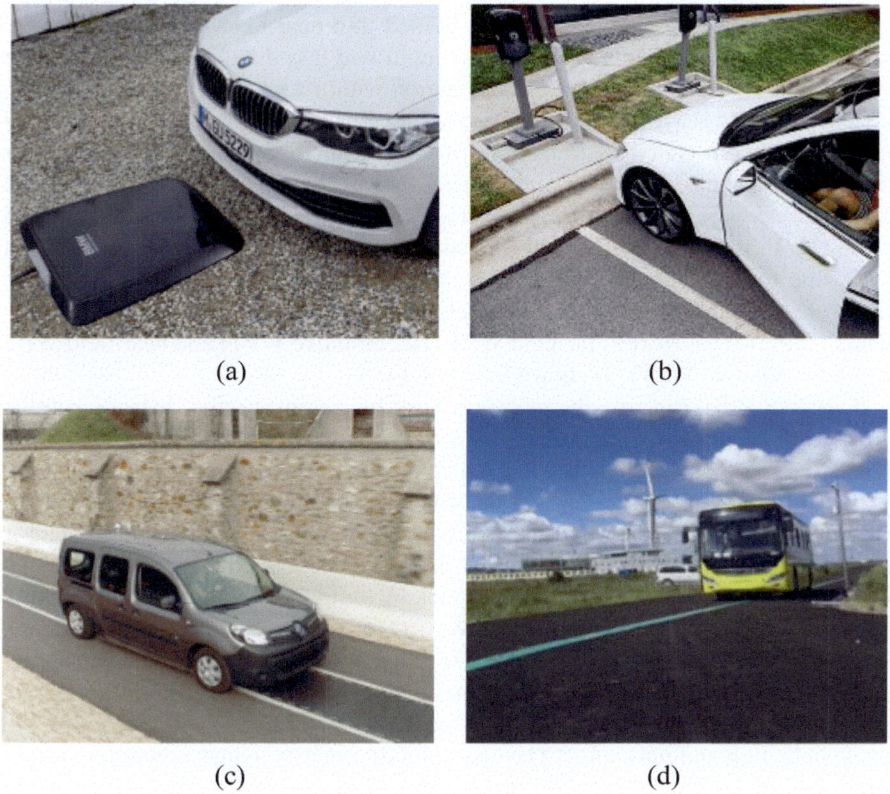

Fig. 1.12 WPT applications for electric vehicles. **a** BMW wireless charging cars. **b** Plugless wireless charging. **c** Qualcomm dynamic wireless charging. **d** State grid mobile wireless charging

Railway Research Institute (KRRI) developed and manufactured a 1 MW WPT system for high-speed trains that successfully ran at 10 km/h. In addition, in order to solve the interoperability problem between dynamic WPT and static WPT, the Korea KAIST team proposed the sixth-generation OLEVs (Online electric vehicles) technology using a coreless wireless power supply [72].

1.1.3 Main Classifications and Characteristics

Throughout the development of WPT technology, many types of WPT technologies have emerged, and their basic principles and structures can be described in Fig. 1.13. The power transmitter and receiver are separated by a distance d, and the characteristics of the electromagnetic wave are determined by its wavelength λ or frequency f. D is the maximum size of the power transmitter and receiver. The behavior of electromagnetic waves is defined by Maxwell's equations, which can be simplified when certain conditions are true, and can be divided into near-field and far-field cases. For transmitters and receivers with a maximum size less than half of the operating wavelength ($D < \lambda/2$), the near field is a region within a single wavelength ($d < \lambda$), while the far field is a region outside two wavelengths ($d > 2\lambda$), and the middle region between the two is called the transition zone. If the maximum size of the transmitter and receiver is greater than half of the operating wavelength ($D > \lambda/2$), the near field is the region where the transfer distance meets $d < 2D^2/\lambda$, and the rest is the far field [135][41]. In the near field, Maxwell's conclusions can be reduced to Kirchhoff's law, while for the far field, RF analysis and optics-based equations are necessary. Therefore, according to the principles of electromagnetics, WPT technologies can be divided into two categories: far-field WPT (i.e., radiative WPT) and near-field WPT (i.e., non-radiative WPT), as shown in Fig. 1.14.

Radiative WPT can usually achieve a very long transmission distance, and its energy carriers include radio waves, microwaves and lasers, etc., but the radio wave mode needs to consider the diffusion and attenuation of electromagnetic waves, and its transmission power is small and the transmission efficiency is low. Microwave and laser methods are a kind of linear directional power transmission, once there is an obstacle to block, it will no longer work, so the above methods are usually used on specific occasions. For non-radiative WPT, including electric-field coupling WPT (EC-WPT) and magnetic-field coupling WPT (MC-WPT). When the transmitter and receiver are inductance coils, they belong to magnetic coupling. When the transmitter and receiver are metal plates, they are electric field coupling, so electric-field coupling is also called capacitive coupling. Besides, the MC-WPT can be divided into two types: induction and resonance. The operating distance of inductive wireless power

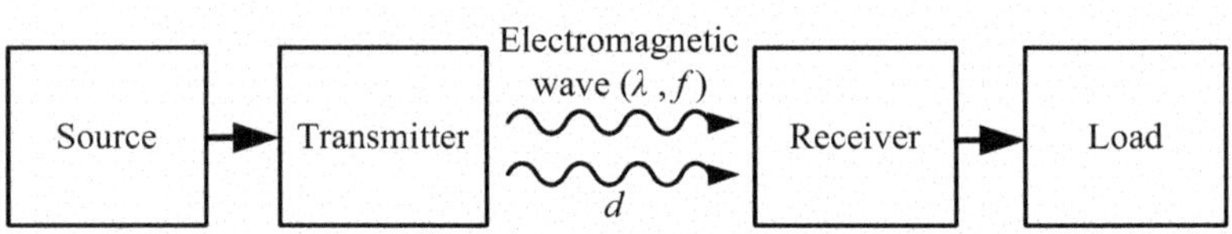

Fig. 1.13 Basic structure diagram of WPT system

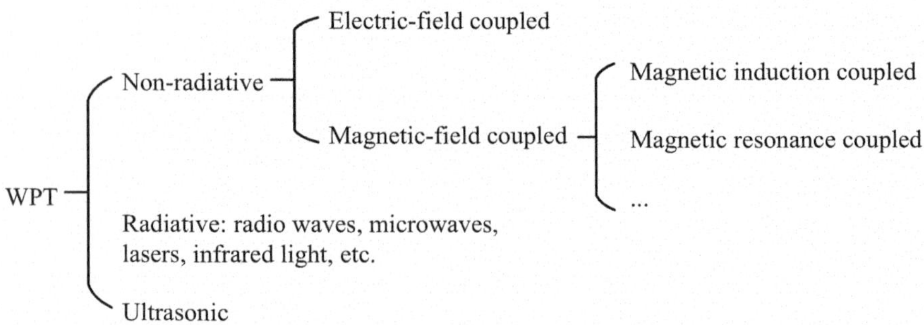

Fig. 1.14 Categories of WPT

transmission (IPT) is usually less than the maximum size of the transmitter or receiver ($d < D$). The working distance of magnetic resonant wireless power transmission (MRWPT) can generally be up to 10 times the maximum size of the transmitter or receiver ($D < d < 10D$).

IPT applies the principle of Faraday electromagnetic induction, which generates an alternating magnetic field by applying a high-frequency alternating current to the transmitting coil, and cross-links to the receiving coil to induce an electromotive force, thereby transmitting electrical energy to the load [73]. The higher the power, the larger the coil is naturally required, and compensation capacitors are often added to reduce the reactive power of the system. As a result, the transfer distance can be increased with the power level, which can range from tens to hundreds of kilowatts. IPT systems can only be transmitted one-to-one and are susceptible to metallic objects, and the operating frequency is generally designed to be slightly different from the resonant frequency of the receiver for power control, and ferrite-enhanced coupling is often used to reduce magnetic flux leakage to improve the transfer efficiency of IPT systems.

MRWPT is based on the principle that resonant objects can be exchanged with high efficiency, and LC resonant circuits are used to achieve efficient wireless power transmission [27]. The operating frequency of the power supply of the system is consistent with the resonant frequency of the transmitter and receiver, so the interaction between the transmitter and receiver is strong, and the interaction with non-resonant objects is very weak. At the same time, due to the resonance, the coil current reaches the maximum, and the coil quality factor Q is high, and the strong interaction between the resonators allows the system to work efficiently and without being affected by obstacles even when the coupling is weak. The energy coupling strength of the MRWPT system mainly depends on the resonant frequency, mutual inductance coefficient and quality factor, and the working frequency is generally designed in the higher-frequency band, so the requirements for the high-frequency inverter link are higher, and wide bandgap semiconductor devices such as SiC or GaN are usually used in high-power occasions.

Capacitively coupled wireless power transfer (CPT) system uses an alternating electric field between the transmitting and receiving plates to generate a displacement current to transfer electrical energy to the load, where the resonant type also uses an LC resonant circuit [74]. Similar to the magnetic coupling method, the current induced by the receiver is proportional to the rate of change of the electric field flux between the two plates. The CPT system is not susceptible to metal objects and has no EMI, but the transmission power is limited by distance and is mainly used in low-power applications.

The ultrasonic WPT system uses sound waves with a frequency greater than 20 kHz, which has the characteristics of strong directionality, can be transmitted through conductors, and does not have the problem of electromagnetic interference, but the overall efficiency is not high due to the large number of energy conversion links [75]. In recent years, the application of near-field WPT technology has become more extensive, and has received great attention, becoming the most promising alternative to wired transmission.

1.2 Existing Problems

The ultimate goal of the application of WPT technology is to obtain a good wireless charging experience, that is, to improve the transfer efficiency, spatial freedom and stability as much as possible while ensuring the load power demand. The wide application of WPT technology has brought challenges to WPT systems, although considerable progress has been made in the research of WPT technology at this stage, it is still urgent to solve the following key problems.

(1) The problem that transfer characteristics are susceptible to changes in transfer distance and misalignment.

Although the transfer distance of MRWPT has been shown to be longer than that of IPT, the transfer efficiency and output power cannot be stable when the transfer distance of the system changes, that is, the performance of the MCRWPT system is very sensitive to the transfer distance. In 2007, the MRWPT system built by MIT can reach about 90% under the condition of transfer distance of 1m, and when the distance is increased to 1.5 m, the efficiency has dropped to about 76%. Moreover, the output power of the system also changes [27]. In practical applications, except for some specific occasions, the transfer distance often changes, such as wireless charging for electric vehicles in the process of moving. Therefore, how to maintain the stability and efficiency of the system when the transfer distance changes is an important problem in WPT technology.

At present, there are four main methods to realize that the WPT system is insensitive to distance changes. The first method is to change the physical structure, by presetting multiple coils with different parameters, and using switches to reconstruct the resonant coils at different distances, so that the transfer characteristics of the system remain unchanged with distance changes within a certain range [3, 76, 77]. The second method is to use dynamic tuning techniques to dynamically adjust the parameters of the components through capacitor arrays and other methods to achieve the stability of output characteristics under the condition of distance change [3, 78, 79]. The third method is to track the working frequency at which the power supply voltage and the supply current or the supply voltage and the secondary side current are in phase, so as to obtain the characteristics that the output power or voltage remains constant with the change of distance, but the transfer efficiency and output power are unstable when the resonant frequency of the transmitter or receiver is offset by interference [80–82]. The fourth method is to use a parity-time (PT) symmetric autonomous circuit, but it can only operate under the condition that the resonant frequencies of the transmitter and receiver are not disturbed [39, 83, 84].

(2) The problem that transfer characteristics are sensitive to changes in load.

In practical applications, depending on the load, it is often necessary for the WPT system to have the characteristics of constant current, constant voltage, or constant power output when the load changes, such as the constant current power supply of LED, battery, and supercapacitor [85]. However, conventional WPT systems are difficult to ensure a constant output when the load changes. In addition, considering the limited space of most of the receiving devices, it is necessary to make the receiving structure of the system as simple as possible. To this end, a lot of work has been done by numerous scholars.

Taking the constant current output as an example, there are four main methods to realize the constant current output of the WPT system. The first method is to design the compensation topology, which has the disadvantage of limited control accuracy [86, 87], for example, the series-series compensation topology can achieve accurate constant current output when the internal resistance of transmitter is zero, but the internal resistance is not zero in practical application. The second method is to adopt a closed-loop control method on the receiving device, which requires additional circuits, such as DC-DC converters, etc., which will greatly increase the weight, volume, and cost of the receiving device [35, 88, 89]. The third method is to only control the transmitter in a closed-loop manner, but bilateral wireless communication between the transmitter and receiver is required to regulate the output current, and the system is prone to instability and failure when the wireless communication is interfered with [90, 91]. The fourth method is to adopt a transmitter-only control strategy, which does not require bilateral wireless communication, but generally requires high-frequency sampling circuits and complex parameter recognition algorithms to obtain the parameters on the receiving side [34, 92]. Due to the shortcomings of the above method, such as the weight increase on the receiving side, the limited control accuracy and the instability, further research is still needed to overcome the load sensitivity of the WPT system.

(3) The problem that transfer characteristics are sensitive to changes in resonant frequency.

The MC-WPT system is generally composed of two or more resonant circuits, and the transfer efficiency and transfer power of the system are very sensitive to the resonant frequency of each circuit, including the resonant frequency of the transmitting circuit and the resonant frequency of the receiving circuit. However, the resonant frequency of a circuit can easily be shifted by interference in practice, such as in an electromagnetic environment generated by a metal object, or under non-resistive load conditions [93–96]. At present, there are few studies on the sensitivity of the transmission performance of MC-WPT systems to the resonant frequency of transmitting or receiving circuits. Ferrites can be used to shield the receiver coil to avoid external interference, but it increases the weight of the receiver and cannot overcome the interference of non-resistive loads on the resonant frequency [97]. When the resonant frequency of the receiver side changes, as long as the information of the resonant frequency of the receiver can be obtained, the resonant frequency of the transmitting circuit can be retuned, so that the two resonators can resonate at the same frequency [98], but it is difficult to obtain the required information when the receiving circuit is disturbed. Therefore, it is a practical need and a major challenge to study WPT systems that can maintain high efficiency and stable output power under the condition of resonant frequency interference.

1.3 Introduction of PT-Symmetric Concept to WPT

The original WPT technology mainly used the principle of electromagnetic induction, and its analysis was similar to that of a transformer, and the Tesla coil, invented by N. Tesla to generate high voltages, was essentially a series resonant transformer. In early experiments, researchers found that capacitive tuning can improve power transfer [99, 100], but they have not been able to elucidate the reasons for this, and they have not realized the significance of increasing the resonant frequency to increase the efficiency of power transmission. Until 2007, the MIT research team compared the high-quality factor LC harmonic oscillator (Resonator or Harmonic Oscillator) to an optical microcavity with a whispering gallery mode, and used the energy $|\boldsymbol{a}_{1,2}|^2$ carried by the optical mode $\boldsymbol{a}_{1,2}$ to characterize the energy stored by the LC harmonic oscillators on both sides of the receiver and transmitter, so that the efficient energy transfer between the weakly coupled LC harmonic oscillators can be transformed into the problem of efficient coupling of the optical modes of the two microcavities. The physical diagram of the MRWPT system is shown in Fig. 1.8, and the equivalent circuit is shown in Fig. 1.15. To analyze the transmission characteristics of the MRWPT system shown in Fig. 1.15, André Kurs et al. developed the following coupling model [27, 101]:

$$\begin{cases} \frac{d\boldsymbol{a}_1}{dt} = (-\Gamma_{10} + j\omega_1)\boldsymbol{a}_1 - j\kappa_{12}\boldsymbol{a}_2 + \boldsymbol{F}_1 \\ \frac{d\boldsymbol{a}_2}{dt} = -j\kappa_{12}\boldsymbol{a}_1 + \left[-(\Gamma_{20} + \Gamma_L) + j\omega_2\right]\boldsymbol{a}_2 \end{cases}, \tag{1.1}$$

where $\boldsymbol{F}_1$ is the driving mode related to the AC source u_S, and $\boldsymbol{F}_1 \propto e^{j\omega_S t}$. ω_S is the angular frequency of the AC source. The transmitting side inductor L_1 and capacitor C_1 form a linear LC harmonic oscillator, which is coupled with the corresponding linear LC harmonic oscillator on the receiving side through mutual inductance M_{12}. $\omega_n = 1/\sqrt{L_n C_n}$ ($n = 1$ is labeled as the transmitting side, $n = 2$ is labeled as the receiving side) represents the resonant angle frequency of the linear LC harmonic oscillator. κ is the coupling rate between the harmonic oscillator on the transmitting side and the receiving side, which is related to the mutual inductance M_{12}. Γ_{10} and Γ_{20} are the intrinsic loss rates related to the internal resistance R_1 and R_2 of the harmonic oscillator on the transmitting side and the receiving side, respectively, i.e., $\Gamma_{10} = R_1/(2L_1)$ and $\Gamma_{20} = R_2/(2L_2)$. $\Gamma_L = R_L/(2L_2)$ is the loss rates related to the load resistance R_L.

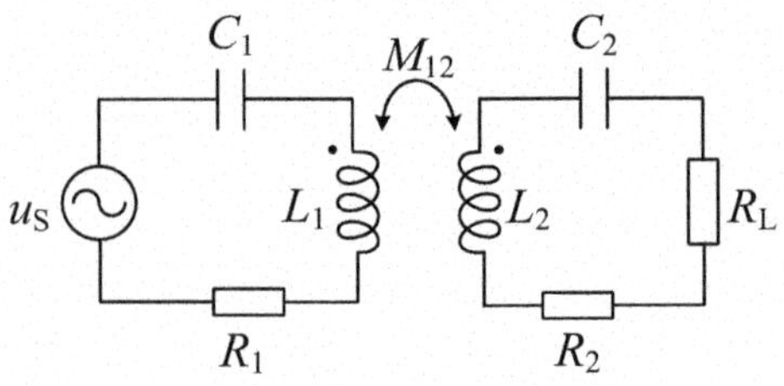

Fig. 1.15 Equivalent circuit of MRWPT system

As can be seen from (1.1), the MRWPT system is a linear system, and the AC source u_S corresponds to the mode $\boldsymbol{F}_1$, which rotates freely at a constant amplitude $|\boldsymbol{F}_1|$ and a constant rate ω_S. Through model analysis, André Kurs et al. further concluded that the energy exchange efficiency between modes is the highest under the resonance condition (i.e., $\omega_S = \omega_1 = \omega_2$), and the maximum efficiency is proportional to the product of the coupling coefficient and quality factor, that is, the conditions for efficient transmission by using strong magnetic resonance are [27, 101]:

$$\omega_S = \omega_1 = \omega_2, \ \eta \propto \frac{\kappa}{\sqrt{\Gamma_1 \Gamma_2}}. \tag{1.2}$$

Strong magnetic resonance not only means that the power supply frequency and harmonic oscillator frequency are consistent, but also requires the system to have a strong coupling rate and a weak loss rate, so even if the coupling coefficient is very small, increasing the operating frequency of the system and reducing the internal resistance of the harmonic oscillator (using a self-resonant open-loop coil) can achieve medium and long-distance efficient transmission with a transmission distance several times larger than the coil size. The proposal of MRWPT breaks through the limitation of short-distance transmission based on the principle of electromagnetic induction, and its resonance property ensures that the power transmission process is not interfered by non-resonant objects.

Although MRWPT has high transfer efficiency, it is sensitive to transfer distance changes and has harsh resonance conditions [80, 102]. When the transfer distance is less than a certain critical value, the output power of the MRWPT system will decrease dramatically. When the transfer distance is small, the power splitting phenomenon occurs in MRWPT. Under the effect of power splitting, the output power of the MRWPT system changes drastically, and it is difficult to keep the output power stable. Moreover, once the resonant frequency of the linear harmonic oscillator is shifted, the transfer efficiency of the system will drop dramatically.

In order to solve the problem that the transfer efficiency of MRWPT system decreases sharply with distance, the research team of Stanford University published a WPT system based on a nonlinear parity-time (PT) symmetric circuit in Nature in 2017 [39], and the circuit schematic diagram is shown in Fig. 1.16. The system adopts a parallel-parallel topology, with a parallel circuit composed of nonlinear negative resistance R_N, inductor L_1, and capacitor C_1 on the transmitting side, and a parallel circuit composed of inductor L_2, capacitor C_2, and load R_L on the receiving side. The negative resistor R_N is composed of an arithmetic amplification circuit and is controlled by the voltage of the capacitor C_1 in the harmonic oscillator on the transmitting side, so the nonlinear gain [39] is used to replace the driving mode $\boldsymbol{F}_1$ in (1.1), and the following coupling mode model is established:

$$\begin{cases} \frac{d\boldsymbol{a}_1}{dt} = [-\Gamma_{10} + g_1(|\boldsymbol{a}_1|) + j\omega_1]\boldsymbol{a}_1 - j\kappa \boldsymbol{a}_2 \\ \frac{d\boldsymbol{a}_2}{dt} = -j\kappa \boldsymbol{a}_1 + [-(\Gamma_{20} + \Gamma_L) + j\omega_2]\boldsymbol{a}_2 \end{cases}, \tag{1.3}$$

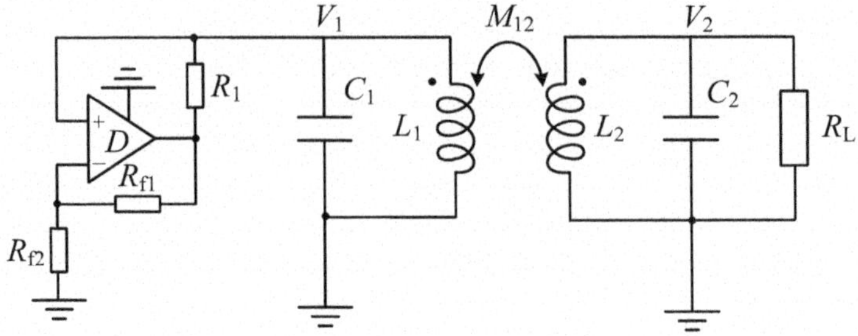

Fig. 1.16 PT-symmetric WPT system proposed by Stanford University

where g_1 is nonlinear gain rate related to the nonlinear resistance R_N, Ref. [39] further states that the system will eventually operate stably at a certain negative resistance value of the R_N to feed energy into the system. Moreover, if the system satisfies the resonant matching $\omega_1 = \omega_2$, the system can automatically operate in the PT-symmetric state within a certain transfer distance.

The concept of PT symmetry is derived from quantum physics, which refers to the invariance of a system under the inversion of the parity inverse (P: $\boldsymbol{a}_1 \leftrightarrow \boldsymbol{a}_2$) and time (T: $t \rightarrow -t, j \rightarrow -j$) inverse evolution. When the system is in a PT-symmetric state, the eigenvalues of the system are all pure imaginary numbers, which means that despite the loss of the system, the eigenstate changes of the mode are conservative and do not decay or enhance with time, and the energy distribution of the mode on both the transmitting and receiving sides is also mirror-symmetrical. However, when the system is in the broken PT-symmetric state, the conservatism is destroyed, and the eigenstate amplitude of the model increases exponentially or decays exponentially. When the system satisfies $\omega_1 = \omega_2$, the negative resistance region of R_N limits the voltage of capacitance C_2 not to exceed its saturation value, thus also restricts the amplitude of the transmitter side mode $|\boldsymbol{a}_1|$ from increasing indefinitely, so for the strong coupled region ($\kappa_{12} > \Gamma_{20} + \Gamma_L$), the system can run stably in the PT-symmetric state, at this time, the nonlinear gain corresponding to R_N is stable at the saturation value, and the system satisfies the PT-symmetric condition:

$$g_{1,\text{sat}} - \Gamma_{10} = \Gamma_{20} + \Gamma_L,\ \omega_1 = \omega_2, \tag{1.4}$$

When the nonlinear PT-symmetric circuit is applied to the WPT system, the system will show that the energy of the transmitter side and the receiver side modes is always equal in the range of the transfer distance satisfying $\kappa_{12} > \Gamma_{20} + \Gamma_L$, so that the transfer efficiency and output power of the system are constant and independent of the transfer distance. therefore, compared with the inductive, resonant, and microwave WPT systems, the PT-symmetric WPT system has a simple circuit structure, and can maintain constant power and efficiency in a wide transfer range, which greatly enhances the practicability of WPT, and is a new research direction and breakthrough with broad prospects for WPT technology.

Since 2017, when Prof. Shanhui Fan first proposed a PT-symmetric MC-WPT system as shown in Fig. 1.16, researchers have begun to try to optimize the system. Prof. Bo Zhang of South China University of Technology used a power electronic converter to construct a nonlinear negative resistor to expand the power level of the PT-symmetric MC-WPT system to the kilowatt level, and introduced PT symmetry into the electric-field coupling WPT system and the electromagnetic field hybrid coupling WPT system, which greatly expanded the application scope and prospect of the WPT system [83, 103–105]. The above WPT systems are all second-order PT-symmetric systems, and in the strong coupling region, the operating frequency of the system follows the intrinsic frequency of the PT-symmetric WPT system, which must vary with the coupling coefficient. M. Sakhdari et al. and the team of Professor Li Yunhui of Tongji University conducted independent research and found that the third-order PT-symmetric WPT system, which is composed of the transmitter oscillator, the relay lossless oscillator, and the receiver oscillator, has an intrinsic frequency of PT-symmetric state independent of the coupling rate, which can overcome the shortcomings of the variable frequency operation of the second-order PT-symmetric system [106, 107]. However, this system requires a very small loss of the relay harmonic oscillator, and the coupling rate between the transmitter-relay oscillator and the relay-receiver oscillator are always equal, which increases the difficulty of the system implementation.

1.4 Summary

This chapter first introduces the development of the WPT system, which can be traced back to the discovery of electromagnetic waves, expounds the research status of the application of the WPT system and the classification of the WPT system, and analyzes the problems that need to be solved urgently in the WPT system. Then, the introduction of the concept of PT symmetry into the WPT system is elaborated. Finally, the chapter arrangement of this book is described.

References

1. Na W, Park J, Lee C, Park K, Kim J, Cho S (2018) Energy-efficient mobile charging for wireless power transfer in internet of things networks. IEEE Internet Things J 5(1):79–92
2. Lee H, Lee KJ, Kim H, Lee I (2018) Wireless information and power exchange for energy-constrained device-to-device communications. IEEE Internet Things J 5(4):3175–3185
3. Ikeda T, Hasaba R, Tanaka Y, Tani H, Shinohara N (2022) Application of wireless power transfer to wearable IoT devices. Wirel Power Week (WPW) Bordeaux, France 218–222
4. Li Y, Chen Y, Chen CS, Wang Z, Zhu YH (2018) Charging while moving: deploying wireless chargers for powering wearable devices. IEEE Trans Veh Technol 67(12):11575–11586
5. Kim J, Kim DH, Choi J, Kim KH, Park YJ (2015) Free-positioning wireless charging system for small electronic devices using a bowl-shaped transmitting coil. IEEE Trans Microw Theory Tech 63(3):791–800

6. Liu H et al (2018) Dynamic wireless charging for inspection robots based on decentralized energy pickup structure. IEEE Trans Industr Inf 14(4):1786–1797
7. Manoufali M, Bialkowski K, Mohammed B, Abbosh A (2018) Wireless power link based on inductive coupling for brain implantable medical devices. IEEE Antennas Wirel Propag Lett 17(1):160–163
8. Yan Z, Song B, Zhang Y, Zhang K, Mao Z, Hu Y (2019) A rotation-free wireless power transfer system with stable output power and efficiency for autonomous underwater vehicles. IEEE Trans Power Electron 34(5):4005–4008
9. Insights GM (2022) Wireless charging market by technology and industry vertical—global opportunity analysis and industry forecast 2021–2030. Available: https://www.prnewswire.com/news-releases/insights-on-the-wireless-charging-global-market-to-2030-featuring-qualcomm-sony-and-texas-instruments-among-others-301464012.html
10. Chu J (2008) 10 breakthrough technologies 2008: wireless power. https://www.technologyreview.com/10-breakthrough-technologies/2008/
11. Tesla N (1891) Experiments with alternate currents of very high frequency and their application to methods of artificial illumination. In: Transactions of the American Institute of Electrical Engineers, vol VIII, no 1, pp 266–319
12. Wheeler LP (1943) II—Tesla's contribution to high frequency. Electr Eng 62(8):355–357
13. Hutin M, Leblanc M (1894) Transformer system for electric railways. U.S. Patent: 527857
14. Brown WC (1966) The microwave powered helicopter. J Microw Power 1(1):1–20
15. Brown WC, Eves EE (1992) Beamed microwave power transmission and its application to space. IEEE Trans Microw Theory Tech 40(6):1239–1250
16. Shinohara N (2014) Wireless power transfer via radiowaves. Wiley-ISTE, Hoboken
17. Glaser PE (1973) Method and apparatus for converting solar radiation to electrical power, U.S. Patent: 3781647
18. Rim CT, Mi C (2017) Wireless power transfer for electric vehicles and mobile devices. John Wiley & Sons, New York
19. Landis GA (1991) Space power by ground-based laser illumination. IEEE Aerosp Electron Syst Mag 6(11):3–7
20. Jin K, Zhou W (2019) Wireless laser power transmission: a review of recent progress. IEEE Trans Power Electron 34(4):3842–3859
21. Coltman JW (2002) The transformer [historical overview]. IEEE Ind Appl Mag 8(1):8–15
22. Boys JT, Covic GA (2015) The inductive power transfer story at the University of Auckland. IEEE Circ Syst Mag 15(2):6–27
23. Hui SYR, Ho WWC (2005) A new generation of universal contactless battery charging platform for portable consumer electronic equipment. IEEE Trans Power Electron 20(3):620–627
24. Agarwal K, Jegadeesan R, Guo YX, Thakor NV (2017) Wireless power transfer strategies for implantable bioelectronics. IEEE Rev Biomed Eng 10:136–161
25. Greatbatch W, Holmes CF (1991) History of implantable devices. IEEE Eng Med Biol Mag 10(3):38–41
26. Cochran GVB, Johnson MW, Kadaba MP, Vosburgh F, Ferguson-Pell MW, Palmeiri VR (1985) Piezoelectric internal fixation devices: a new approach to electrical augmentation of osteogenesis. J Orthop Res 3(4):508–513
27. Kurs A, Karalis A, Moffatt R, Joannopoulos JD, Fisher P, Soljačić M (2007) Wireless power transfer via strongly coupled magnetic resonances, vol 317, no 5834, pp 83–86
28. Choi BH, Park BC, Lee JH (2015) Near-field beamforming loop array for selective wireless power transfer. IEEE Microw Wirel Compon Lett 25(11):748–750
29. Choi BH, Thai VX, Lee ES, Kim JH, Rim CT (2016) Dipole-coil-based wide-range inductive power transfer systems for wireless sensors. IEEE Trans Industr Electron 63(5):3158–3167
30. Zhang W, Mi CC (2016) Compensation topologies of high-power wireless power transfer systems. IEEE Trans Veh Technol 65(6):4768–4778
31. Li S, Mi CC (2015) Wireless power transfer for electric vehicle applications. IEEE J Emerg Sel Top Power Electron 3(1):4–17

32. Kan T, Nguyen TD, White JC, Malhan RK, Mi CC (2017) A new integration method for an electric vehicle wireless charging system using LCC compensation topology: analysis and design. IEEE Trans Power Electron 32(2):1638–1650
33. Dai X, Sun Y (2014) An accurate frequency tracking method based on short current detection for inductive power transfer system. IEEE Trans Industr Electron 61(2):776–783
34. Su YG, Zhang HY, Wang ZH, Hu AP, Chen L, Sun Y (2015) Steady-state load identification method of inductive power transfer system based on switching capacitors. IEEE Trans Power Electron 30(11):6349–6355
35. Li Z, Zhu C, Jiang J, Song K, Wei G (2017) A 3 kW wireless power transfer system for sightseeing car supercapacitor charge. IEEE Trans Power Electron 32(5):3301–3316
36. Yin N et al (2012) Analysis of wireless energy transmission for implantable device based on coupled magnetic resonance. IEEE Trans Magn 48(2):723–726
37. Tan L et al (2020) Power stability optimization design of three-dimensional wireless power transmission system in multi-load application scenarios. IEEE Access 8:91843–91854
38. Yan K, Chen Q, Hou J, Ren X, Ruan X (2014) Self-oscillating contactless resonant converter with phase detection contactless current transformer. IEEE Trans Power Electron 29(8):4438–4449
39. Assawaworrarit S, Yu X, Fan S (2017) Robust wireless power transfer using a nonlinear parity-time-symmetric circuit. Nature 546(7658):387–390
40. Jiang Y, Zhang B (2020) A fractional-order wireless power transfer system insensitive to resonant frequency. IEEE Trans Power Electron 35(5):5496–5505
41. Hamam RE, Karalis A, Joannopoulos JD, Soljačić M (2009) Efficient weakly-radiative wireless energy transfer: an EIT-like approach. Ann Phys 324(8):1783–1795
42. Ra'Di Y et al (2018) On-site wireless power generation. IEEE Trans Antennas Propag 66(8):4260–4268
43. Krasnok AE, Baranov DG, Generalov A, Li S, Alú A (2017) Coherently enhanced wireless power transfer. Phys Rev Lett 120(14):143901
44. Mao F, Lu Y, Martins RP (2019) A reconfigurable cross-connected wireless-power transceiver for bidirectional device-to-device wireless charging. IEEE J Solid-State Circ 54(9):2579–2589
45. Sondhi K et al (2019) Flexible screen-printed coils for wireless power transfer using low-frequency magnetic fields. J Micromech Microeng 29(8):084006
46. Hua R, Hu AP (2019) Modeling and analysis of inductive power transfer system with passive matrix power repeater. IEEE Trans Industr Electron 66(6):4406–4413
47. Nagendra GR, Covic GA, Boys JT (2017) Sizing of inductive power pads for dynamic charging of evs on ipt highways. IEEE Trans Transp Electrification 3(2):405–417
48. Pearce MGS, Covic GA, Boys JT (2019) Robust Ferrite-Less Double D Topology for Roadway IPT Applications. IEEE Trans Power Electron 34(7):6062–6075
49. Tejeda A, Kim S, Lin FY, Covic GA, Boys JT (2019) A hybrid solenoid coupler for wireless charging applications. IEEE Trans Power Electron 34(6):5632–5645
50. Choi BG, Sohn YH, Lee ES, Han SH, Kim HR, Rim CT (2019) Coreless transmitting coils with conductive magnetic shield for wide-range ubiquitous IPT. IEEE Trans Power Electron 34(3):2539–2552
51. Jeong SY, Park JH, Hong GP, Rim CT (2019) Autotuning control system by variation of self-inductance for dynamic wireless EV charging with small air gap. IEEE Trans Power Electron 34(6):5165–5174
52. Thai VX, Jang GC, Jeong SY, Park JH, Kim YS, Rim CT (2020) Symmetric sensing coil design for the blind-zone free metal object detection of a stationary wireless electric vehicles charger. IEEE Trans Power Electron 35(4):3466–3477
53. Beh TC, Kato M, Imura T, Oh S, Hori Y (2013) Automated impedance matching system for robust wireless power transfer via magnetic resonance coupling. IEEE Trans Industr Electron 60(9):3689–3698
54. Jiwariyavej V, Imura T, Hori Y (2015) Coupling coefficients estimation of wireless power transfer system via magnetic resonance coupling using information from either side of the system. IEEE J Emerg Sel Top Power Electron 3(1):191–200

55. Koh KE, Beh TC, Imura T, Hori Y (2014) Impedance matching and power division using impedance inverter for wireless power transfer via magnetic resonant coupling. IEEE Trans Ind Appl 50(3):2061–2070
56. Cheng C et al (2020) A multiload inductive power transfer repeater system with constant load current characteristics. IEEE J Emerg Select Top Power Electron 8(4):3533–3541
57. Lu F, Zhang H, Hofmann H, Su W, Mi CC (2018) A dual-coupled LCC-compensated IPT system with a compact magnetic coupler. IEEE Trans Power Electron 33(7):6391–6402
58. Zhang Y, Yan Z, Liang Z, Li S, Mi CC (2020) A high-power wireless charging system using LCL-N topology to achieve a compact and low-cost receiver. IEEE Trans Power Electron 35(1):131–137
59. Arteaga JM, Aldhaher S, Kkelis G, Kwan C, Yates DC, Mitcheson PD (2019) Dynamic capabilities of multi-mhz inductive power transfer systems demonstrated with batteryless drones. IEEE Trans Power Electron 34(6):5093–5104
60. Lawson J, Yates DC, Mitcheson PD (2019) High Q coil measurement for inductive power transfer. IEEE Trans Microw Theory Tech 67(5):1962–1973
61. Pacini A, Costanzo A, Aldhaher S, Mitcheson PD (2017) Load- and position-independent moving MHz WPT system based on GaN-distributed current sources. IEEE Trans Microw Theory Tech 65(12):5367–5376
62. Colak K, Asa E, Bojarski M, Czarkowski D, Onar OC (2015) A novel phase-shift control of semibridgeless active rectifier for wireless power transfer. IEEE Trans Power Electron 30(11):6288–6297
63. Miller JM, Onar OC, Chinthavali M (2015) Primary-side power flow control of wireless power transfer for electric vehicle charging. IEEE J Emerg Sel Top Power Electron 3(1):147–162
64. Onar OC, Chinthavali M, Campbell SL, Seiber LE, White CP (2019) Vehicular integration of wireless power transfer systems and hardware interoperability case studies. IEEE Trans Ind Appl 55(5):5223–5234
65. Pries J, Galigekere VPN, Onar OC, Su GJ (2020) A 50 kW three-phase wireless power transfer system using bipolar windings and series resonant networks for rotating magnetic fields. IEEE Trans Power Electron 35(5):4500–4517
66. Sasatani T, Chabalko MJ, Kawahara Y, Sample AP (2017) Multimode quasistatic cavity resonators for wireless power transfer. IEEE Antennas Wirel Propag Lett 16:2746–2749
67. Hui SYR (2018) Technical and safety challenges in emerging trends of near-field wireless power transfer industrial guidelines. IEEE Electromagn Compat Mag 7(1):78–86
68. Witricity (2020) Automative solution. https://witricity.com/products/automotive/
69. Galigekere VP et al (2018) Design and implementation of an optimized 100 kW stationary wireless charging system for EV battery recharging. In: 2018 IEEE energy conversion congress and exposition (ECCE), pp 3587–3592
70. Chae SH, Jeong C, Lim SH (2018) Simultaneous wireless information and power transfer for internet of things sensor networks. IEEE Internet Things J 5(4):2829–2843
71. Orekan T, Zhang P (2018) Underwater wireless power transfer: smart ocean energy converters. Springer
72. Mi CC, Buja G, Choi SY, Rim CT (2016) Modern advances in wireless power transfer systems for roadway powered electric vehicles. IEEE Trans Industr Electron 63(10):6533–6545
73. Mahesh A, Chokkalingam B, Mihet-Popa L (2021) Inductive wireless power transfer charging for electric vehicles. IEEE Access 9:137667–137713
74. Erel MZ, Bayindir KC, Aydemir MT, Chaudhary SK, Guerrero JM (2022) A comprehensive review on wireless capacitive power transfer technology: fundamentals and applications. IEEE Access 10:3116–3143
75. Roes MGL, Duarte JL, Hendrix MAM, Lomonova EA (2013) Acoustic energy transfer: a review. IEEE Trans Industr Electron 60(1):242–248
76. Kim J, Jeong J (2015) Range-adaptive wireless power transfer using multiloop and tunable matching techniques. IEEE Trans Industr Electron 62(10):6233–6241
77. Lee G, Waters BH, Shin YG, Smith JR, Park WS (2016) A reconfigurable resonant coil for range adaptation wireless power transfer. IEEE Trans Microw Theory Tech 64(2):624–632

78. Aldhaher S, Luk PCK, Whidborne JF (2014) Electronic tuning of misaligned coils in wireless power transfer systems. IEEE Trans Power Electron 29(11):5975–5982
79. Tian J, Hu AP (2017) A DC-voltage-controlled variable capacitor for stabilizing the ZVS frequency of a resonant converter for wireless power transfer. IEEE Trans Power Electron 32(3):2312–2318
80. Sample AP, Meyer DT, Smith JR (2011) Analysis, experimental results, and range adaptation of magnetically coupled resonators for wireless power transfer. IEEE Trans Industr Electron 58(2):544–554
81. Gati E, Kampitsis G, Manias S (2017) Variable frequency controller for inductive power transfer in dynamic conditions. IEEE Trans Power Electron 32(2):1684–1696
82. Yan Z, Wu L, Baoyun W (2018) High-efficiency coupling-insensitive wireless power and information transmission based on the phase-shifted control. IEEE Trans Power Electron 33(9):7821–7831
83. Zhou J, Zhang B, Xiao W, Qiu D, Chen Y (2019) Nonlinear parity-time-symmetric model for constant efficiency wireless power transfer: application to a drone-in-flight wireless charging platform. IEEE Trans Industr Electron 66(5):4097–4107
84. Assawaworrarit S, Fan S (2020) Robust and efficient wireless power transfer using a switch-mode implementation of a nonlinear parity-time symmetric circuit. Nat Electron 3(5):273–279
85. Buja G, Bertoluzzo M, Mude KN (2015) Design and experimentation of WPT charger for electric city car. IEEE Trans Industr Electron 62(12):7436–7447
86. Qu X, Jing Y, Han H, Wong SC, Tse CK (2017) Higher order compensation for inductive-power-transfer converters with constant-voltage or constant-current output combating transformer parameter constraints. IEEE Trans Power Electron 32(1):394–405
87. Vu VB, Tran DH, Choi W (2018) Implementation of the constant current and constant voltage charge of inductive power transfer systems with the double-sided LCC compensation topology for electric vehicle battery charge applications. IEEE Trans Power Electron 33(9):7398–7410
88. Matsumoto H, Shibako Y, Neba Y (2018) Contactless power transfer system for AGVs. IEEE Trans Industr Electron 65(1):251–260
89. Li Z, Song K, Jiang J, Zhu C (2018) Constant current charging and maximum efficiency tracking control scheme for supercapacitor wireless charging. IEEE Trans Power Electron 33(10):9088–9100
90. Cai H, Shi L, Li Y (2014) Harmonic-based phase-shifted control of inductively coupled power transfer. IEEE Trans Power Electron 29(2):594–602
91. Berger A, Agostinelli M, Vesti S, Oliver JA, Cobos JA, Huemer M (2015) A wireless charging system applying phase-shift and amplitude control to maximize efficiency and extractable power. IEEE Trans Power Electron 30(11):6338–6348
92. Song K, Li Z, Jiang J, Zhu C (2018) Constant current/voltage charging operation for series-series and series-parallel compensated wireless power transfer systems employing primary-side controller. IEEE Trans Power Electron 33(9):8065–8080
93. Qing X, Chen ZN (2007) Proximity effects of metallic environments on high frequency RFID reader antenna: study and applications. IEEE Trans Antennas Propag 55(11):3105–3111
94. Redman-White W, Kennedy H, Bodnar R, Lee T (2017) Adaptive tuning of large-signal resonant circuits using phase-switched fractional capacitance. IEEE Trans Circ Syst II Express Briefs 64(9):1072–1076
95. Zhang P, Yang Q, Zhang X, Li Y, Li Y (2017) Comparative study of metal obstacle variations in disturbing wireless power transmission system. IEEE Trans Magn 53(6):1–4
96. Li B, Xiang G, Li C, Zang K (2017) Detuning mechanism and frequency tracking algorithm for wireless power transmission system. J Eng Sci Technol Rev 10:100–108
97. Tan L, Li J, Chen C, Yan C, Guo J, Huang X (2016) Analysis and performance improvement of WPT systems in the environment of single non-ferromagnetic metal plates. Energies 9:576
98. Yu X, Skauli T, Skauli B, Sandhu S, Catrysse PB, Fan S (2013) Wireless power transfer in the presence of metallic plates: experimental results. AIP Adv 3(6):062102
99. Hui SYR (2016) Past, present and future trends of non-radiative wireless power transfer. CPSS Trans Power Electron Appl 1(1):83–91

100. Tesla N (1999) High frequency oscillators for electro-therapeutic and other purposes. Proc IEEE 87(7):1282
101. Moffatt RA (2009) Wireless transfer of electric power. Massachusetts Institute of Technology, Cambridge
102. Niu WQ, Chu JX, Gu W, Shen AD (2013) Exact analysis of frequency splitting phenomena of contactless power transfer systems. IEEE Trans Circ Syst I Regul Pap 60(6):1670–1677
103. Wu L, Zhang B, Zhou J (2020) Efficiency improvement of the parity-time-symmetric wireless power transfer system for electric vehicle charging. IEEE Trans Power Electron 35(11):12497–12508
104. Qiu D et al (2024) A robust parity-time-symmetric hybrid wireless power transfer system with extended coupling range. Int J Circ Theory Appl 1–16
105. Gu W, Qiu D, Shu X, Zhang B, Xiao W, Chen Y (2023) A constant output capacitive wireless power transfer system based on parity-time symmetric. IEEE Trans Circ Syst II Express Briefs 70(7):2585–2589
106. Sakhdari M, Hajizadegan M, Chen P-Y (2020) Robust extended-range wireless power transfer using a higher-order PT-symmetric platform. Phys Rev Res 2:013152
107. Zeng C et al (2020) High-order parity-time symmetric model for stable three-coil wireless power transfer. Phys Rev Appl 13:034054

Chapter 2
PT-Symmetric Principle and Circuits

The concept of PT symmetry, which originated from the field of quantum mechanics, was proposed in the study of quantum operators and their energy spectra, and has been intensively investigated in quantum field theory, optics, acoustics [1–4]. In quantum mechanics, Hamiltonian H plays a very important role, which generally acts on a system and is an observable physical quantity that reflects the total energy of the system. Meanwhile, the measured value of the observables is equal to the eigenvalue of its corresponding Hamiltonians [5, 6]. If the physical quantity can be observed, the measured value of the observables must be real, that is, the eigenvalue of Hamiltonian H corresponding to the observables must be real. It is generally believed that the Hamiltonian H of a system is Hermitian, so that the eigenvalues of the Hamiltonian H are all real, otherwise its eigenvalues will appear complex. However, in 1988, Bender and Boettcher discovered that Hermiticity is not a necessary condition for Hamiltonian H to have real eigenvalues [7–9]. In fact, a whole class of Hamiltonians can have real eigenvalues without being Hermitian, if they are PT-symmetric. When the characteristic equation of a system is invariant under the combined operation of parity (P) reversal and time (T) reversal operations, the PT-symmetric condition is satisfied. Remarkably, there is a critical value of the parameter in such systems [10]. When the parameter that controls the degree of non-Hermiticity is less than the critical value, the system is in the PT-symmetric region and all eigenvalues are real [11]. Conversely, beyond this critical threshold, the system enters the PT-symmetric broken regime with complex eigenvalues. This critical value is called the exceptional point (EP), where both the eigenvalues and the corresponding eigenstates of the system coalesce [12–14]. These discoveries extend people's attention from Hermitian systems to non-Hermitian systems.

B. Zhang et al., *Parity-Time Symmetric Wireless Power Transfer*, CPSS Power Electronics Series, https://doi.org/10.1007/978-981-96-3149-0_2

2.1 Principle

2.1.1 Parity Reversal

Parity (P) reversal is defined by spatial inversion, also known as spatial reflection. Classically, P is the reflection of position vectors through the origin, that is, in three dimension, it refers to the simultaneous flip in the sign of all three spatial coordinates or a point reflection [15, 16].

Taking a single continuous coordinate as an example, P denotes a reflection $x \rightarrow -x$ across one spatial axis. Quantities depending on odd powers of x or its derivatives are consequently inverted as well. In fact, P reversal can contain coordinate transformations on three spatial axes, and only coordinate transformation under one of them is discussed here.

In the field of electronics, the wavelengths in the system are long, so spatial considerations can be eliminated, and spatial symmetry is reduced to a matter of circuit topology. Therefore, P reversal is defined by exchanging the labels of the left side and the right side in the circuit configuration [17]. For this reason, P-reversal symmetry of circuit systems is much easier to achieve than that in a real spatial system.

In Hamiltonian dynamics, P reversal takes

$$j \rightarrow j,\ \hat{x} \rightarrow -\hat{x},\ \hat{p} \rightarrow -\hat{p}, \tag{2.1}$$

where $\hat{x}$ and $\hat{p}$ are position and momentum operators respectively, j represents the imaginary number, and H is P-symmetric if $[H, P] = 0$ [7, 18].

2.1.2 Time Reversal

Time (T) reversal refers to inverting the flow of time, and can be viewed as a mathematical operation that replaces the expression for time with its negative in formulas or equations so that they describe an event in which time runs backward or all the motions are reversed [17]. This backward propagation is achieved by the operation of $t \rightarrow -t$, and time derivatives $\frac{d^n}{dt^n}$ of odd order are also inverted by this process.

In electronics, T reversal effectively reverses the sign of resistive impedances, while the sign of reactive impedances remains unchanged. Applying T to Ohm's law of $U(t) = I(t)Z$ yields

$$U(-t) = -I(-t)Z, \tag{2.2}$$

where the sign of I is reversed because the current is the time derivative of charge.

For resistors, it satisfies $U = IR$, then the T-reversal statement can be further written as

$$Ue^{j\omega(-t)} = -Ie^{j\omega(-t)}R. \tag{2.3}$$

For the negative resistor $-R$, it satisfies $U = -IR$ and its T-reversal statement is

$$Ue^{j\omega(-t)} = Ie^{j\omega(-t)}R. \tag{2.4}$$

For reactive components, such as inductors and capacitors, taking an additional derivative causes the negative sign to cancel out. This characteristic is related to the fact that resistors are power-dissipating components. Energy is dissipated in a resistor in normal time, and when time is reversed, energy is drawn out from the resistor.

In Hamiltonian dynamics, T reversal takes

$$j \to -j,\ \hat{x} \to \hat{x},\ \hat{p} \to -\hat{p}. \tag{2.5}$$

Here, H is T-symmetric if $[H, P] = 0$ [7].

Therefore, H is PT-symmetric when it satisfies $[H, PT] = 0$, $[H, P] \neq 0$ and $[H, T] \neq 0$.

2.1.3 PT Symmetry

Systems which are invariant under the combined action of the P and T-reversal operations, but not under P or T reversal individually, are PT-symmetric. System exhibiting PT symmetry are generally realized by adding balanced energy gain and energy loss mechanisms. When the gain and loss mechanisms have the same magnitude but opposite signs, the system is conservative, in which the energy flows steadily through the mechanisms to balance. However, when the strength of this balanced non-conservation factor increases continuously, there will be a spontaneous transition from energy conservative to a non-conservative state. This phenomenon actually corresponds to the spontaneous PT symmetry breaking of the solution of the symmetric system, and the transition of the system from PT-symmetric to PT-symmetric breaking is also called exceptional point (EP) [19].

The existence of the PT-symmetric region and the conservation of the system are of great significance. It demonstrates that Hamiltonians with PT symmetry can have real eigenvalues even if they are not Hermitian. However, this contradicts the normal criterion of quantum mechanics, which states that the measured energies of a physical system can be real only if the Hamiltonian is Hermitian, which is obviously inconsistent with the properties of PT-symmetric systems. The Hamiltonian of the PT-symmetric system is the first known non-Hermitian Hamiltonian with this property.

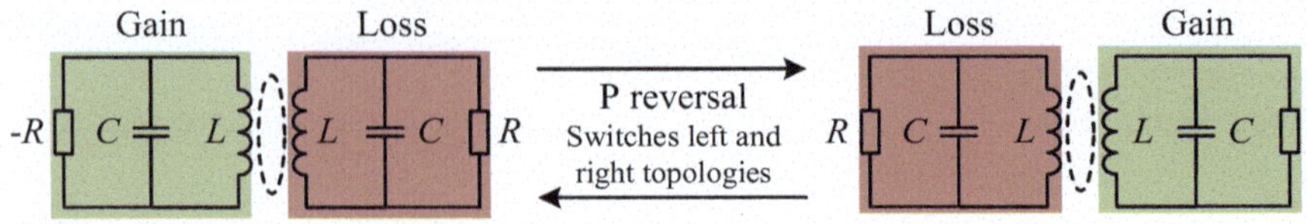

Fig. 2.1 P reversal

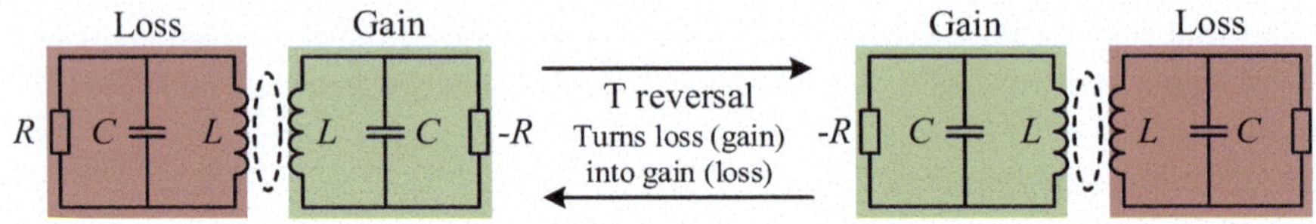

Fig. 2.2 T reversal

2.2 Representation of PT Symmetry in Circuits

2.2.1 *Topological Symmetry*

Compared with other application fields, the application of PT symmetry in circuits has an obvious advantage, that is, in the circuit, the wavelength of electromagnetic waves is much larger than the size of the circuit. This limitation allows all considerations of parity or spatial symmetry to be reduced to the symmetry of the circuit topology described by Kirchhoff's laws. Topological symmetry can be easily achieved as long as the system has the desired node topologies and appropriate circuit parameters. Specifically, P inversion of the circuit can be realized by swapping the indices corresponding to the gain side ($n = 1$) and loss side ($n = 2$) [17], that is, $n = 1 \leftrightarrow n = 2$, which is characterized by topological symmetry in the circuit, as shown in Fig. 2.1.

Resistors, inductors, and capacitors are the most basic components in circuits, in which resistors become negative resistors that produces power after T inversion, due to its natural characteristic of dissipating power. Inductors and capacitors remain unchanged after T inversion, because the relationship between voltage and current is related to the time derivative, and the same Kirchhoff relationship is restored after T inversion, as shown in Fig. 2.2.

2.2.2 *Negative Resistance and Negative Impedance*

The simplest conception of a resistor is defined by Ohm's law for a DC voltage. Generally speaking, according to Ohm's law, the potential difference (i.e., voltage U) across a resistor is directly proportional to the current I flowing through it. The current in a resistor flows from high potential to low potential, which is defined as the

Fig. 2.3 Negative resistor

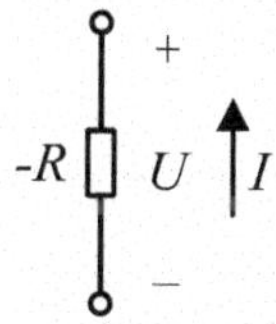

reference positive direction of the current, so the I-V curve of a resistor is a straight line through the origin with a positive slope $1/R$. This leads to the mathematical equation of $R = U/I$, where R is the resistance in ohms (Ω), U is the voltage in volts (V), and I is the current in amperes (A). For an AC voltage, Ohm's law for a resistor is also preserved. The AC impedance of an ideal resistor is real and independent of operating frequency, which means that the AC current I flowing through the resistor remains in phase with the AC voltage U across the resistor at each instant.

As mentioned above, the negative resistor is a regular resistor with a negative value, in which the current and voltage are inversely proportional to each other, as shown in Fig. 2.3. The negative resistor also obeys Ohm's law, but its current flows backwards, from low potential to high potential, and its I-U curve is a negatively sloped line. In other words, a normal circuit with a resistor following Ohm's law has a current drop when the voltage decreases, whereas in the case of negative resistor, the current increases with a voltage drop. For the AC voltage, the current flowing out of the negative resistor remains in phase with the voltage across the negative resistor for each cycle.

In addition to tunnel diodes and Gunn diodes, which have a negative resistance effect, the typical negative resistor is an active component that relies on other circuit components and power supplies to implement.

Similarly, negative impedance, like positive impedance, has two components: resistance and reactance. The reactance component can be either positive or negative, but the resistance component is always negative at some frequency in the range from zero to infinity. Therefore, the negative impedance can be written as

$$Z = -R + jX = |Z|\angle\theta, \tag{2.6}$$

where $-R$ is the negative resistance, X is the inductive or capacitive reactance, θ is the impedance angle with the range of $[\pi/2, \pi]$ and $[-\pi/2, -\pi]$.

2.3 PT-Symmetric Coupled RLC Oscillators

2.3.1 RLC Oscillators

The RLC oscillators, as shown in Fig. 2.4, including parallel RLC oscillator and series RLC oscillator, are the simple and well-understood devices for exploring the behavior of PT-symmetric systems [16, 20]. Notice that the results of this section extend to both positive and negative resistors. Both L and C are considered non-negative. Here, $G = 1/R$ is the conductance of the resistor R.

According to the reference direction of the current in Fig. 2.4a, the equation of motion for the parallel RLC oscillator can be given by KCL and VCR, as follows:

$$i_{\mathrm{G}}(t) + i_{\mathrm{C}}(t) + i_{\mathrm{L}}(t) = 0, \tag{2.7}$$

$$i_{\mathrm{C}}(t) = C\frac{du(t)}{dt}, \tag{2.8}$$

$$i_{\mathrm{G}}(t) = Gu(t), \tag{2.9}$$

$$u(t) = L\frac{di_{\mathrm{L}}(t)}{dt}. \tag{2.10}$$

Then, (2.7) can be rewritten as

$$\frac{d^2u(t)}{dt^2} + \frac{G}{C}\frac{du(t)}{dt} + \omega_0^2 u(t) = 0, \tag{2.11}$$

where $\omega_0 = 1/\sqrt{LC}$ is the natural angular frequency of the parallel RLC oscillator.

Changing the variable from time t to the phase $\varphi = \omega_0 t$, (2.11) can be further written as

$$\frac{d^2u(\varphi)}{d\varphi^2} + 2\gamma\frac{du(\varphi)}{d\varphi} + u(\varphi) = 0, \tag{2.12}$$

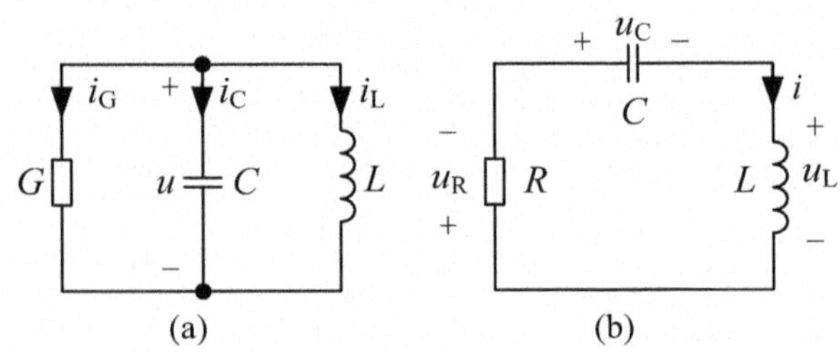

Fig. 2.4 RLC oscillators. **a** Parallel, **b** Series

where $\gamma = \frac{G}{2}\sqrt{\frac{L}{C}}$.

Assuming that $u = U_0 e^{j\omega\varphi}$, the solutions of (2.11) can be derived as

$$u(\varphi) = U_0 e^{\left(-\gamma \pm j\sqrt{1-\gamma^2}\right)\varphi}, \tag{2.13}$$

which have both a periodic oscillating part and an exponential part in the under-damped region $0 < |\gamma| < 1$. When γ is positive, the amplitude of the periodic oscillating part decays with time, and when γ is negative, the amplitude of the periodic oscillating part grows with time. As $|\gamma|$ increases toward one, the admittance of the resistor becomes very large, causing the rate of decay or growth to become faster than the rate of the periodic oscillating part. The sign of γ is the sign of resistor R.

At the natural frequency ω_0, the impedances of inductor and capacitor are equal in magnitude, that is

$$|j\omega_0 L| = \frac{1}{|j\omega_0 C|} = \sqrt{\frac{L}{C}}, \tag{2.14}$$

but the phases of inductive impedance and capacitive impedance are opposite, resulting in the impedances of the inductor and capacitor canceling each other, which means that the resonance occurs in the parallel RLC oscillator. The admittance of a parallel RLC oscillator is pure resistance, that is, $Y(\omega_0) = G$. Therefore, for a given voltage, the energy dissipated in the resistor is proportional to its conductance.

Moreover, in the resonant RLC circuits, the quality factor Q is defined as the ratio of the reactive power of the capacitor or inductor to the average power of the resistor at resonance, that is

$$Q = \frac{\omega_0 C}{G} = \frac{1}{G}\sqrt{\frac{C}{L}} = \frac{1}{2\gamma}. \tag{2.15}$$

The series RLC oscillator is the dual of the parallel RLC oscillator, so it can be analyzed in a similar way to the parallel RLC oscillator, which is not repeated here.

2.3.2 *Coupled RLC Oscillators and PT Symmetry*

The coupled RLC oscillator is composed of a pair of RLC units coupled through different media, which can be divided into three types: capacitive coupling, inductive coupling, and hybrid inductive-capacitive coupling. Meanwhile, the inductor and capacitor in each pair of RLC units can be connected in series or parallel, as shown in Figs. 2.5, 2.6 and 2.7. Here, the RLC unit connected to the negative resistor is defined as the primary side, which has the gain property and provides power. The other side with a positive resistor is called the secondary side, which has a lossy property and dissipates power. It is noted that the coupling is provided by a mutual inductance M or mutual capacitance C_M.

The coupled RLC oscillators are the simplest circuit implementation to demonstrate PT symmetry. Without loss of generality, this section takes the coupled RLC oscillators shown in Fig. 2.5 as examples to analyze the PT symmetry, which lays a foundation for the research of the PT-symmetric wireless power transfer system.

According to the circuit theory, the state equations of coupled RLC oscillators can be established by KCL, KVL, and VCR. For the coupled RLC oscillator with series connection of Fig. 2.5a, the circuit equations are obtained by KVL and VCR, as follows:

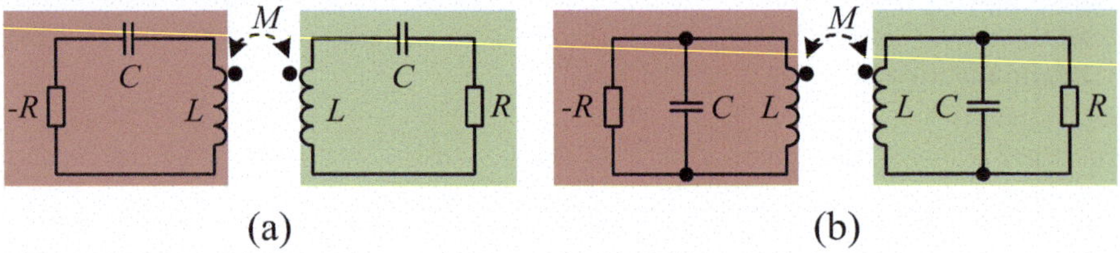

Fig. 2.5 Coupled RLC oscillator with inductive coupling. **a** Series connection, **b** Parallel connection

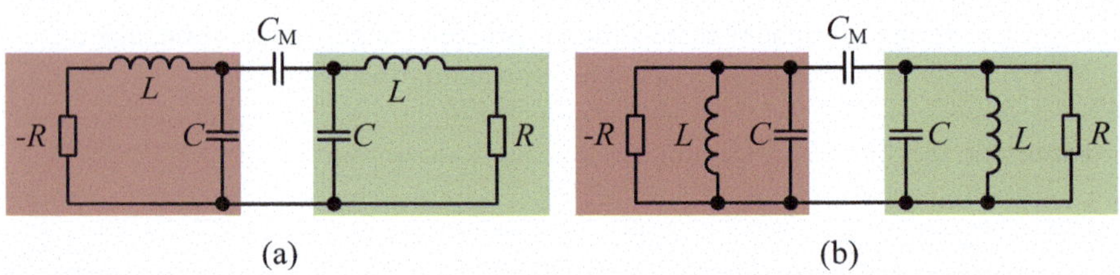

Fig. 2.6 Coupled RLC oscillator with capacitive coupling. **a** Series connection, **b** Parallel connection

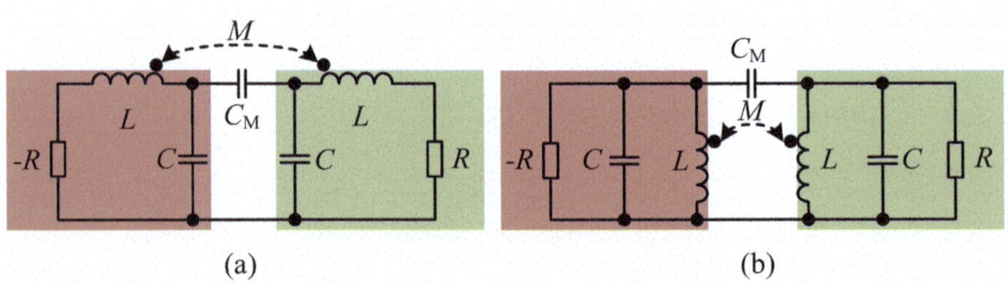

Fig. 2.7 Coupled RLC oscillator with hybrid inductive-capacitive coupling. **a** Series connection, **b** Parallel connection

$$\begin{cases} L\dfrac{di_1}{dt} + M\dfrac{di_2}{dt} - Ri_1 + u_1 = 0 \\ i_1 = C\dfrac{du_1}{dt} \\ L\dfrac{di_2}{dt} + M\dfrac{di_1}{dt} + Ri_2 + u_2 = 0 \\ i_2 = C\dfrac{du_2}{dt} \end{cases}, \tag{2.16}$$

where the subscripts correspond to the primary ($n = 1$) and secondary ($n = 2$) sides, i_1 and i_2 are currents flowing through the inductors of the primary and secondary sides, u_1 and u_2 are voltages across the capacitors of the primary and secondary sides, respectively.

Taking the charges $q_1 = Cu_1$ and $q_2 = Cu_2$ of the capacitors of the primary and secondary sides as state variables, which can be used to represent the energy stored in primary and secondary sides, respectively, (2.16) can be further rewritten into the form of state equations, as follows:

$$\begin{cases} \dfrac{d^2q_1}{dt^2} = \dfrac{R}{L(1-k^2)}\dfrac{dq_1}{dt} - \dfrac{\omega_0^2}{1-k^2}q_1 + \dfrac{Rk}{L(1-k^2)}\dfrac{dq_2}{dt} + \dfrac{\omega_0^2 k}{1-k^2}q_2 \\ \dfrac{d^2q_2}{dt^2} = -\dfrac{R}{L(1-k^2)}\dfrac{dq_2}{dt} - \dfrac{\omega_0^2}{1-k^2}q_2 - \dfrac{Rk}{L(1-k^2)}\dfrac{dq_1}{dt} + \dfrac{\omega_0^2 k}{1-k^2}q_1 \end{cases}, \tag{2.17}$$

where $k = M/L$ is the coupling coefficient, it is used to characterize the strength of the inductive coupling. $\omega_0 = 1/\sqrt{LC}$ is the natural angular frequency of the primary and secondary RLC oscillators, respectively.

Based on the representation of PT symmetry in circuits, P-reversal, T-reversal, and PT-reversal transformations of the coupled RLC circuit in Fig. 2.5a are performed as follows:

(1) P reversal of the coupled RLC circuit is realized by swapping the subscripts 1 and 2 in state equations, i.e., $q_1 \leftrightarrow q_2$, then the state equations of (2.17) become

$$\begin{cases} \dfrac{d^2q_2}{dt^2} = \dfrac{R}{L(1-k^2)}\dfrac{dq_2}{dt} - \dfrac{\omega_0^2}{1-k^2}q_2 + \dfrac{Rk}{L(1-k^2)}\dfrac{dq_1}{dt} + \dfrac{\omega_0^2 k}{1-k^2}q_1 \\ \dfrac{d^2q_1}{dt^2} = -\dfrac{R}{L(1-k^2)}\dfrac{dq_1}{dt} - \dfrac{\omega_0^2}{1-k^2}q_1 - \dfrac{Rk}{L(1-k^2)}\dfrac{dq_2}{dt} + \dfrac{\omega_0^2 k}{1-k^2}q_2 \end{cases}, \tag{2.18}$$

By comparing (2.17) and (2.18), it can be found that the coefficients of the first derivatives of q_1 and q_2 with respect to time have changed after the P-reversal operation, which means that the system characteristics are different from the original system. Consequently, P reversal of the coupled RLC circuit is essentially equivalent to swapping the space positions of the primary and secondary sides, while the fact that the negative resistor -R provides power and the positive resistor R dissipates power does not change.

(2) T reversal means running backward in time, i.e., $t \to -t$. Applying T reversal to (2.17) yields

$$\begin{cases} \dfrac{d^2q_1}{dt^2} = -\dfrac{R}{L(1-k^2)}\dfrac{dq_1}{dt} - \dfrac{\omega_0^2}{1-k^2}q_1 - \dfrac{Rk}{L(1-k^2)}\dfrac{dq_2}{dt} + \dfrac{\omega_0^2 k}{1-k^2}q_2 \\ \dfrac{d^2q_2}{dt^2} = \dfrac{R}{L(1-k^2)}\dfrac{dq_2}{dt} - \dfrac{\omega_2^2}{1-k^2}q_2 + \dfrac{Rk}{L(1-k^2)}\dfrac{dq_1}{dt} + \dfrac{\omega_0^2 k}{1-k^2}q_1 \end{cases}, \tag{2.19}$$

From (2.17) and (2.19), it can be seen that the negative resistor -R turns to positive resistor R, indicating that the original gain term in the coupled RLC circuit becomes loss term when time flows backward, and the positive resistor R turns to negative resistor $-R$, indicating that the original loss term in the coupled RLC circuit becomes gain term when time flows backward, which means that the characteristics of the original system are changed. Thus, T reversal reversals the direction of energy flow.

(3) PT-reversal transformation is a combined P-reversal (i.e., $q_1 \leftrightarrow q_2$) and T-reversal (i.e., $t \to -t$) transformations. Under the PT-reversal transformation, the state equations can be written as

$$\begin{cases} \dfrac{d^2q_2}{dt^2} = -\dfrac{R}{L(1-k^2)}\dfrac{dq_2}{dt} - \dfrac{\omega_0^2}{1-k^2}q_2 - \dfrac{Rk}{L(1-k^2)}\dfrac{dq_1}{dt} + \dfrac{\omega_0^2 k}{1-k^2}q_1 \\ \dfrac{d^2q_1}{dt^2} = \dfrac{R}{L(1-k^2)}\dfrac{dq_1}{dt} - \dfrac{\omega_2^2}{1-k^2}q_1 + \dfrac{Rk}{L(1-k^2)}\dfrac{dq_2}{dt} + \dfrac{\omega_0^2 k}{1-k^2}q_2 \end{cases}. \tag{2.20}$$

It can be found that the state Eq. (2.20) is completely consistent with the original Eq. (2.17), which represents that the characteristics of the system remain unchanged after the PT-reversal operation.

According to the PT symmetric principle, systems whose state equations are invariant under the combined P and T-reversal transformations are PT symmetric. Therefore, the coupled RLC circuit shown in Fig. 2.5a is PT symmetric.

Similarly, for the coupled RLC oscillator with parallel connection of Fig. 2.5b, the state equations are

$$\begin{cases} \dfrac{d^2q_1}{dt^2} = \dfrac{1}{RC}\dfrac{dq_1}{dt} - \dfrac{\omega_0^2}{1-k^2}q_1 + \dfrac{\omega_0^2 k}{1-k^2}q_2 \\ \dfrac{d^2q_2}{dt^2} = -\dfrac{1}{RC}\dfrac{dq_2}{dt} - \dfrac{\omega_0^2}{1-k^2}q_2 + \dfrac{\omega_0^2 k}{1-k^2}q_1 \end{cases}. \tag{2.21}$$

Then, applying PT-reversal transformation to (2.21) yields

$$\begin{cases} \dfrac{d^2q_2}{dt^2} = -\dfrac{1}{RC}\dfrac{dq_2}{dt} - \dfrac{\omega_0^2}{1-k^2}q_2 + \dfrac{\omega_0^2 k}{1-k^2}q_1 \\ \dfrac{d^2q_1}{dt^2} = \dfrac{1}{RC}\dfrac{dq_1}{dt} - \dfrac{\omega_0^2}{1-k^2}q_1 + \dfrac{\omega_0^2 k}{1-k^2}q_2 \end{cases}, \tag{2.22}$$

it can be seen that (2.22) and (2.21) are identical under the PT-reversal transformation, which demonstrates that the coupled RLC oscillator shown in Fig. 2.5b is PT symmetry.

Moreover, from the state Eqs. (2.21) and (2.17) of coupled RLC circuits with different topologies, it can also be found that the analysis of PT symmetry of coupled RLC oscillators by circuit theory is quite straightforward and easy to understand. However, the form of state equations of different circuit topologies is not the same, and the derivation is complicated. In order to solve this issue, a common approximate method, which is called coupled-mode theory, is proposed to simplify the mathematical model of complicated coupled systems weakly coupled by multiple RLC units.

In the field of physics, coupled-mode theory is used to study the general law of coupling between two or more electromagnetic wave modes, and intuitively reveal the flow process of energy between modes from the perspective of energy [21]. In particular, coupled-mode equations have a general formalism [22–24], regardless of the topology of each mode, as follows:

$$\frac{d\boldsymbol{a}_{\mathrm{m}}(t)}{dt} = (j\omega_{\mathrm{m}} - \Gamma_{\mathrm{m}})\boldsymbol{a}_{\mathrm{m}}(t) + \sum_{\mathrm{n}\neq\mathrm{m}} j\kappa_{\mathrm{mn}}\boldsymbol{a}_{\mathrm{n}}(t) + \boldsymbol{F}_{\mathrm{m}}(t), \tag{2.23}$$

where the indices denote the different RLC oscillator objects. The variables $\boldsymbol{a}_{\mathrm{m}}(t) = \sqrt{\frac{L_{\mathrm{m}}}{2}}i_{\mathrm{m}} + j\sqrt{\frac{C_{\mathrm{m}}}{2}}u_{\mathrm{m}}$ are modes, which are defined so that the energy stored in object m is $|\boldsymbol{a}_{\mathrm{m}}(t)|^2$. $\boldsymbol{F}_{\mathrm{m}}(t)$ are driving terms, which can be included in the first term to the right of (2.23) when it depends on $\boldsymbol{a}_{\mathrm{m}}(t)$. ω_{m} is the natural angular frequency of the isolated object m. Γ_{m} is the intrinsic decay rate of the object m, which is caused by absorption and radiated losses. $\kappa_{\mathrm{mn}} = \kappa_{\mathrm{mn}}$ are coupling rates between the oscillator objects indicated by the subscripts.

Therefore, the coupled-mode equations of coupled RLC oscillators shown in Figs. 2.5, 2.6 and 2.7 have a unified form, that is

$$\begin{cases} \dfrac{d\boldsymbol{a}_1}{dt} = (j\omega_1 + g_1)\boldsymbol{a}_1 + j\kappa\boldsymbol{a}_2 \\ \dfrac{d\boldsymbol{a}_2}{dt} = (j\omega_2 - \Gamma_2)\boldsymbol{a}_2 + j\kappa\boldsymbol{a}_1 \end{cases}. \tag{2.24}$$

Applying PT-reversal operation to (2.24) yields

$$\begin{cases} \dfrac{d\boldsymbol{a}_2}{dt} = (j\omega_1 - g_1)\boldsymbol{a}_2 + j\kappa\boldsymbol{a}_1 \\ \dfrac{d\boldsymbol{a}_1}{dt} = (j\omega_2 + \Gamma_2)\boldsymbol{a}_1 + j\kappa\boldsymbol{a}_2 \end{cases}. \tag{2.25}$$

Then, according to the PT-symmetric principle, (2.25) and (2.24) are identical under the following PT-symmetric conditions:

$$\omega_1 = \omega_2,\ g_1 = \Gamma_2, \tag{2.26}$$

It is worth noting that the coefficients of the coupled-mode equations corresponding to different topologies are different. For the RLC series topology, the expressions of the coefficients are $g_1 = \Gamma_2 = \frac{R}{2L}$, while the RLC parallel topology has $g_1 = \Gamma_2 = \frac{G}{2C}$. Furthermore, the coupling rate of inductive coupling is $\kappa = \frac{\sqrt{\omega_1\omega_2}M}{2\sqrt{L_1L_2}}$, while the coupling rate of capacitive coupling is $\kappa = \frac{\sqrt{\omega_1\omega_2}C_{\mathrm{m}}}{2\sqrt{C_1C_2}}$.

From above, it can be observed that the coupled RLC circuits with a symmetric topology and matched parameters are PT-symmetric.

2.4 Realization Method of Negative Resistors

In practice, the independent components of negative resistor have not yet been discovered, and the existing negative resistors are mainly constructed by other components. At present, there are two main realization methods of negative resistor, one is achieved by an operational amplifier, and the other is realized by a controlled power electronic inverter.

2.4.1 Operational Amplifier

Negative impedance converter (NIC) is a circuit that uses an operational amplifier to achieve negative impedance [25–27]. By replacing its impedances with resistances, a realization circuit of the negative resistor can be obtained [28, 29], as shown in Fig. 2.8. The output characteristics of the proposed negative resistor are described in Fig. 2.9. Here, U_{sat} is the saturation voltage of the operational amplifier, and $A = 1 + R_{\mathrm{f1}}/R_{\mathrm{f2}}$ is amplification factor (or gain factor) of the operational amplifying circuit.

From Fig. 2.9b, it can be seen that the nonlinear resistor behaves as a negative resistor under $|u_{\mathrm{NR}}| < U_{\mathrm{sat}}$, that is

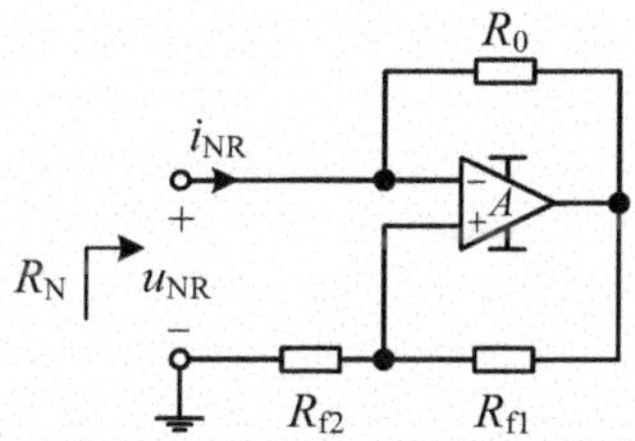

Fig. 2.8 Negative resistor realized by an operational amplifier

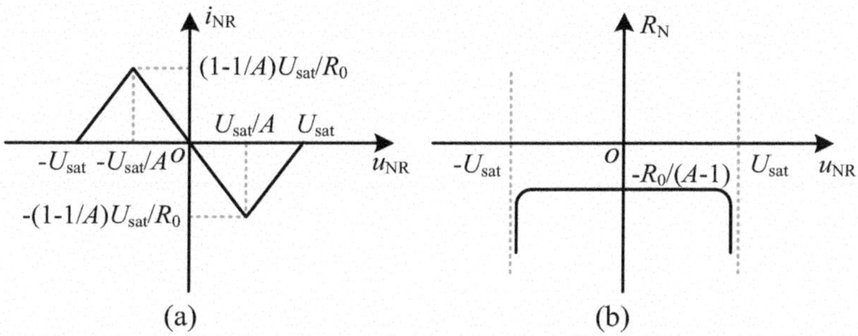

Fig. 2.9 Output characteristics of negative resistor R_N. **a** Voltage versus current, **b** Negative resistance versus voltage

$$R_N = \frac{u_{NR}}{i_{NR}} = \begin{cases} \dfrac{u_{NR}}{u_{NR} - U_{sat}} R_0, \; u_{NR} > \dfrac{U_{sat}}{A} \\ -\dfrac{R_{f2}}{R_{f1}} R_0, \; |u_{NR}| < \dfrac{U_{sat}}{A} \\ \dfrac{u_{NR}}{u_{NR} + U_{sat}} R_0, \; u_{NR} < -\dfrac{U_{sat}}{A} \end{cases} \tag{2.27}$$

Hence, the required negative resistance can be obtained by reasonably designing the values of positive resistors R_0, R_{f1}, and R_{f2}.

2.4.2 *Power Electronic Inverter*

Due to the limitation of the power level of the operational amplifier, the realization method of negative resistor with an operational amplifier can only apply to the mW-level power requirements, which is difficult to apply in the actual wireless power transfer system [28, 30–32]. Inspired by this, the realization method of a negative resistor using a power electronic inverter is proposed, which can drastically improve the power capacity and maximum efficiency compared with an operational amplifier.

According to the output characteristics of the negative resistor, that is, the output voltage and current are in phase when selecting a non-associated direction, there are two main ways to realize negative resistor by controlling the power electronic inverter. One method is to sample the phase signals of the output voltage u_{NR} and current i_{NR} of the power electronic inverter, and send them to a controller (such as phase-locked-loop, PLL) for phase comparison to generate PWM signals of different frequencies, so as to continuously track the operating frequencies that makes the phase difference between voltage u_{NR} and current i_{NR} zero, thereby finally realizing the negative resistance characteristic. This control approach is also known as zero-phase-angle (ZPA) control [33–35]. In addition, based on different external circuits, the power electronic inverter can adopt a voltage source inverter (VSI) or current source inverter (CSI), as shown in Fig. 2.10.

The other method is to use the zero-crossing detection of the output current of VSI or output voltage of CSI to generate driving signals, which are used to drive the corresponding switches, so that the output voltage and current of the inverter remain in phase at any time, as shown in Fig. 2.11. This control approach is also called self-oscillation control [36, 37]. Significantly, inverters that realize the negative resistor can adopt full-bridge, half-bridge, class D, and class E inverters [28, 38–40].

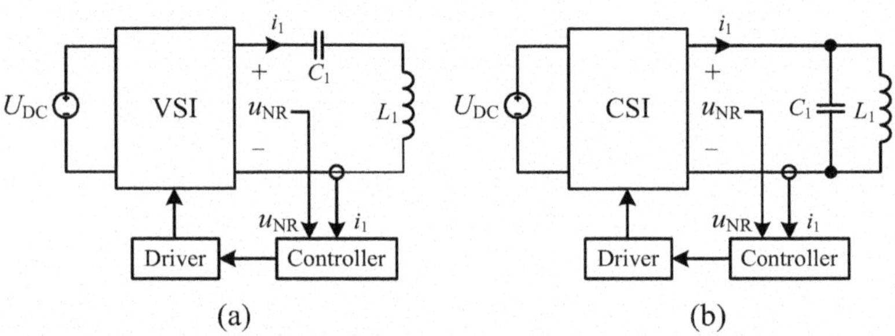

Fig. 2.10 Realization method of the negative resistor by the power electronic inverter with ZPA control. **a** VSI, **b** CSI

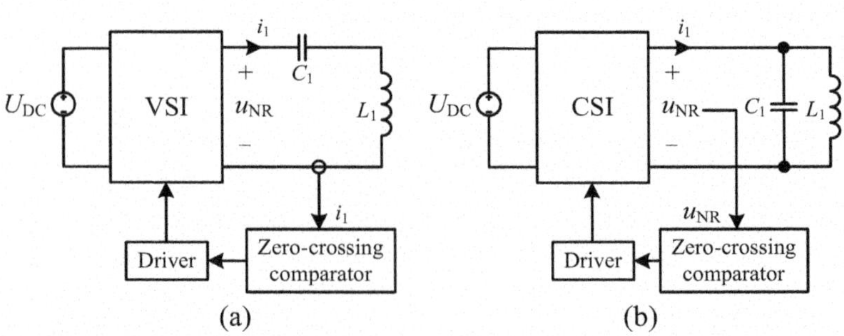

Fig. 2.11 Realization method of the negative resistor by the power electronic inverter with self-oscillation control. **a** VSI, **b** CSI

As can be seen from Figs. 2.10 and 2.11, the negative resistor constructed by the power electronic inverter is equivalent to a power supply whose output voltage and current are controlled to be in the same phase.

Taking the class D inverter with self-oscillation control as an example, a schematic of the negative resistor is given in Fig. 2.12 [28, 41]. Here, the negative resistor can be regarded as a nonlinear controlled voltage source, and the relationship between the DC input voltage U_{DC} and the output voltage u_{NR} of the class D inverter can be expressed by

$$u_{NR} = -\frac{\mathrm{sign}(i_1) + 1}{2} U_{DC}, \tag{2.28}$$

where sign() is the sign function.

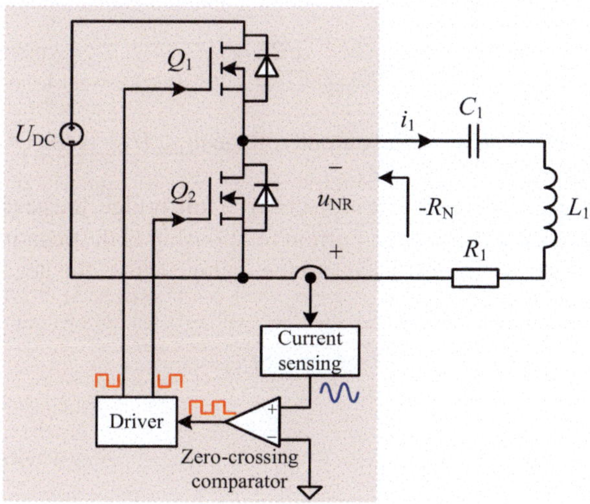

Fig. 2.12 Schematic of the negative resistor using a class D inverter

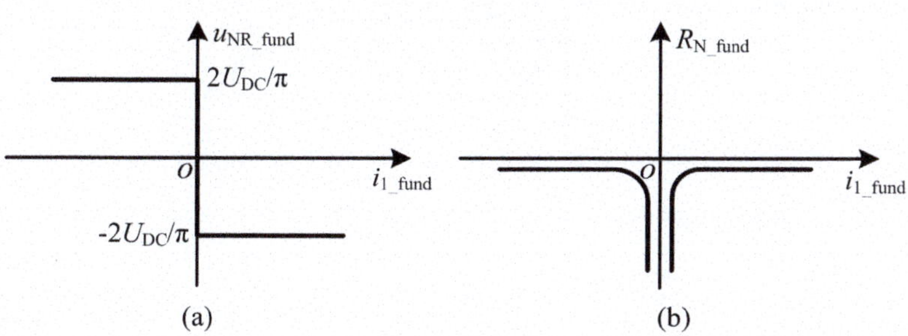

Fig. 2.13 Characteristic curves of the fundamental wave component of the nonlinear controlled voltage source. **a** Voltage-current, **b** Negative resistance-current

Based on (2.28), the characteristic curves of the fundamental wave component of the nonlinear controlled voltage source can be drawn, as shown in Fig. 2.13. It can be seen that the output voltage of the nonlinear controlled voltage source is naturally limited to the range $[-2U_{DC}/\pi, 2U_{DC}/\pi]$, and the negative resistance R_N has a nonlinear negative correlation with the current i_1 (Fig. 2.12).

In Fig. 2.12, the control circuit uses the current of the transmitting coil as a feedback signal to generate the driving signals of the switches Q_1 and Q_2, and the voltage-current characteristic of negative resistance is achieved by a zero-crossing comparator, maintaining a 180° phase difference between the output voltage and the current of the inverter, as shown in Fig. 2.14. Therefore, the output of the class D inverter can always be equivalent as a negative resistance. Meanwhile, the control strategy without any communication requirement has a very short transient response, which can satisfy the demand of electric vehicles. In addition, the fundamental component of the output voltage of the inverter can be derived as

$$U_{NR} = \frac{\sqrt{2}U_{DC}}{\pi}, \tag{2.29}$$

where U_{NR} is the RMS value of the output voltage u_{NR}, U_{DC} is the DC input voltage of the class D inverter.

Similarly, if the class D inverter is replaced by a full-bridge, the negative resistance can still be achieved, but the voltage-current relationship of the negative resistance in the steady state is shown in Fig. 2.15, and the voltage across the negative resistance is [39].

$$U_{NR} = \frac{2\sqrt{2}U_{DC}}{\pi}. \tag{2.30}$$

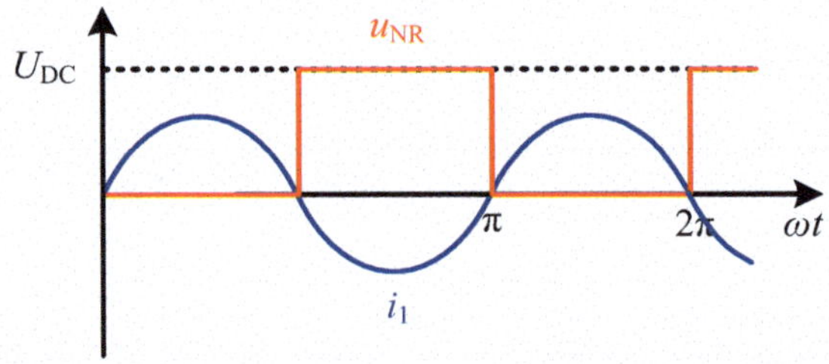

Fig. 2.14 The voltage-current relationship of the negative resistance constructed by the class D inverter

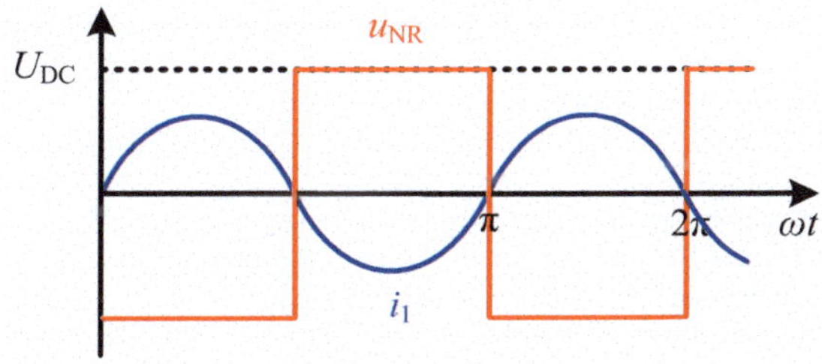

Fig. 2.15 The voltage-current relationship of the negative resistance constructed by the full-bridge inverter

2.5 Summary

This chapter first introduces the PT-symmetric principle, including P reversal, T reversal, and PT symmetry. Then, the representation of PT symmetry in circuits is described, and the concepts of negative resistance and negative impedance are stated. On this basis, the coupled RLC circuits with PT symmetry are introduced and analyzed in depth. Finally, two realization methods of a negative resistor based on the operational amplifier and power electronic inverter are elaborated.

References

1. Fleury R, Sounas DL, Alù A (2016) Parity-time symmetry in acoustics: theory, devices, and potential applications. IEEE J Sel Top Quantum Electron 22(5):121–129
2. Regensburger A, Bersch C, Miri M-A, Onishchukov G, Christodoulides DN, Peschel U (2012) Parity-time synthetic photonic lattices. Nature 488(7410):167–171
3. Zyablovsky AA, Vinogradov AP, Pukhov AA, Dorofeenko AV, Lisyansky AA (2014) PT-symmetry in optics. Phys Usp 57(11):1063–1082
4. Huang C, Ye F, Kartashov YV, Malomed BA, Chen X (2014) PT symmetry in optics beyond the paraxial approximation. Opt Lett 39(18):5443–5446
5. Meyur S, Debnath S (2006) PT-symmetry, pseudo-hermiticity: the real spectra of non-Hermitian Hamiltonians. Acta Physica Polonica 37(6):1697–1700
6. Mostafazadeh A, Batal A (2004) Physical aspects of pseudo-Hermitian and PT-symmetric quantum mechanics. J Phys A: Math Gen 37(48):11645–11679
7. Bender CM, Boettcher S, Meisinger PN (1999) PT-symmetric quantum mechanics. J Math Phys 40(5):2201–2229
8. Bender CM (2007) Making sense of non-Hermitian Hamiltonians. Rep Prog Phys 70(6):947–1018
9. Mostafazadeh A (2003) Pseudo-hermiticity and generalized PT- and CPT-symmetries. J Math Phys 44(3):974–989
10. Bender CM, Boettcher S (1998) Real spectra in non-Hermitian Hamiltonians having PT symmetry. Phys Rev Lett 80(24):5243–5246
11. El-Ganainy R, Makris KG, Khajavikhan M, Musslimani ZH, Rotter S, Christodoulides DN (2018) Non-Hermitian physics and PT symmetry. Nat Phys 14(1):11–19
12. Hodaei H et al (2017) Enhanced sensitivity at higher-order exceptional points. Nature 548(7666):187–191

13. Chen P-Y, El-Ganainy R (2019) Exceptional points enhance wireless readout. Nat Electron 2(8):323–324
14. Sakhdari M, Hajizadegan M, Zhong Q, Christodoulides DN, El-Ganainy R, Chen PY (2019) Experimental observation of PT symmetry breaking near divergent exceptional points. Phys Rev Lett 123(19):193901
15. Gan WS (2021) Time reversal acoustics. Springer Singapore, Singapore, pp 49–57
16. Lin Z, Schindler J, Ellis FM, Kottos T (2012) Experimental observation of the dual behavior of PT-symmetric scattering. Phys Rev A 85(5):2012
17. Schindler J, Lin Z, Lee JM, Ramezani H, Ellis FM, Kottos T (2012) PT-symmetric electronics. J Phys A Math Theor 45(44)
18. Bender CM, Brody DC, Jones HF (2002) Complex extension of quantum mechanics. Phys Rev Lett 89(27):270401
19. Hassan AU, Hodaei H, Miri MA, Khajavikhan M, Christodoulides DN (2015) Nonlinear reversal of the PT-symmetric phase transition in a system of coupled semiconductor microring resonators. Phys Rev A 92(6)
20. Schindler J, Li A, Zheng MC, Ellis FM, Kottos T (2011) Experimental study of active LRC circuits with PT symmetries. Phys Rev A 84(4):040101
21. Kiani M, Ghovanloo M (2012) The circuit theory behind coupled-mode magnetic resonance-based wireless power transmission. IEEE Trans Circ Syst I Regul Papers 59(9):2065–2074
22. Haus HA (1984) Waves and fields in optoelectronics. Prentice-Hall, New Jersey
23. Kurs A, Karalis A, Moffatt R, Joannopoulos JD, Fisher P, Soljačić M (2007) Wireless power transfer via strongly coupled magnetic resonances. Science 317(5834):83–86
24. Haus HA, Huang W (2002) Coupled-mode theory. Proc IEEE 79(10):1505–1518
25. Assawaworrarit S, Yu X, Fan S (2017) Robust wireless power transfer using a nonlinear parity-time-symmetric circuit. Nature 546(7658):387–390
26. Durukan S, Yilmaz A, Keser M (2022) Analysis, design and realization of negative impedance converter circuit with current feedback operational amplifier. Int J Wirel Microw Technol 12(2):1–10
27. Vincelj L, Hrabar S (2021) Influence of load resonance on operation of negative impedance converter circuits. In: International symposium ELMAR Zadar, Croatia, pp 39–42
28. Zhou J, Zhang B, Xiao W, Qiu D, Chen Y (2019) Nonlinear parity-time-symmetric model for constant efficiency wireless power transfer: application to a drone-in-flight wireless charging platform. IEEE Trans Industr Electron 66(5):4097–4107
29. Liu G, Zhang B (2019) Analytical model of a 25–50 m robust single-wire electric-field coupling power transfer system using a limiter. IEEE Trans Circ Syst II Express Briefs 66(6):978–982
30. Assawaworrarit S, Fan S (2020) Robust and efficient wireless power transfer using a switch-mode implementation of a nonlinear parity-time symmetric circuit. Nat Electron 3(5):273–279
31. Shu X, Zhang B (2018) Single-wire electric-field coupling power transmission using nonlinear parity-time-symmetric model with coupled-mode theory. Energies 11(3):532
32. Hou Y, Lin M, Chen W, Yang X (2018) Parity-time-symmetric wireless power transfer system using switch-mode nonlinear gain element. In: 2018 IEEE international power electronics and application conference and exposition (PEAC) Shenzhen, China, pp 1–5
33. Liu Y, Madawala UK, Mai R, He Z (2020) Zero-phase-angle controlled bidirectional wireless EV charging systems for large coil misalignments. IEEE Trans Power Electron 35(5):5343–5353
34. Liu W, Tang H (2007) Analysis of voltage source inductive coupled power transfer systems based on zero phase angle resonant control method. In: 2007 2nd IEEE conference on industrial electronics and applications, Harbin, China, pp 1873–1877
35. Chwei-Sen W, Covic GA, Stielau OH (2001) General stability criterions for zero phase angle controlled loosely coupled inductive power transfer systems. In: IECON'01 27th annual conference of the IEEE industrial electronics society (Cat. No.37243), Denver, CO, USA, 2001, pp 1049–1054
36. Wang X, Liu S, Li Q, Xu H (2017) Self-oscillating control method and topology analysis for parallel resonant contactless power transfer systems. In: 2017 IEEE PELS workshop on emerging technologies: wireless power transfer (WoW) Chongqing, China, pp 1–6

37. Namadmalan A, Moghani JS (2012) Self-oscillating switching technique for current source parallel resonant induction heating systems. J Power Electron 12(6):851–858
38. Zhu H, Zhang B, Wu L (2020) Output power stabilization for wireless power transfer system employing primary-side-only control. IEEE Access 8:63735–63747
39. Zhang Z, Zhang B (2020) Omnidirectional and efficient wireless power transfer system for logistic robots. IEEE Access 8:13683–13693
40. Wu L, Zhang B, Jiang Y (2023) Robust parity-time-symmetric WPT system with reduced switching-frequency and improved step-down conversion ratio. IEEE Trans Transp Electrification 9(2):2090–2103
41. Wu L, Zhang B, Zhou J (2020) Efficiency improvement of the parity-time-symmetric wireless power transfer system for electric vehicle charging. IEEE Trans Power Electron 35(11):12497–12508

Chapter 3
PT-Symmetric Magnetic-Field Coupled Wireless Power Transfer System

Magnetic-field coupled wireless power transfer (MC-WPT) is a mechanism that consists of at least a pair of LC resonators with magnetically coupled, in which energy is fed from an independent AC power source to the transmitting LC resonator, and transferred to the receiving LC resonator through the magnetic fields coupling between the resonators, and then transferred to the load. However, conventional MC-WPT systems are susceptible to variation of operating conditions, such as transfer distance, resonator frequency shift, load disturbance, etc. Fortunately, the introduction of PT symmetry into wireless power transfer systems can solve these issues to some extent. Based on the PT-symmetric principle and circuit described in Chap. 2, this chapter first introduces the basic structure of the PT-symmetric MC-WPT system, then establishes the circuit model and coupled-mode model of the system respectively, as well as gives the PT-symmetric conditions, analyzes the basic characteristics of the system in detail, including frequency characteristic, voltage and current gain characteristics, output power and transfer efficiency characteristics, and finally build an experimental prototype to verify the correctness of the theoretical analysis.

3.1 Basic Structure

The PT-symmetric MC-WPT system is mainly composed of the negative resistor, magnetic coupling mechanism, transmitting and receiving LC compensation networks, and the load equipment, as shown in Fig. 3.1. The implementation of the negative resistor has been detailed in Sect. 2.4, which will not be repeated here. The magnetic coupling mechanism consists of the transmitting coil and receiving coil, which are coupled by the magnetic field represented by mutual inductance. The LC compensation networks have series (S), parallel (P), and the combination of S and P.

B. Zhang et al., *Parity-Time Symmetric Wireless Power Transfer*, CPSS Power Electronics Series, https://doi.org/10.1007/978-981-96-3149-0_3

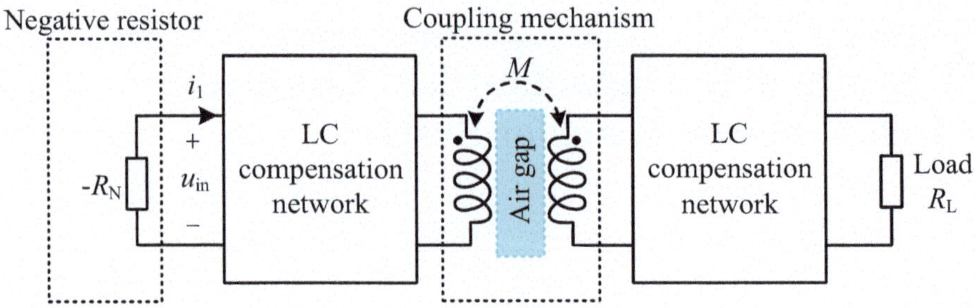

Fig. 3.1 System structure of the PT-symmetric MC-WPT system

Generally, MC-WPT systems have four basic topologies, including SS, PP, SP and PS compensation topologies [1, 2], in which SS and PP compensation topologies are exact PT-symmetric in terms of topological symmetry, SP and PS compensation topologies are PT-symmetric in the case of approximate modeling by coupled-mode theory. The equivalent circuits of PT-symmetric MC-WPT systems are shown in Fig. 3.1, wherein the negative resistor $-R_N$ is used as a power supply to provide energy for the system, and the phase difference between its output voltage u_{in} and current i_1 is always zero. L_1, C_1, R_1 are inductance, compensation capacitance, internal resistance of the transmitting coil, respectively. L_2, C_2, R_2 are inductance, compensation capacitance, internal resistance of the receiving coil, respectively. M is the mutual inductance between transmitting and receiving coils. i_1 and i_2 are currents flowing through the transmitting and receiving coils, respectively. u_1 and u_2 are voltages across the compensation capacitance C_1 and C_2, respectively. In particular, the coupling coefficient of the mutual inductance is defined as $k = \mathrm{M}/\sqrt{L_1L_2}$, and the natural resonant frequencies of the transmitting and receiving coils are defined as $\omega_1=1/\sqrt{L_1C_1}$ and $\omega_2 = 1/\sqrt{L_2C_2}$, respectively.

3.2 System Modeling and PT Symmetry

When the dynamic model of the MC-WPT system remains invariant after a combined P and T inversion transformation, the MC-WPT system can be treated as PT-symmetric, and the energy stored in the transmitting and receiving resonators will maintain equal in the exact PT-symmetric region, typically leading to a constant transfer efficiency, which also reveals that not all topologies satisfy PT symmetry. Considering that there are two most mainstream approaches to model the PT-symmetric MC-WPT system, including coupled-mode theory (CMT) and circuit theory (CT) [3, 4], we would analyze the PT symmetry and fundamental characteristics of the basic PT-symmetric MC-WPT topologies shown in Fig. 3.2 from different modeling perspectives.

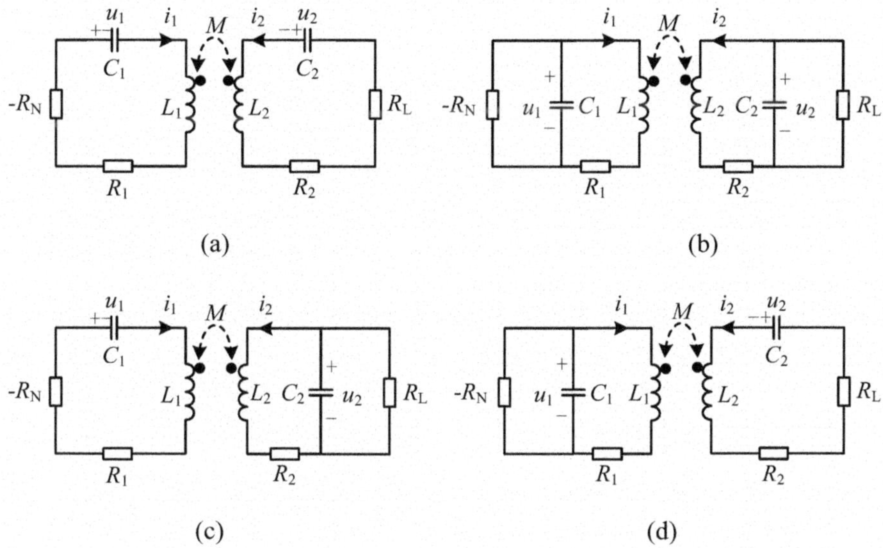

Fig. 3.2 Equivalent circuits of PT-symmetric MC-WPT systems. **a** SS compensation topology, **b** PP compensation topology, **c** SP compensation topology, **d** PS compensation topology

3.2.1 Circuit Model and Its PT Symmetry

Based on the CT, the dynamics of the SS compensation topology can be fully described by the following state equations:

$$\begin{bmatrix} \frac{di_1}{dt} \\ \frac{du_1}{dt} \\ \frac{di_2}{dt} \\ \frac{du_2}{dt} \end{bmatrix} = \begin{bmatrix} \frac{L_2(R_N-R_1)}{L_1L_2-M^2} & -\frac{L_2}{L_1L_2-M^2} & \frac{M(R_2+R_L)}{L_1L_2-M^2} & \frac{M}{L_1L_2-M^2} \\ \frac{1}{C_1} & 0 & 0 & 0 \\ -\frac{M(R_N-R_1)}{L_1L_2-M^2} & \frac{M}{L_1L_2-M^2} & -\frac{L_1(R_2+R_L)}{L_1L_2-M^2} & -\frac{L_1}{L_1L_2-M^2} \\ 0 & 0 & \frac{1}{C_2} & 0 \end{bmatrix} \begin{bmatrix} i_1 \\ u_1 \\ i_2 \\ u_2 \end{bmatrix}. \quad (3.1)$$

Then, the steady-state model under the CT can be obtained by phasor method, that is

$$\begin{bmatrix} -R_N + R_1 + j\left(\omega L_1 - \frac{1}{\omega C_1}\right) & j\omega M \\ j\omega M & R_2 + R_L + j\left(\omega L_2 - \frac{1}{\omega C_2}\right) \end{bmatrix} \begin{bmatrix} \dot{I}_1 \\ \dot{I}_2 \end{bmatrix} = 0, \quad (3.2)$$

where $\dot{I}_1$ and $\dot{I}_2$ are the phasor forms of i_1 and i_2.

Meanwhile, (3.1) can be rewritten for the charges $q_n = C_n u_n$ $(n = 1, 2)$ in the form:

$$\begin{cases} \dfrac{d^2q_1}{dt^2} = \dfrac{(R_N - R_1)}{(1-k^2)L_1}\dfrac{dq_1}{dt} - \dfrac{\omega_1^2}{1-k^2}q_1 + \dfrac{k(R_2+R_L)}{(1-k^2)\sqrt{L_1L_2}}\dfrac{dq_2}{dt} + \dfrac{\omega_2^2 k}{1-k^2}\sqrt{\dfrac{L_2}{L_1}}q_2 \\ \dfrac{d^2q_2}{dt^2} = -\dfrac{(R_2+R_L)}{(1-k^2)L_2}\dfrac{dq_2}{dt} - \dfrac{\omega_2^2}{1-k^2}q_2 - \dfrac{k(R_N-R_1)}{(1-k^2)\sqrt{L_1L_2}}\dfrac{dq_1}{dt} + \dfrac{\omega_1^2 k}{1-k^2}\sqrt{\dfrac{L_1}{L_2}}q_1 \end{cases}. \tag{3.3}$$

Based on the definition of PT symmetry, (3.3) should be invariant under a combined P inversion (i.e., $n = 1 \leftrightarrow n = 2$) and T inversion (i.e., $t \to -t$ transformation, then the transformed equation from (3.3) can be described as follows

$$\begin{cases} \dfrac{d^2q_2}{dt^2} = -\dfrac{(R_N - R_1)}{(1-k^2)L_1}\dfrac{dq_2}{dt} - \dfrac{\omega_1^2}{1-k^2}q_2 - \dfrac{k(R_2+R_L)}{(1-k^2)\sqrt{L_1L_2}}\dfrac{dq_1}{dt} + \dfrac{\omega_2^2 k}{1-k^2}\sqrt{\dfrac{L_2}{L_1}}q_1 \\ \dfrac{d^2q_1}{dt^2} = \dfrac{(R_2+R_L)}{(1-k^2)L_2}\dfrac{dq_1}{dt} - \dfrac{\omega_2^2}{1-k^2}q_1 + \dfrac{k(R_N-R_1)}{(1-k^2)\sqrt{L_1L_2}}\dfrac{dq_2}{dt} + \dfrac{\omega_1^2 k}{1-k^2}\sqrt{\dfrac{L_1}{L_2}}q_2 \end{cases}. \tag{3.4}$$

By comparing (3.3) and (3.4), if they are required to be equivalent, the constraint conditions for the circuit parameters can be obtained as $L_1 = L_2$, $C_1 = C_2$, and $R_N - R_1 = R_2 + R_L$, which are also called PT-symmetric conditions. It can be seen that if the circuit model of the SS compensation topology satisfies the PT symmetry, the circuit parameters of the transmitter and receiver have to be symmetric.

Similarly, for the PP compensation topology shown in Fig. 3.2b, the circuit equation can be expressed as

$$\begin{bmatrix} \frac{di_1}{dt} \\ \frac{du_1}{dt} \\ \frac{di_2}{dt} \\ \frac{du_2}{dt} \end{bmatrix} = \begin{bmatrix} -\frac{L_2R_1}{L_1L_2-M^2} & \frac{L_2}{L_1L_2-M^2} & \frac{MR_2}{L_1L_2-M^2} & -\frac{M}{L_1L_2-M^2} \\ -\frac{1}{C_1} & \frac{1}{R_NC_1} & 0 & 0 \\ \frac{MR_1}{L_1L_2-M^2} & -\frac{M}{L_1L_2-M^2} & -\frac{L_1R_2}{L_1L_2-M^2} & \frac{L_1}{L_1L_2-M^2} \\ 0 & 0 & -\frac{1}{C_2} & -\frac{1}{R_LC_2} \end{bmatrix} \begin{bmatrix} i_1 \\ u_1 \\ i_2 \\ u_2 \end{bmatrix}. \tag{3.5}$$

Substituting $q_n = C_nu_n$ ($n = 1, 2$) into (3.5), (3.5) can be rewritten as

$$\begin{cases} \dfrac{d^2q_1}{dt^2} = \left[\dfrac{1}{R_{\mathrm{N}}C_1} - \dfrac{R_1}{(1-k^2)L_1}\right]\dfrac{dq_1}{dt} - \dfrac{\omega_1^2}{1-k^2}\left(1-\dfrac{R_1}{R_{\mathrm{N}}}\right)q_1 \\ \qquad + \dfrac{kR_2}{(1-k^2)\sqrt{L_1L_2}}\dfrac{dq_2}{dt} + \dfrac{\omega_2^2k}{1-k^2}\sqrt{\dfrac{L_2}{L_1}}\left(1+\dfrac{R_2}{R_{\mathrm{L}}}\right)q_2 \\ \dfrac{d^2q_2}{dt^2} = -\left[\dfrac{1}{R_{\mathrm{L}}C_2} + \dfrac{R_2}{(1-k^2)L_2}\right]\dfrac{dq_2}{dt} - \dfrac{\omega_2^2}{1-k^2}\left(1+\dfrac{R_2}{R_{\mathrm{L}}}\right)q_2 \\ \qquad + \dfrac{kR_1}{(1-k^2)\sqrt{L_1L_2}}\dfrac{dq_1}{dt} + \dfrac{\omega_1^2k}{1-k^2}\sqrt{\dfrac{L_1}{L_2}}\left(1-\dfrac{R_1}{R_{\mathrm{N}}}\right)q_1. \end{cases} \tag{3.6}$$

Then, the transformed equation after PT transformation from (3.6), we have

$$\begin{cases} \dfrac{d^2q_2}{dt^2} = -\left[\dfrac{1}{R_{\mathrm{N}}C_1} - \dfrac{R_1}{(1-k^2)L_1}\right]\dfrac{dq_2}{dt} - \dfrac{\omega_1^2}{1-k^2}\left(1-\dfrac{R_1}{R_{\mathrm{N}}}\right)q_2 \\ \qquad - \dfrac{kR_2}{(1-k^2)\sqrt{L_1L_2}}\dfrac{dq_1}{dt} + \dfrac{\omega_2^2k}{1-k^2}\sqrt{\dfrac{L_2}{L_1}}\left(1+\dfrac{R_2}{R_{\mathrm{L}}}\right)q_1 \\ \dfrac{d^2q_1}{dt^2} = \left[\dfrac{1}{R_{\mathrm{L}}C_2} + \dfrac{R_2}{(1-k^2)L_2}\right]\dfrac{dq_1}{dt} - \dfrac{\omega_2^2}{1-k^2}\left(1+\dfrac{R_2}{R_{\mathrm{L}}}\right)q_1 \\ \qquad - \dfrac{kR_1}{(1-k^2)\sqrt{L_1L_2}}\dfrac{dq_2}{dt} + \dfrac{\omega_1^2k}{1-k^2}\sqrt{\dfrac{L_1}{L_2}}\left(1-\dfrac{R_1}{R_{\mathrm{N}}}\right)q_2 \end{cases} \tag{3.7}$$

Therefore, the PT symmetry can be obtained in the PP compensation topology with the constraint conditions that $L_1 = L_2$, $C_1 = C_2$, $R_1 = R_2 = 0$, and $R_{\mathrm{N}} = R_{\mathrm{L}}$. It is worth noting that the PP compensation topology requires not only the circuit parameters to be symmetric, but also the internal resistances of the transmitting and receiving coils to be zero. In addition, considering that the internal resistance is generally small in practical MC-WPT system, the PP compensation topology can be approximated as being PT-symmetric.

For the SP compensation topology, the state equation can be described as

$$\begin{cases} \dfrac{d^2q_1}{dt^2} = \dfrac{R_N - R_1}{L_1\left(1-k^2\right)}\dfrac{dq_1}{dt} - \dfrac{\omega_1^2}{1-k^2}q_1 \\ \qquad - \dfrac{kR_2}{(1-k^2)\sqrt{L_1L_2}}\dfrac{dq_2}{dt} - \dfrac{\omega_2^2 k}{1-k^2}\sqrt{\dfrac{L_2}{L_1}}\left(\dfrac{R_2}{R_L}+1\right)q_2 \\ \dfrac{d^2q_2}{dt^2} = -\left[\dfrac{1}{R_LC_2} + \dfrac{R_2}{(1-k^2)L_2}\right]\dfrac{dq_2}{dt} - \dfrac{\omega_2^2}{1-k^2}\left(\dfrac{R_2}{R_L}+1\right)q_2 \\ \qquad + \dfrac{k(R_N - R_1)}{(1-k^2)\sqrt{L_1L_2}}\dfrac{dq_1}{dt} - \dfrac{\omega_1^2 k}{1-k^2}\sqrt{\dfrac{L_1}{L_2}}q_1 \end{cases}. \tag{3.8}$$

After PT transformation, we have

$$\begin{cases} \dfrac{d^2q_2}{dt^2} = -\dfrac{R_N - R_1}{L_1\left(1-k^2\right)}\dfrac{dq_2}{dt} - \dfrac{\omega_1^2}{1-k^2}q_2 \\ \qquad + \dfrac{kR_2}{(1-k^2)\sqrt{L_1L_2}}\dfrac{dq_1}{dt} - \dfrac{\omega_2^2 k}{(1-k^2)}\sqrt{\dfrac{L_2}{L_1}}\left(\dfrac{R_2}{R_L}+1\right)q_1 \\ \dfrac{d^2q_1}{dt^2} = +\left[\dfrac{1}{R_LC_2} + \dfrac{R_2}{(1-k^2)L_2}\right]\dfrac{dq_1}{dt} - \dfrac{\omega_2^2}{1-k^2}\left(\dfrac{R_2}{R_L}+1\right)q_1 \\ \qquad - \dfrac{k(R_N - R_1)}{(1-k^2)\sqrt{L_1L_2}}\dfrac{dq_2}{dt} - \dfrac{\omega_1^2 k}{1-k^2}\sqrt{\dfrac{L_1}{L_2}}q_2 \end{cases}. \tag{3.9}$$

Meanwhile, for PS compensation topology, the state equation can be deduced as

$$\begin{cases} \dfrac{d^2q_1}{dt^2} = \left[\dfrac{1}{R_NC_1} - \dfrac{R_1}{(1-k^2)L_1}\right]\dfrac{dq_1}{dt} + \dfrac{\omega_1^2}{1-k^2}\left(\dfrac{R_1}{R_N}-1\right)q_1 \\ \qquad - \dfrac{k(R_2+R_L)}{(1-k^2)\sqrt{L_1L_2}}\dfrac{dq_2}{dt} - \dfrac{\omega_2^2 k}{1-k^2}\sqrt{\dfrac{L_2}{L_1}}q_2 \\ \dfrac{d^2q_2}{dt^2} = -\dfrac{(R_2+R_L)}{(1-k^2)L_2}\dfrac{dq_2}{dt} - \dfrac{\omega_2^2}{1-k^2}q_2 \\ \qquad - k\sqrt{\dfrac{L_1}{L_2}}\dfrac{R_1}{(1-k^2)L_1}\dfrac{dq_1}{dt} + \dfrac{\omega_1^2 k}{1-k^2}\sqrt{\dfrac{L_1}{L_2}}\left(\dfrac{R_1}{R_N}-1\right)q_1 \end{cases}. \tag{3.10}$$

After PT transformation, we have

$$\begin{cases} \dfrac{d^2q_2}{dt^2} = -\left[\dfrac{1}{R_N C_1} - \dfrac{R_1}{(1-k^2)L_1}\right]\dfrac{dq_2}{dt} + \dfrac{\omega_1^2}{1-k^2}\left(\dfrac{R_1}{R_N} - 1\right)q_2 \\ \quad + \dfrac{k(R_2+R_L)}{(1-k^2)\sqrt{L_1L_2}}\dfrac{dq_1}{dt} - \dfrac{\omega_2^2 k}{1-k^2}\sqrt{\dfrac{L_2}{L_1}}q_1 \\ \dfrac{d^2q_1}{dt^2} = \dfrac{(R_2+R_L)}{(1-k^2)L_2}\dfrac{dq_1}{dt} - \dfrac{\omega_2^2}{1-k^2}q_1 \\ \quad + \dfrac{kR_1}{(1-k^2)\sqrt{L_1L_2}}\dfrac{dq_2}{dt} + \dfrac{\omega_1^2 k}{1-k^2}\sqrt{\dfrac{L_1}{L_2}}\left(\dfrac{R_1}{R_N} - 1\right)q_2 \end{cases} \tag{3.11}$$

From (3.8–3.11), it can be found that the state equations of the SP and PS compensation topologies cannot kept invariant after PT transformation based on CT, so SP and PS compensation topologies do not satisfy PT symmetry.

In a word, from the perspective of CT, only SS and PP compensation topologies satisfy PT symmetry. For the SS compensation topology, the state equation of (3.3) can be rewritten as the following Liouvillian formalism when $\tau = \omega_0 t$, $\mathbf{\Psi} = (q_1, q_2, \dot{q}_1, \dot{q}_2)^T$, $L_1 = L_2 = L$, $C_1 = C_2 = C$ and R_N-$R_1 = R_2 + R_L = R$ [5, 6]

$$L\mathbf{\Psi} = \frac{d\mathbf{\Psi}}{d\tau}\ L = \begin{bmatrix} 0 & 0 & 1 & 0 \\ 0 & 0 & 0 & 1 \\ -\frac{1}{1-k^2} & \frac{k}{1-k^2} & \frac{1}{\gamma(1-k^2)} & \frac{k}{\gamma(1-k^2)} \\ \frac{k}{1-k^2} & -\frac{1}{1-k^2} & -\frac{k}{\gamma(1-k^2)} & -\frac{1}{\gamma(1-k^2)} \end{bmatrix}, \tag{3.12}$$

where $\omega_0 = \omega_1 = \omega_2 = 1/\sqrt{LC}$, $\gamma = R^{-1}\sqrt{L/C}$, q_n and $\dot{q}_n$ correspond to the charge and displacement current of the compensation capacitors in the transmitter ($n = 1$) and receiver ($n = 2$), respectively.

The PT-symmetric Hamiltonian H can be written as $H = j\mathrm{L}$ ($H\mathbf{\Psi} = jd\mathbf{\Psi}/d\tau$), then the characteristic equation of the WPT system can be given by

$$\lambda^4 - \lambda^2\frac{2\gamma^2 - 1}{\gamma^2(1-k^2)} + \frac{1}{1-k^2} = 0. \tag{3.13}$$

By solving (3.13), four eigenfrequencies on the time scale t can be derived as follows

$$\omega_{e1,\,e3} = \pm\omega_0\sqrt{\frac{2\gamma^2 - 1 + \sqrt{(2\gamma^2-1)^2 - 4\gamma^4(1-k^2)}}{2\gamma^2(1-k^2)}}, \tag{3.14}$$

$$\omega_{e2,\,e4} = \pm\omega_0\sqrt{\frac{2\gamma^2 - 1 - \sqrt{(2\gamma^2-1)^2 - 4\gamma^4(1-k^2)}}{2\gamma^2(1-k^2)}}, \tag{3.15}$$

where $\omega_{e1\sim e4}$ are also called operating angular frequencies of the WPT system.

From (3.14) and (3.15), it can be seen that the SS compensation topology of PT-symmetric MC-WPT system has two operating regions, when the coupling coefficient k satisfies the critical condition shown in Table 3.1, all operating frequencies are real. While as the coupling coefficient k decreases, the operating frequencies become complex conjugate pairs.

Similarly, the results of PP compensation topology can also be obtained by the same way, that is, the PT-symmetric Hamiltonian H of PP compensation topology can be given by (3.6) as below

$$H = \begin{bmatrix} 0 & 0 & j & 0 \\ 0 & 0 & 0 & j \\ -j\frac{1}{1-k^2} & j\frac{k}{1-k^2} & j\gamma & 0 \\ j\frac{k}{1-k^2} & -j\frac{1}{1-k^2} & 0 & -j\gamma \end{bmatrix}. \tag{3.16}$$

Hence, the characteristic equation of the PP compensation topology can be expressed by

$$\lambda^4 - \lambda^2\frac{2-\gamma^2(1-k^2)}{1-k^2} + \frac{1}{1-k^2} = 0. \tag{3.17}$$

Then, the eigenfrequencies can be derived as

$$\omega_{e1,\,e3} = \pm\omega_0\sqrt{\frac{2-\gamma^2(1-k^2)+\sqrt{\left[2-\gamma^2(1-k^2)\right]^2-4(1-k^2)}}{2(1-k^2)}}, \tag{3.18}$$

$$\omega_{e2,\,e4} = \pm\omega_0\sqrt{\frac{2-\gamma^2(1-k^2)-\sqrt{\left[2-\gamma^2(1-k^2)\right]^2-4(1-k^2)}}{2(1-k^2)}}. \tag{3.19}$$

Table 3.1 gives the comparison of SS, PP, SP, and PS compensation topologies based on CT, it can be seen that SS and PP compensation topologies are PT-symmetric

Table 3.1 Comparison of SS, PP, SP and PS compensation topologies based on CT

Structure	PT symmetry	Conditions of PT symmetry	PT-symmetric region
SS	Satisfied	$L_1 = L_2, C_1 = C_2, \omega_1 = \omega_2 = \omega_0, R_N - R_1 = R_2 + R_L$	$\sqrt{2\left(1-\sqrt{1-k^2}\right)} \geq \frac{R_2+R_L}{\omega_0 L_2}$
PP	Satisfied	$L_1 = L_2, C_1 = C_2, R_1 = R_2 = 0, \omega_1 = \omega_2 = \omega_0, R_N = R_L$	$\frac{\sqrt{1-k^2}}{\sqrt{2\left(1-\sqrt{1-k^2}\right)}} \geq \frac{1}{\omega_0 R_L C_2}$
SP	Not satisfied	–	–
PS	Not satisfied	–	–

under certain parameter constraints known as PT-symmetric conditions, whereas SP and PS compensation topologies cannot satisfy PT symmetry in the circuit model analysis.

3.2.2 Coupled-Mode Model and Its PT Symmetry

CMT is a framework to analyze energy exchange between two or more resonators that are coupled to each other, its general formalism is simple and can ignore the specific topology of the resonator [7]. Meanwhile, CMT can also reduce the order of the differential equation by half due to its modeling from the perspective of energy coupling, thus simplifying the analysis. However, CMT is accurate only if the coupling strength $k = M/\sqrt{L_1L_2}$ of the mutual inductance is small, and coil quality factors $Q_{10} = \omega_0 L_1/R_1$ and $Q_{20} = \omega_0 L_2/R_2$ are large, which means that the internal resistances R_1 and R_2 are small [3, 8, 9]. Fortunately, the MC-WPT system are generally satisfied the above two conditions, so CMT is a useful tool for analyzing the physical mechanism of energy transmission in a PT-symmetric WPT system.

Based on the coupled-mode theory [8–11], the coupled modes are defined as.

$$\boldsymbol{a}_n = \sqrt{\frac{L_n}{2}} i_n + j\sqrt{\frac{C_n}{2}} u_n = A_n e^{j(\omega t + \theta_n)}, n = 1, 2 \tag{3.20}$$

where the complex variables $\boldsymbol{a}_n$ represent the energy modes, wherein the energy stored in the transmitting and receiving resonators are $|\boldsymbol{a}_1|^2$ and $|\boldsymbol{a}_2|^2$, respectively. The variables A_n and θ_n denote the amplitudes and phases of the modes, respectively. Both A_n and θ_n are slowly varying variables on the time scale $\tau = \omega t$, that is, $dA_n/dt << \omega$ and $d\theta_n/dt << \omega$.

According to (3.20), by representing i_n and u_n with A_n and θ_n, there is

$$\begin{cases} u_n = \sqrt{\dfrac{2}{C_n}} A_n \cos(\omega t + \theta_n) \\ i_n = \sqrt{\dfrac{2}{L_n}} A_n \sin(\omega t + \theta_n) \\ \dfrac{du_n}{dt} = \sqrt{\dfrac{2}{C_n}} \dfrac{dA_n}{dt} \cos(\omega t + \theta_n) - \sqrt{\dfrac{2}{C_n}} A_n \left(\omega + \dfrac{d\theta_n}{dt}\right) \sin(\omega t + \theta_n) \\ \dfrac{di_n}{dt} = \sqrt{\dfrac{2}{L_n}} \dfrac{dA_n}{dt} \sin(\omega t + \theta_n) + \sqrt{\dfrac{2}{L_n}} A_n \left(\omega + \dfrac{d\theta_n}{dt}\right) \cos(\omega t + \theta_n) \end{cases} . \tag{3.21}$$

By substituting (3.21) into the state Eq. (3.1) of the SS compensation topology, the dynamics of the coupled modes of SS compensation topology can be derived.

Assuming that the slowly varying variables A_n and θ_n are constant during a self-oscillating period, the high-frequency terms which vary at several times of operating frequency f can be eliminated by the averaging method [12–14]. Thus, only considering low-frequency characteristics and ignoring small high-frequency ripples, the approximate model related to A_n and θ_n can be derived as follows:

$$\frac{dA_1}{dt} = -\frac{L_1L_2}{L_1L_2 - M^2}\frac{R_1 - R_N}{2L_1}A_1 + \frac{M\sqrt{L_1L_2}}{L_1L_2 - M^2}\frac{(R_2 + R_L)A_2}{2L_2}\cos(\theta_1 - \theta_2) + \frac{M\sqrt{L_1L_2}}{L_1L_2 - M^2}\frac{A_2}{2\sqrt{L_2C_2}}\sin(\theta_2 - \theta_1), \tag{3.22}$$

$$A_1\left(\omega + \frac{d\theta_1}{dt}\right) = \left(\frac{L_1L_2}{L_1L_2 - M^2} + 1\right)\frac{A_1}{2\sqrt{L_1C_1}} - \frac{M\sqrt{L_1L_2}}{L_1L_2 - M^2}\frac{A_2}{2\sqrt{L_2C_2}}\cos(\theta_2 - \theta_1) - \frac{M\sqrt{L_1L_2}}{L_1L_2 - M^2}\frac{(R_2 + R_L)A_2}{2L_2}\sin(\theta_1 - \theta_2), \tag{3.23}$$

$$\frac{dA_2}{dt} = \frac{M\sqrt{L_1L_2}}{L_1L_2 - M^2}\frac{R_1 - R_N}{2L_1}A_1\cos(\theta_1 - \theta_2) + \frac{M\sqrt{L_1L_2}}{L_1L_2 - M^2}\frac{A_1}{2\sqrt{L_1C_1}}\sin(\theta_1 - \theta_2) - \frac{L_1L_2}{L_1L_2 - M^2}\frac{(R_2 + R_L)A_2}{2L_2}, \tag{3.24}$$

$$A_2\left(\omega + \frac{d\theta_2}{dt}\right) = \left(\frac{L_1L_2}{L_1L_2 - M^2} + 1\right)\frac{A_2}{2\sqrt{L_2C_2}} - \frac{M\sqrt{L_1L_2}}{L_1L_2 - M^2}\frac{A_1}{2\sqrt{L_1C_1}}\cos(\theta_1 - \theta_2) - \frac{M\sqrt{L_1L_2}}{L_1L_2 - M^2}\frac{(R_1 - R_N)A_1}{2L_1}\sin(\theta_2 - \theta_1). \tag{3.25}$$

By selecting complex variables $\boldsymbol{a}_n$ as the new state variables, the state space can be described by.

$$\frac{d\boldsymbol{a}_n}{dt} = \frac{dA_n}{dt}e^{j(\omega t + \theta_n)} + jA_n\left(\omega + \frac{d\theta_n}{dt}\right)e^{j(\omega t + \theta_n)}, n = 1, 2. \tag{3.26}$$

By substituting (3.22–3.25) into (3.26), the dynamics of the coupled modes can be written as follows:

$$\frac{d\boldsymbol{a}_1}{dt} = \left[\frac{g_0 - \tau_{10}}{1 - k^2} + j\frac{\omega_1}{2}\left(1 + \frac{1}{1 - k^2}\right)\right]\boldsymbol{a}_1 + \left[\frac{k(\tau_{20} + \tau_L)}{1 - k^2} - j\frac{\omega_2 k}{2(1 - k^2)}\right]\boldsymbol{a}_2, \tag{3.27}$$

$$\frac{d\boldsymbol{a}_2}{dt} = \left[\frac{k(\tau_{10} - g_0)}{1 - k^2} - j\frac{\omega_1 k}{2(1 - k^2)}\right]\boldsymbol{a}_1 + \left[j\frac{\omega_2}{2}\left(1 + \frac{1}{1 - k^2}\right) - \frac{\tau_{20} + \tau_L}{1 - k^2}\right]\boldsymbol{a}_2, \tag{3.28}$$

where $\omega_1 = 1/\sqrt{L_1C_1}$ and $\omega_2 = 1/\sqrt{L_2C_2}$ are defined as natural resonant frequencies of transmitting and receiving resonators. $\tau_{10} = R_1/(2L_1)$ and $\tau_{20} = R_2/(2L_2)$ are defined as intrinsic loss rates of transmitter and receiver, which are determined by intrinsic quality factors of transmitting and receiving resonators Q_{10} and Q_{20}, that is, $\tau_{10} = \omega_1/(2Q_{10})$ and $\tau_{20} = \omega_2/(2Q_{20})$. $\tau_L = R_L/(2L_2)$ is defined as the loss rate due to the load resistance. $k = M/\sqrt{L_1L_2}$ is the coupling coefficient due to the mutual inductance. $g_0 = R_N/(2L_1)$ is defined as the gain coefficient.

For the mid-range MC-WPT system, the intrinsic quality factors are generally high, and natural resonant frequencies of the transmitting and receiving resonators are usually close to giving rise to a complete energy transfer, i.e., $\omega_1 = \omega_2$ [9]. Besides, the PT-symmetric conditions also require matched resonators, i.e., $\omega_1 = \omega_2$. Therefore, the case of $\omega_1 = \omega_2 = \omega_0$ is analyzed in the following. Then, the terms τ_{10}/ω_0, $(\tau_{10} + \tau_L)/\omega_0$, k and G_0/ω_0 are nearly of the same order with the small parameter ε, so higher-order terms $o(\varepsilon^2)$ such as $k\tau_{10}/\omega_0$ and k^2 have little effect on the variations of the state variables on the time scale $\tau = \omega_0 t$. By neglecting the higher-order terms $o(\varepsilon^2)$ in the state Eqs. (3.27) and (3.28), the simplified coupled-mode model can be further presented as

$$\frac{d}{dt}\begin{bmatrix} \boldsymbol{a}_1 \\ \boldsymbol{a}_2 \end{bmatrix} = \begin{bmatrix} j\omega_0 + g_0 - \tau_{10} & -j\kappa \\ -j\kappa & j\omega_0 - (\tau_{20} + \tau_L) \end{bmatrix}\begin{bmatrix} \boldsymbol{a}_1 \\ \boldsymbol{a}_2 \end{bmatrix} \tag{3.29}$$

where $\kappa = \omega_0 k/2$ is defined as the energy coupling rate between $\boldsymbol{a}_1$ and $\boldsymbol{a}_2$.

Similarly, for PP, SP and PS compensation topologies, the coupled-mode equations with the same form can be derived by the above analysis method, wherein the relationship between the gain rate g_{10}, intrinsic loss rate τ_{10} and τ_{20}, loss rate τ_L and the circuit parameters are expressed in Table 3.2.

Table 3.2 Comparison of SS, PP, SP and PS compensation topologies based on CMT

Structure	Total gain rate g	Total loss rate τ	Conditions of PT-symmetry	PT-symmetric region
SS	$g_0 = R_N/(2L_1)$, $\tau_{10} = R_1/(2L_1)$	$\tau_{20} = R_2/(2L_2)$, $\tau_L = R_L/(2L_2)$	$\omega_1 = \omega_2 = \omega_0$, $g_0 - \tau_{10} = \tau_{20} + \tau_L$	$\kappa \geq \tau_{20} + \tau_L$
PP	$g_0 = 1/(2R_NC_1)$, $\tau_{10} = R_1/(2L_1)$	$\tau_{20} = R_2/(2L_2)$, $\tau_L = 1/(2R_LC_2)$	$\omega_1 = \omega_2 = \omega_0$, $g_0 - \tau_{10} = \tau_{20} + \tau_L$	$\kappa \geq \tau_{20} + \tau_L$
SP	$g_0 = R_N/(2L_1)$, $\tau_{10} = R_1/(2L_1)$	$\tau_{20} = R_2/(2L_2)$, $\tau_L = 1/(2R_LC_2)$	$\omega_1 = \omega_2 = \omega_0$, $g_0 - \tau_{10} = \tau_{20} + \tau_L$	$\kappa \geq \tau_{20} + \tau_L$
PS	$g_0 = 1/(2R_NC_1)$, $\tau_{10} = R_1/(2L_1)$	$\tau_{20} = R_2/(2L_2)$, $\tau_L = R_L/(2L_2)$	$\omega_1 = \omega_2 = \omega_0$, $g_0 - \tau_{10} = \tau_{20} + \tau_L$	$\kappa \geq \tau_{20} + \tau_L$

When the coupled-mode model (3.29) enters the steady state, the rotational angular frequency, amplitude and phase difference between modes are constant. If the steady-state value of the mode is defined as $\tilde{\boldsymbol{a}}_n (n = 1, 2)$, the steady-state model under the CMT can be expressed by

$$\begin{bmatrix} j(\omega-\omega_1)-g_0+\tau_{10} & j\kappa \\ j\kappa & j(\omega-\omega_2)+\tau_{20}+\tau_{\mathrm{L}} \end{bmatrix}\begin{bmatrix} \tilde{\boldsymbol{a}}_1 \\ \tilde{\boldsymbol{a}}_2 \end{bmatrix} = 0 \tag{3.30}$$

where the steady-state rotational angular frequency ω of the mode corresponds to the operating angular frequency of the system.

According to the definition of PT symmetry, the coupled-mode Eq. (3.29) remains invariant after a combined P ($\boldsymbol{a}_1 \leftrightarrow \boldsymbol{a}_2$) and T ($t \rightarrow -t, j \rightarrow -j$) transformation, the PT-symmetric conditions can be obtained as shown in Table 3.2. From Table 3.2, it can be known that SS, PP, SP and PS compensation topologies all satisfy PT symmetry, revealing that the coupled-mode model broadens the constraints of the PT-symmetric MC-WPT system under reasonable approximation. Compared with the circuit model, the coupled-mode model does not require strict topological symmetry of the MC-WPT system [11].

Assuming that the eigen-solution forms of (3.29) is $[\boldsymbol{a}_1, \boldsymbol{a}_2]^T = [1, \boldsymbol{a}_{20}]^T e^{j\omega t}$, the characteristic equation of the PT-symmetric MC-WPT system can be derived as

$$\begin{aligned} &\omega^2 + \omega\big[j(g_0 - \tau_{10} - \tau_{20} - \tau_{\mathrm{L}}) - (\omega_1 + \omega_2)\big] - \kappa^2 \\ &+ (g_0 - \tau_{10} + j\omega_1)(\tau_{20} + \tau_{\mathrm{L}} - j\omega_2) = 0. \end{aligned} \tag{3.31}$$

By solving (3.31), the eigenfrequencies of the PT-symmetric MC-WPT system can be obtained as

$$\omega = \frac{\omega_1 + \omega_2}{2} - j\frac{g_0 - \tau_{10} - \tau_{20} - \tau_{\mathrm{L}}}{2} \pm \sqrt{\kappa^2 - \left(\frac{g_0 - \tau_{10} + \tau_{20} + \tau_{\mathrm{L}}}{2} - j\frac{\omega_1 - \omega_2}{2}\right)^2}. \tag{3.32}$$

When PT-symmetric conditions g_0-$\tau_{10} = \tau_{20} + \tau_{\mathrm{L}}$ and $\omega_1 = \omega_2 = \omega_0$ are satisfied, and $\kappa \geq \tau_{20} + \tau_{\mathrm{L}}$, the eigenfrequencies are purely real, there are

$$\omega_{\mathrm{e1,e2}} = \omega_0 \pm \sqrt{\kappa^2 - (\tau_{20} + \tau_{\mathrm{L}})^2}. \tag{3.33}$$

3.2.3 *Comparison Between Circuit Model and Coupled-Mode Model*

Without loss of generality, taking SS compensation topology as an example, the numerical simulating waveforms of the transmitting and receiving currents obtained by the circuit model (3.1) and coupled-mode model (3.29) are depicted in Fig. 3.3. Here, the relationship between the transmitting and receiving currents i_1 and i_2 in the circuit model and the modes $\boldsymbol{a}_1$ and $\boldsymbol{a}_2$ in the coupled-mode model is $i_n = (\boldsymbol{a}_n + \boldsymbol{a}_n^*)/\sqrt{2L_n}(n = 1, 2)$ [13]. The natural resonant frequency is $f_0 = \omega_0/(2\pi) =$ 1 MHz, and other circuit parameters are listed in Table 3.3 in Sect. 3.4.2.

From Fig. 3.3, it can be seen that the transient currents of the circuit model and coupled-mode model almost overlap when the coupling coefficient is small, but when the coupling is strengthened, there is a certain error between them, which is mainly reflected in the larger deviation of the phase of them. Nonetheless, the amplitudes of the currents in the steady state remain nearly equal.

According to the definition of the modes, the relationship between the circuit phasor and the steady state value of the mode can be expressed by $I_n = \sqrt{2/L_n}\boldsymbol{a}_n$, ($n$ = 1, 2). Then, substituting $I_n = \sqrt{2/L_n}\boldsymbol{a}_n$ into the steady-state circuit model (3.2), we have

$$\begin{bmatrix} \tau_{10} - g_{10} + j\frac{1}{2}\left(\omega - \frac{\omega_1^2}{\omega}\right) & j\omega\frac{k}{2} \\ j\omega\frac{k}{2} & \tau_{20} + \tau_{\mathrm{L}} + j\frac{1}{2}\left(\omega - \frac{\omega_2^2}{\omega}\right) \end{bmatrix} \begin{bmatrix} \tilde{\boldsymbol{a}}_1 \\ \tilde{\boldsymbol{a}}_2 \end{bmatrix} = 0. \tag{3.34}$$

By comparing (3.30) and (3.34), it can be seen that the following approximations are made from the steady-state circuit model to the coupled-mode model:

$$\frac{1}{2}\left(\omega - \frac{\omega_n^2}{\omega}\right) \approx \omega - \omega_n, n = 1, 2, \tag{3.35}$$

Table 3.3 Circuit parameters of the experimental prototype

Description	Symbol	Values
Inductance of transmitting coil	L_1	41.76 μH
Inductance of receiving coil	L_2	17.84 μH
Compensation capacitance of transmitter	C_1	605 pF
Compensation capacitance of receiver	C_2	1.4 nF
Internal resistance of transmitter	R_1	0.796 Ω
Internal resistance of receiver	R_2	0.349 Ω
Load resistance	R_{L}	10 Ω
Natural resonant frequency of transmitter	f_1	1 MHz
Natural resonant frequency of receiver	f_2	1 MHz
DC input voltage	U_{DC}	36 V

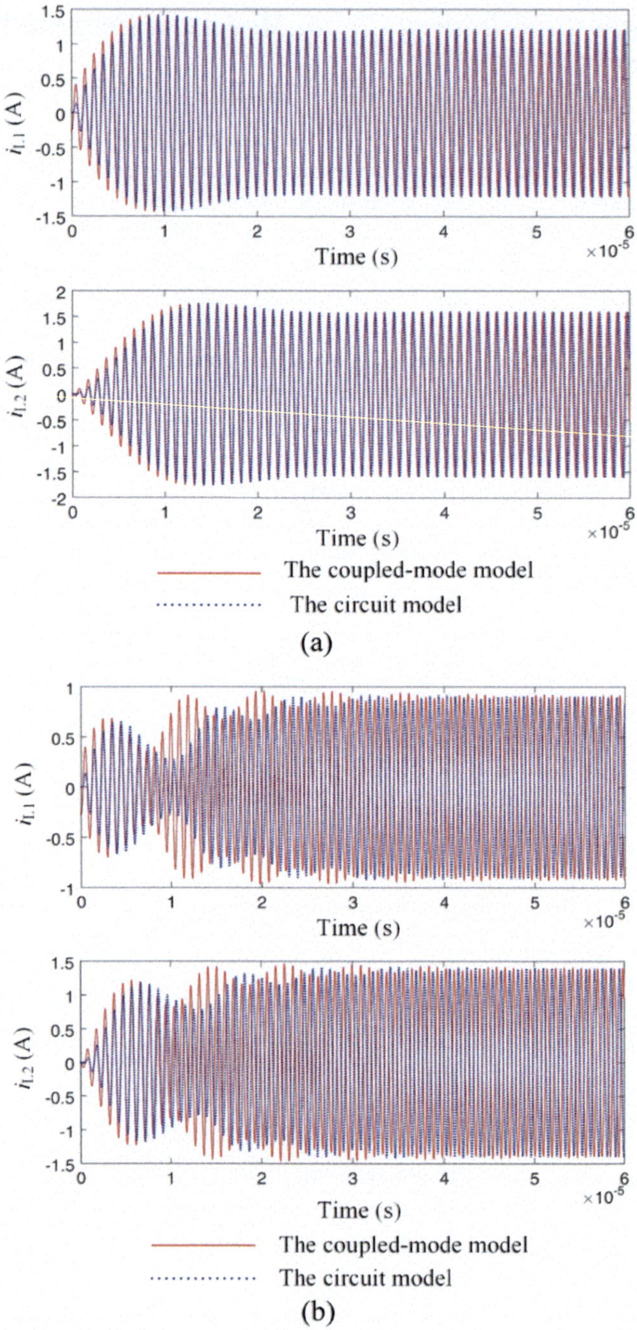

Fig. 3.3 Simulated waveforms of the transmitting and receiving currents obtained by circuit model and coupled-mode model under **a** the weak coupling coefficient $k = 0.03$, **b** the strong coupling coefficient $k = 0.3$

$$\omega\frac{k}{2} \approx \kappa = \frac{\omega_1 + \omega_2}{2}\frac{k}{2}. \tag{3.36}$$

Therefore, the coupled-mode model is an approximate model, which is suitable for the case of identical natural resonant frequencies of coils and weak coupling coefficients. Nevertheless, the second-order coupled-mode model reduces the order by half compared with the fourth-order circuit model, which not only reveals the wireless power transfer mechanism but also greatly simplifies the analysis of the PT-symmetric MC-WPT system. Considering that the circuit model can more accurately analyze the influence of parameter variations on the transfer characteristics of the MC-WPT system, it is necessary to choose the appropriate analysis model when solving different types of issues.

3.3 Fundamental Characteristic Analysis

3.3.1 *Frequency Characteristic*

For the steady-state circuit model shown in (3.2), it has nontrivial solutions only if the coefficient matrix has a zero determinant [15]

$$\left[-R_N + R_1 + j\left(\omega L_1 - \frac{1}{\omega C_1}\right)\right]\left[R_2 + R_L + j\left(\omega L_2 - \frac{1}{\omega C_2}\right)\right] + \omega^2 M^2 = 0. \tag{3.37}$$

The real and imaginary parts of (3.37) can be separated as

$$\left(\frac{-R_N + R_1}{L_1}\right)\left(\frac{R_2 + R_L}{L_2}\right) - \omega^2\left(1 - \frac{\omega_0^2}{\omega^2}\right)^2 + \omega^2 k^2 = 0, \tag{3.38}$$

$$\left(1 - \frac{\omega_0^2}{\omega^2}\right)\left(\frac{-R_N + R_1}{L_1} + \frac{R_2 + R_L}{L_2}\right) = 0. \tag{3.39}$$

From (3.38) and (3.39), it can be found that the operating frequency of the SS compensation topology can be divided into two regions:

(1) Exact PT-symmetric state.

First, in this case, the following PT-symmetric conditions are automatically satisfied.

$$L_1 = L_2,\ C_1 = C_2,\ R_N - R_1 = R_2 + R_L. \tag{3.40}$$

Second, the operating frequency of SS compensation topology of PT-symmetric MC-WPT system can be derived as

$$\omega_{o1} = \frac{\omega_0}{\sqrt{2(1-k^2)}}\sqrt{2-\left(\frac{R_2+R_L}{\omega_0 L_2}\right)^2+\sqrt{\left[2-\left(\frac{R_2+R_L}{\omega_0 L_2}\right)^2\right]^2-4(1-k^2)}}, \tag{3.41}$$

$$\omega_{o2} = \frac{\omega_0}{\sqrt{2(1-k^2)}}\sqrt{2-\left(\frac{R_2+R_L}{\omega_0 L_2}\right)^2-\sqrt{\left[2-\left(\frac{R_2+R_L}{\omega_0 L_2}\right)^2\right]^2-4(1-k^2)}}, \tag{3.42}$$

where the operating frequencies are consistent with the eigenfrequencies mentioned earlier.

Third, in order to ensure that the operating frequency is purely real, the strong coupling region is needed, that is

$$k \geq k_C = \sqrt{1-\frac{1}{4}\left[2-\left(\frac{R_2+R_L}{\omega_0 L_2}\right)^2\right]^2}, \tag{3.43}$$

where k_C is the critical coupling coefficient.

(2) Broken PT-symmetric state.

In this case, the coupling coefficient satisfies $k < k_C$, and there is only one solution for operating frequency at

$$\omega_{o3} = \omega_0. \tag{3.44}$$

In addition, the negative resistance can be derived from (3.44) and (3.38) as

$$R_N = \frac{\omega_0^2 k^2 L_1 L_2}{R_2+R_L} + R_1. \tag{3.45}$$

Therefore, the frequency characteristics of the PT-symmetric MC-WPT system with SS compensation topology based on CT are summarized as

$$\omega = \begin{cases} \omega_0, & k < k_c \\ \frac{\omega_0}{\sqrt{2(1-k^2)}}\sqrt{2-\left(\frac{R_2+R_L}{\omega_0 L_2}\right)^2 \pm \sqrt{\left[2-\left(\frac{R_2+R_L}{\omega_0 L_2}\right)^2\right]^2-4(1-k^2)}}, & k \geq k_c \end{cases}. \tag{3.46}$$

Similarly, for the steady-state coupled-mode model described in (3.30), the operating frequency ω in the steady state can be determined by the characteristic Eq. (3.31). Separating the real and imaginary parts of (3.31), we get

$$(\omega - \omega_1)(\omega - \omega_2) + (g_0 - \tau_{10})(\tau_{20} + \tau_L) - \kappa^2{=}0, \tag{3.47}$$

$$g_0 - \tau_{10} = \frac{\omega - \omega_1}{\omega - \omega_2}(\tau_{20} + \tau_L). \tag{3.48}$$

Substituting the definitions of g_0, τ_{10}, τ_{20}, τ_L and κ into (3.47) and (3.48), the steady solutions can be derived as follows:

(1) Exact PT-symmetric state.

First, in this case, the following PT-symmetric conditions are automatically satisfied.

$$\omega_1 = \omega_2 = \omega_0, \frac{R_N - R_1}{2L_1} = \frac{R_2 + R_L}{2L_2}. \tag{3.49}$$

Second, the operating frequency of SS compensation topology of PT-symmetric MC-WPT system can be derived as

$$\omega_{o1} = \omega_0 + \sqrt{\left(\frac{\omega_0 k}{2}\right)^2 - \left(\frac{R_2 + R_L}{2L_2}\right)^2}, \tag{3.50}$$

$$\omega_{o2} = \omega_0 - \sqrt{\left(\frac{\omega_0 k}{2}\right)^2 - \left(\frac{R_2 + R_L}{2L_2}\right)^2}. \tag{3.51}$$

Third, in order to ensure that the operating frequency is purely real, the strong coupling region is needed, that is

$$k \geq k_C = \frac{R_2 + R_L}{\omega_0 L_2}. \tag{3.52}$$

(2) Broken PT-symmetric state.

In this case, $k < k_C$ and only one solution for operating frequency exists at ω_0. By substituting $\omega = \omega_0$ into (3.47), the negative resistance can be obtained, as consistent with (3.45).

Therefore, the frequency characteristics of the PT-symmetric MC-WPT system with SS compensation topology based on CMT are summarized as

$$\omega = \begin{cases} \omega_0 & k < k_C \\ \omega_0 \pm \sqrt{\left(\frac{\omega_0 k}{2}\right)^2 - \left(\frac{R_2+R_L}{2L_2}\right)^2}, & k \geq k_C \end{cases}. \tag{3.53}$$

From the above analysis, it can be found that whether the circuit model or the coupled-mode model, the system has two real solutions of operating frequency in the PT-symmetric region, and the negative resistance automatically balances the load resistance and internal resistance of coils. When the coefficient decreases to $k = k_C$,

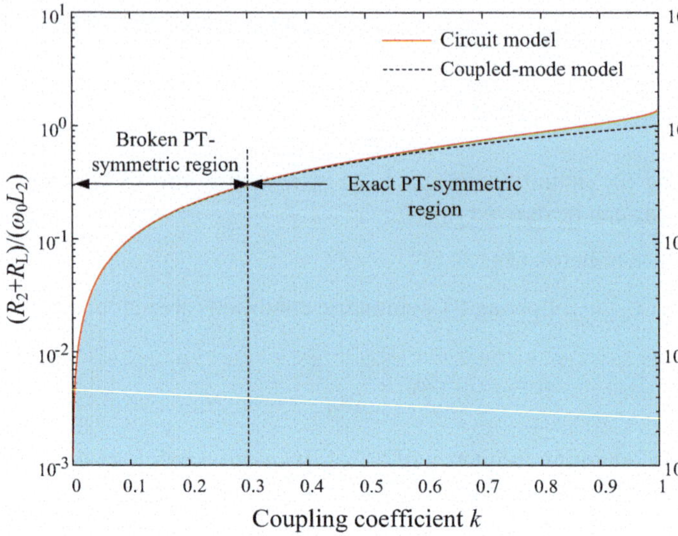

Fig. 3.4 The boundary of exact PT-symmetric region

the bifurcation of the operating frequency degenerates and the operating frequencies become the same at the so-called exceptional point (EP, corresponding to $\omega_{o1} = \omega_{o2} = \omega_{o3}$) [16, 17]. Then, in the broken PT-symmetric region ($k < k_C$), the negative resistance no longer balances the load resistance and internal resistance of coils, and the system operates at the natural resonator frequency ω_0.

Moreover, from (3.43) and (3.52), the exact and broken PT-symmetric regions defined by the loaded quality factor $Q_L = \omega_0 L_2/(R_2 + R_L)$ can be drawn, as shown in Fig. 3.4, it can be found that the critical coupling coefficients of circuit model and coupled-mode model are almost equal when the loaded quality factor $Q_L = \omega_0 L_2/(R_2 + R_L)$ is high (here, $Q_L > 10$).

Only one of steady solutions ω_{o1}, ω_{o2}, ω_{o3} is stable for the definite parameters of the PT-symmetric MC-WPT system. In order to analyze the stability of multi-stable solutions in the strong coupled region $k \geq k_C$, the coupled-mode model of (3.29) can be used to analyze the stability of steady-state values. According to Lyapunov's theory, the stability of the steady-state solution can be determined by the eigenvalues of the Jacobian matrix via linearization near the steady-state solution [13]. By applying perturbation ρ_n around the steady-state solution $\tilde{\boldsymbol{a}}_n$, that is $\boldsymbol{a}_n = \tilde{\boldsymbol{a}}_n + \rho_n e^{j\omega t}$, then the equation of perturbation can be obtained as

$$\begin{cases} \dfrac{d\rho_1}{dt} = \left[j(\omega_0 - \omega) - \tau_{10} + g_{10}\right]\rho_1 - \dfrac{g_{10}\tilde{\boldsymbol{a}}_1^2}{|\tilde{\boldsymbol{a}}_1|^2}\rho_1^* - j\kappa\rho_2 \\ \dfrac{d\rho_2}{dt} = -j\kappa\rho_1 + \left[j(\omega_0 - \omega) - (\tau_{20} + \tau_L)\right]\rho_2 \end{cases}, \tag{3.54}$$

where g_{10} can be written as $g_{10} = G_{10}/(2|\tilde{\boldsymbol{a}}_1|)$.

Recall $\tilde{\boldsymbol{a}}_{\mathrm{n}} = \tilde{A}_{\mathrm{n}} e^{j(\omega t + \tilde{\theta}_{\mathrm{n}})}$ and let $\rho_{\mathrm{n}} = x_{\mathrm{n}} + \mathrm{j} y_{\mathrm{n}}$, then substitute each steady-state solution of (3.50), (3.51), (3.44) into (3.54), the corresponding Jacobian matrix can be derived as

$$J = \begin{bmatrix} \frac{G_{10}}{2\tilde{A}_1}\left(1 - \cos 2\tilde{\theta}_1\right) - \tau_{10} & (\omega - \omega_0) - \frac{G_{10}}{2\tilde{A}_1}\sin 2\tilde{\theta}_1 & 0 & \kappa \\ (\omega - \omega_0) - \frac{G_{10}}{2\tilde{A}_1}\sin 2\tilde{\theta}_1 & \frac{G_{10}}{2\tilde{A}_1}\left(1 - \cos 2\tilde{\theta}_1\right) - \tau_{10} & -\kappa & 0 \\ 0 & \kappa & -(\tau_{20} + \tau_{\mathrm{L}}) & \omega - \omega_0 \\ -\kappa & 0 & \omega_0 - \omega & -(\tau_{20} + \tau_{\mathrm{L}}) \end{bmatrix} \tag{3.55}$$

By substituting the steady-state solutions $\omega_{01}, \omega_{02}, \omega_{03}$ into Jacobian matrix (3.55) and deducing the corresponding Jacobian matrix eigenvalues, it can be easily determined that the eigenvalue of the operating frequency solution $\omega_{03} = \omega_0$ has a positive real part in the strong coupled region $k \geq k_{\mathrm{C}}$, which means that the operating frequency solution $\omega_{03} = \omega_0$ is unstable for $k \geq k_{\mathrm{C}}$. Consequently, in the strong coupling region either ω_{01} or ω_{02} solutions is stable.

Figure 3.5 depicts the curves of the operating frequency versus the coupling coefficient under different mathematical models, it can be seen that the variation trend of operating frequency obtained by circuit model and coupled-mode model is the same, the difference is that the operating frequencies ω_{01} and ω_{02} in exact PT-symmetric region obtained by coupled-mode model is symmetrical about the natural resonant frequency ω_0, while the operating frequencies ω_{01} and ω_{02} obtained by circuit model is asymmetrical. Moreover, when the coupling coefficient is small, the operating frequencies obtained by the circuit model and coupled-mode model are basically identical.

Similarly, the PP compensation topology of PT-symmetric MC-WPT system also has the same frequency characteristics, no tired in words here.

3.3.2 *Voltage and Current Ratio*

From (3.2), the current ratio can be deduced as

$$\frac{I_1}{I_2} = \left|\frac{\dot{I}_1}{\dot{I}_2}\right| = \sqrt{\frac{(R_2 + R_{\mathrm{L}})^2 + \omega^2 L_2^2\left(1 - \frac{\omega_2^2}{\omega^2}\right)^2}{\omega^2 k^2 L_1 L_2}}, \tag{3.56}$$

where I_1 and I_2 are the root-mean-square (RMS) values of $\dot{I}_1$ and $\dot{I}_2$, respectively.

Then, according to $U_{\mathrm{in}} = R_{\mathrm{N}} I_1$ and $U_{\mathrm{o}} = R_{\mathrm{L}} I_2$, the voltage ratio can be obtained as

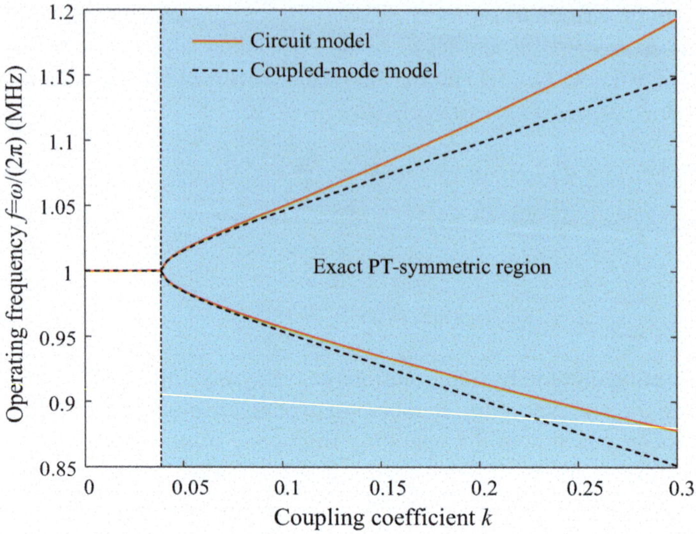

Fig. 3.5 The operating frequency characteristic

$$\frac{U_{\text{in}}}{U_{\text{o}}} = \left|\frac{\dot{U}_{\text{in}}}{\dot{U}_{\text{o}}}\right| = \frac{R_{\text{N}} I_1}{R_{\text{L}} I_2}. \tag{3.57}$$

Therefore, the current and voltage ratio of SS compensation topology can be derived as follows:

(1) Exact PT-symmetric region.

By substituting (3.40) and (3.41) into (3.56), the current ratio can be deduced as

$$\frac{I_1}{I_2} = \left|\frac{\dot{I}_1}{\dot{I}_2}\right| = 1. \tag{3.58}$$

Then, the voltage ratio can be obtained as

$$\frac{U_{\text{in}}}{U_{\text{o}}} = \left|\frac{\dot{U}_{\text{in}}}{\dot{U}_{\text{o}}}\right| = \frac{R_{\text{L}} + R_1 + R_2}{R_{\text{L}}}. \tag{3.59}$$

(2) Broken PT-symmetric region.

In this case, by substituting (3.45) into (3.56) and (3.57), the current and voltage ratios can be obtained as

$$\frac{I_1}{I_2} = \left|\frac{\dot{I}_1}{\dot{I}_2}\right| = \frac{R_2 + R_{\text{L}}}{\omega_0 k \sqrt{L_1 L_2}}, \tag{3.60}$$

$$\frac{U_{\text{in}}}{U_{\text{o}}} = \left|\frac{\dot{U}_{\text{in}}}{\dot{U}_{\text{o}}}\right| = \frac{\omega_0^2 k^2 L_1 L_2 + R_1(R_2 + R_{\text{L}})}{\omega_0 k \sqrt{L_1 L_2} R_{\text{L}}}. \tag{3.61}$$

From the above analysis, we can find that the current and voltage ratios are constant against the variations of coupling coefficient in the exact PT-symmetric region.

3.3.3 Output Power and Transfer Efficiency

According to the definition of output power and transfer efficiency, there are

$$P_{\text{O}} = \begin{cases} I_2^2 R_{\text{L}}, & \text{CT} \\ 2\tau_{\text{L}} |\boldsymbol{a}_2|^2, & \text{CMT} \end{cases}, \tag{3.62}$$

$$\eta = \begin{cases} \frac{I_2^2 R_{\text{L}}}{I_1^2 R_1 + I_2^2 (R_2 + R_{\text{L}})} & \text{CT} \\ \frac{2\tau_{\text{L}} |\boldsymbol{a}_2|^2}{2\tau_{10} |\boldsymbol{a}_1|^2 + 2(\tau_{20} + \tau_{\text{L}}) |\boldsymbol{a}_2|^2} & \text{CMT} \end{cases}. \tag{3.63}$$

Without loss of generality, the use of an inverter to construct the negative resistor is considered here, as described in Sect. 2.4.2, we get

$$U_{\text{in}} = \begin{cases} \frac{\sqrt{2} U_{\text{DC}}}{\pi}, & \text{half-birdge inverter} \\ \frac{2\sqrt{2} U_{\text{DC}}}{\pi}, & \text{full-birdge inverter} \end{cases}, \tag{3.64}$$

$$g_{10} = \frac{U_{\text{in}}}{2\sqrt{L_1} |\boldsymbol{a}_1|}. \tag{3.65}$$

Then, combined with the above analysis of operating frequency characteristic, the output power and transfer efficiency can also be divided into two regions:

(1) Exact PT-symmetric region.

The output power in this region can be derived as

$$P_{\text{O}} = \begin{cases} \frac{L_1 L_2 R_{\text{L}} U_{\text{in}}^2}{L_1^2 (R_2 + R_{\text{L}})^2 + 2L_1 L_2 R_1 (R_2 + R_{\text{L}}) + L_2^2 R_1^2}, & \text{CT} \\ \frac{\tau_{\text{L}} U_{\text{in}}^2}{2L_1 (\tau_{10} + \tau_{20} + \tau_{\text{L}})^2}, & \text{CMT} \end{cases}, \tag{3.66}$$

and the transfer efficiency is

$$\eta = \begin{cases} \frac{L_1 R_{\text{L}}}{L_2 R_1 + L_1 (R_2 + R_{\text{L}})}, & \text{CT} \\ \frac{\tau_{\text{L}}}{\tau_{10} + \tau_{20} + \tau_{\text{L}}}, & \text{CMT} \end{cases}. \tag{3.67}$$

(2) Broken PT-symmetric region.

The output power in this region can be deduced as

$$P_O = \begin{cases} \dfrac{\omega_0^2 k^2 L_1 L_2 R_L U_{in}^2}{\left[\omega_0^2 k^2 L_1 L_2 + R_1(R_2 + R_L)\right]^2}, & \text{CT} \\ \dfrac{\tau_L \kappa^2 U_{in}^2}{2L_1\left[\tau_{10}(\tau_{20} + \tau_L) + \kappa^2\right]^2}, & \text{CMT} \end{cases}, \tag{3.68}$$

and the transfer efficiency is

$$\eta = \begin{cases} \dfrac{\omega_0^2 k^2 L_1 L_2 R_L}{R_1(R_2 + R_L)^2 + \omega_0^2 k^2 L_1 L_2 (R_2 + R_L)}, & \text{CT} \\ \dfrac{\kappa^2 \tau_L}{\left[\tau_{10}(\tau_{20} + \tau_L) + \kappa^2\right](\tau_{20} + \tau_L)}, & \text{CMT} \end{cases}. \tag{3.69}$$

According to the above expressions of output power and transfer efficiency, it can be found that the output power and transfer efficiency based on CT and CMT are consistent, which means that the two models can be regarded as equivalent. Moreover, from (3.66) and (3.67), it can be seen that the variation laws of output power and transfer efficiency with the coupling coefficient are extremely different in the exact and broken PT-symmetric regions. In the exact PT-symmetric region, the output power and transfer efficiency always remain constant in response to the variations of the coupling coefficient, while in the broken PT-symmetric region, the output power and transfer efficiency vary with the coupling coefficient.

Figure 3.6 gives the curves of the output power and transfer efficiency versus the coupling coefficient, it can be demonstrated that the output power and transfer efficiency are independent of the coupling coefficient in the exact PT-symmetric region.

3.4 System Design and Verification

3.4.1 Negative Resistor Implementation

According to Sect. 2.4.2, the negative resistor can be constructed by the power converter. Here, the class D inverter circuit is used to build up the negative resistor, the schematic of the negative resistor circuit is shown in Fig. 3.7a, and the control circuit is shown in Fig. 3.7b. In the experimental prototype, the class D inverter is built based on a GaN E-HEMT evaluation board (GS66508B-EVBDB 650 V). The GaN transistors can operate at high-frequency operation with low power consumption and switching loss, which further improves the overall system efficiency. The two

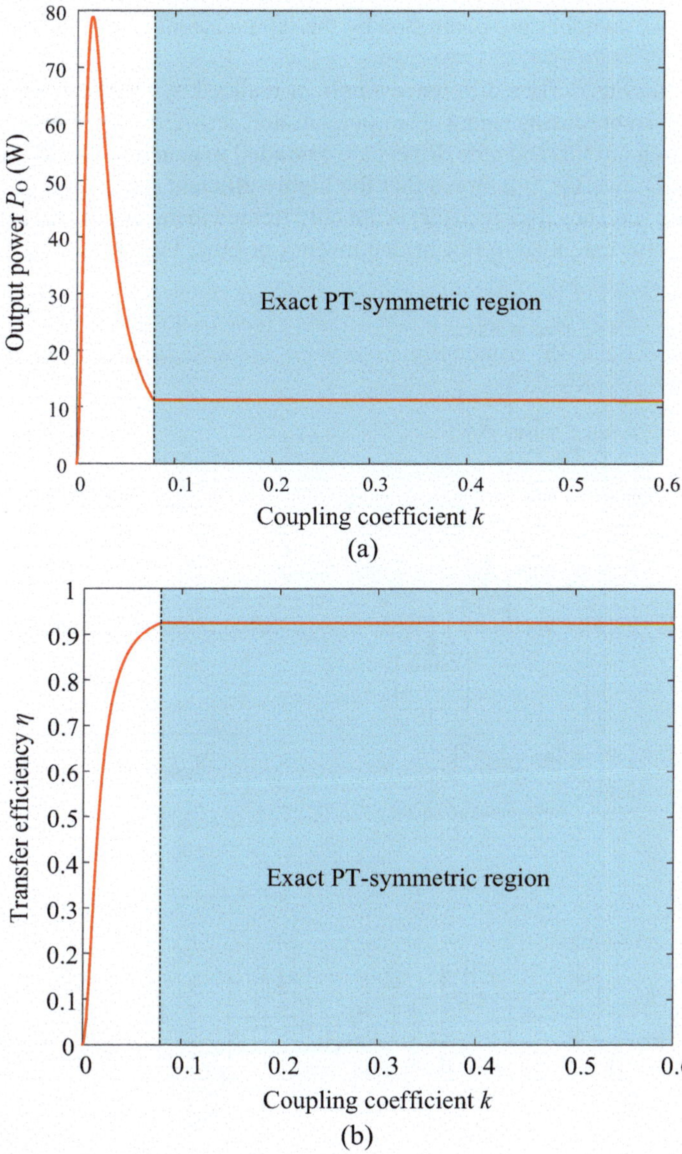

Fig. 3.6 Output power P_O and transfer efficiency η versus coupling coefficient k. **a** Output power. **b** Transfer efficiency

complementary switches are controlled by the input current across the transmitting coil.

As shown in Fig. 3.7b, a difference amplifier realized by OPA2690 is adapted to amplify the current sensing signal. The compensator, zero-crossing comparator, dead time generation circuit, and gate drivers are cascaded to generate the driving signals for the class D inverter. It is noted that the highly efficient realization of nonlinear saturable gain (i.e., negative resistor) is not only limited to the use of class D inverter, but it can be also extended to full-bridge inverter or other high frequency inverter.

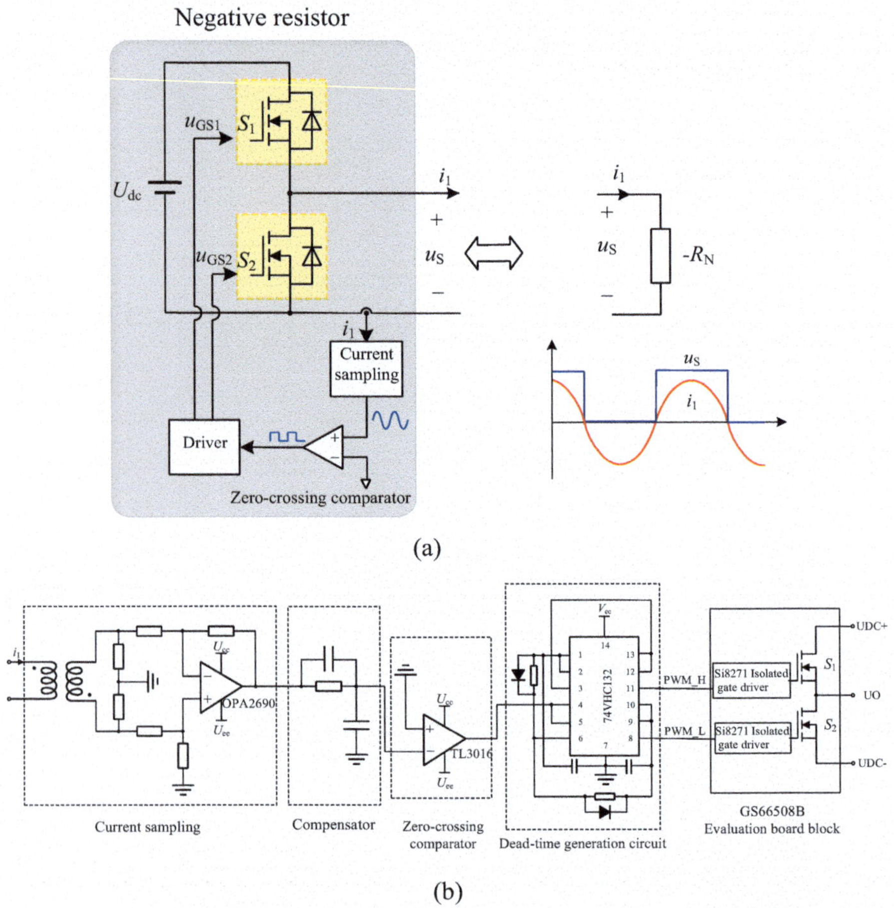

Fig. 3.7 Negative resistance $-R_N$ implementation. **a** The full-bridge inverter realizing $-R_N$. **b** The control circuit of $-R_N$

3.4.2 System Implementation

Figure 3.8 shows the experimental prototype of the PT-symmetric MC-WPT system for UAV. To obtain accurate experimental results, the drone is statically positioned on the platform above the transmitter. The position of the transmitting coil is fixed while the drone with the receiving coil can be adjusted along the *x*-*y*-*z* direction in a three-dimensional way. All circuit parameters of setup components and models are listed in Table 3.3, and the critical coupling coefficient can be calculated as $k_C = 0.092$. In addition, the experimental values of the coupling coefficients were measured by using a precision impedance analyzer (Wayne Kerr 6500B), and the numerical calculation and experimental measurement results are shown in Fig. 3.9.

The current probe and oscilloscope can be used to record the current waveform of the transmitting coil and the receiving coil in the experiment, and read the frequency and RMS value of the current, so that the experimental value of the output power and transmission efficiency of the system can be easily obtained. The experimental values are obtained from $P_o = I_2^2 R_L$ and $\eta = I_2^2 R_L/[I_2^2 R_1 + I_2^2 (R_2 + R_L)]$, where R_1 and R_2 are the internal resistances of the transmitting side and the receiving side respectively, which are measured by the precision impedance analyzer.

As can be seen from Fig. 3.9, when the receiving coil and the transmitter coil are axially aligned and the axial transfer distance varies from 0 to 10 cm, the coupling

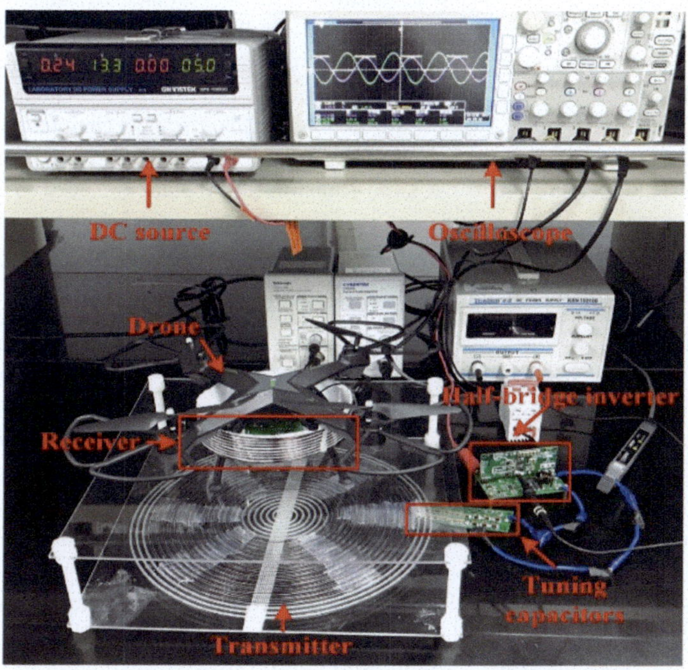

Fig. 3.8 Experimental prototype of the PT-symmetric MC-WPT system

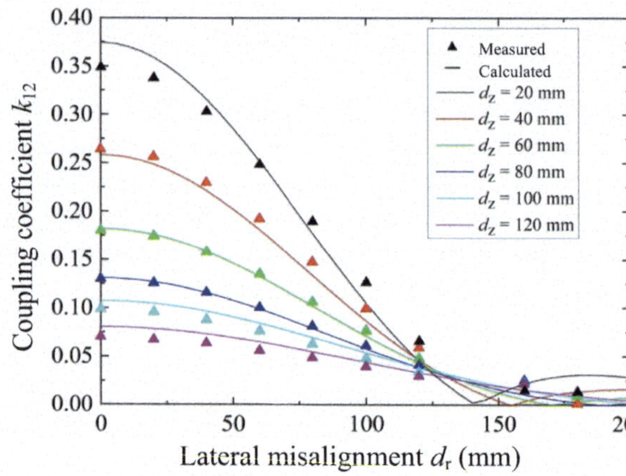

Fig. 3.9 The coupling coefficient k_{12} varies with the axial distance d_z and the lateral misalignment d_r

coefficient between the coils is greater than the critical coupling coefficient. Therefore, the axial transfer distance between the coils varies by 2–16 cm every 2 cm, and the experimentally measured data of the operating frequency, output power, and transfer efficiency of the system are shown in Fig. 3.10. Because the transmitting coil and receiving coil are both axisymmetric, the center-to-center distance d_r can be used to describe the lateral misalignment while d_z denotes the transfer distance. Here, the coaxial aligned cases with various transfer distance d_z, and the lateral misaligned cases against lateral misalignment d_r with various transfer distance d_z are considered. It can be seen from Fig. 3.10 that the system can maintain a constant transfer efficiency of 10 W at 93.6% within the range of 10 cm, which is consistent with the theoretical analysis. When the axial transfer distance changes within the range of 10 cm, the working frequency of the system changes with the change of the transfer distance, and when the axial transmission distance exceeds 10 cm, the working frequency of the system is equal to the natural resonant frequency of the system, which means that the system begins to enter the broken PT-symmetric state, and the output power and efficiency of the system are no longer constant.

When the axial transfer distance between the transmitter and receiving coils is fixed, and the receiver coil is moved along the diameter of the transmitter coil, the experimental values of the output power and efficiency of the system are shown in Fig. 3.11, which are consistent with the theoretical curve. Under different axial transfer distances, even if there is a certain lateral misalignment between the transmitting coil and the receiving coil, the system can obtain constant output power and transfer efficiency, and it can be found that within the range of 10 cm axial transfer distance, the smaller the axial transfer distance, the greater the lateral misalignment of the system. The transmitting and receiving coils are at an axial transfer distance of 8 cm, and the system is still in a PT-symmetric state within the lateral misalignment of 6 cm, which means that the UAV can hover and charge at a height of 8 m on the WPT platform, and can obtain constant power and constant transfer efficiency within 6 cm from the center of the platform.

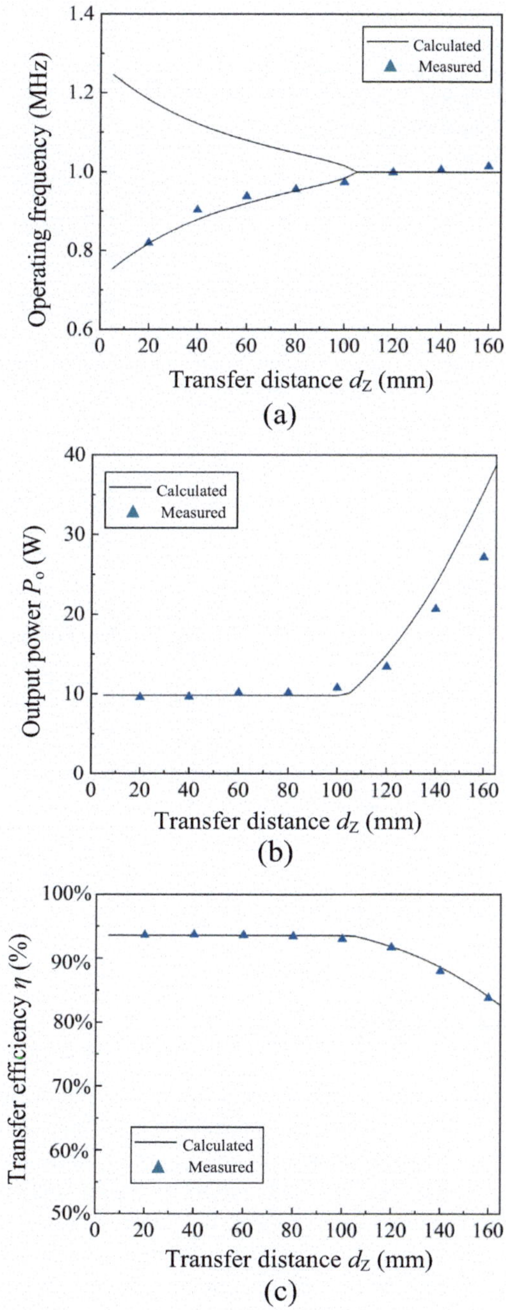

Fig. 3.10 Transfer performance of the prototype for the coaxial aligned case. **a** The output power versus the transfer distance d_Z. **b** The transfer efficiency versus the transfer distance d_Z. **c** The operating frequency versus the transfer distance d_Z

Fig. 3.11 Transfer performance of the experimental prototype with lateral misalignment. **a** output power. **b** Transfer efficiency

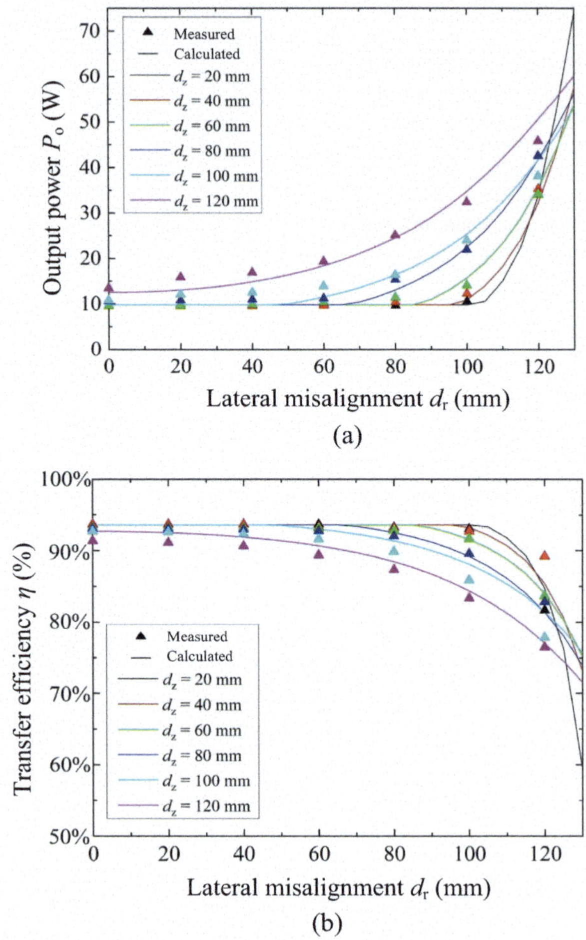

3.5 Summary

This chapter first introduces the basic structure and four kinds of compensated topologies of the PT-symmetric MC-WPT system, including SS, PP, SP, and PP topologies. Then, the circuit model and coupled-mode model of the PT-symmetric MC-WPT system are established, and the comparison between the circuit model and coupled-mode model is discussed. It is demonstrated that the coupled-mode model can be equivalent to the circuit model under certain approximate conditions. Meanwhile, the fundamental characteristics of the PT-symmetric MC-WPT system are analyzed, including operating frequency, voltage and current ratios, output power, and transfer efficiency. Finally, an experimental prototype of the PT-symmetric MC-WPT system is built, and the experimental results show that the PT-symmetric MC-WPT system has constant output power and transfer efficiency in the exact PT-symmetric region.

References

1. Hong H, Yang D, Won S (2017) The analysis for selecting compensating capacitances of two-coil resonant wireless power transfer system. In: 2017 IEEE international conference on energy internet (ICEI), Beijing, China, pp 220–225
2. Shevchenko V, Husev O, Strzelecki R, Pakhaliuk B, Poliakov N, Strzelecka N (2019) Compensation topologies in IPT systems: standards, requirements, classification, analysis, comparison and application. IEEE Access 7:120559–120580
3. Kiani M, Ghovanloo M (2012) The circuit theory behind coupled-mode magnetic resonance-based wireless power transmission. IEEE Trans Circ Syst I Regul Pap 59(9):2065–2074
4. Cheon S, Kim YH, Kang SY, Lee ML, Lee JM, Zyung T (2011) Circuit-model-based analysis of a wireless energy-transfer system via coupled magnetic resonances. IEEE Trans Ind Electron 58(7):2906–2914
5. Schindler J, Li A, Zheng MC, Ellis FM, Kottos TJPRA (2011) Experimental study of active LRC circuits with PT-symmetries. Phys Rev A 84(4):040101
6. Sakhdari M, Hajizadegan M, Li Y, Cheng MMC, Hung JCH, Chen PY (2018) Ultrasensitive, parity-time-symmetric wireless reactive and resistive sensors. IEEE Sens J 18(23):9548–9555
7. Kurs A, Karalis A, Moffatt R, Joannopoulos JD, Fisher P, Soljacic MJS (2007) Wireless power transfer via strongly coupled magnetic resonances. Science 317(5834):317
8. Haus HA (1984) Waves and fields in optoelectronics. Prentice-Hall, Englewood Cliffs, NJ
9. Louisell WH (1960) Coupled mode and parametric electronics. Wiley, New York, NY, USA
10. Assawaworrarit S, Yu X, Fan S (2017) Robust wireless power transfer using a nonlinear parity-time-symmetric circuit. Nature 546(7658):387–390
11. Rong C, Zhang B, Wei Z, Wu L, Shu X (2022) A wireless power transfer system for spinal cord stimulation based on generalized parity-time symmetry condition. IEEE Trans Ind Appl 58(1):1330–1339
12. Sanders JA, Verhulst F (1985) Averaging methods in nonlinear dynamical systems. Springer-Verlag, New York
13. Zhou J, Zhang B, Xiao W, Qiu D, Chen Y (2019) Nonlinear parity-time-symmetric model for constant efficiency wireless power transfer: application to a drone-in-flight wireless charging platform. IEEE Trans Ind Electron 66(5):4097–4107
14. Li H, Wang K, Huang L, Chen W, Yang X (2015) Dynamic modeling based on coupled modes for wireless power transfer systems. IEEE Trans Power Electron 30(11):6245–6253
15. Wu L, Zhang B, Zhou J (2020) Efficiency improvement of the parity-time-symmetric wireless power transfer system for electric vehicle charging. IEEE Trans Power Electron 35(11):12497–12508
16. Hassan AU, Hodaei H, Miri MA, Khajavikhan M, Christodoulides DN (2015) Nonlinear reversal of the PT-symmetric phase transition in a system of coupled semiconductor microring resonators. Phys Rev A 92(6):063807
17. Özdemir ŞK, Rotter S, Nori F, Yang L (2019) Parity-time symmetry and exceptional points in photonics. Nat Mater 18(8):783–798

Chapter 4
Transfer Characteristics of Constant Power Against Distance and Misalignment

At present, the MC-WPT system has been widely used in portable electronic devices, electric vehicles, and kitchen appliances. Stable output power is one of the critical factors to evaluate an MC-WPT system. The PT-symmetric MC-WPT system can achieve inherently constant output power with constant transfer efficiency independent of the coupling coefficient in the exact PT-symmetric region (i.e., the strong coupling region), as described in Sect. 3.3.3. In other words, the PT-symmetric MC-WPT system is especially suitable for domestic kitchen appliances equipped with resistive heating elements (such as toaster, hot water pot, and rice cooker), because these loads are constant and given ahead, resulting in negligible changes in load conditions. Therefore, this chapter mainly considers the variation of the coupling coefficient (including transfer distance and misalignment), and introduces a PT-symmetric MC-WPT system with stable output power in the whole coupling range. Here, the self-oscillating (SO) mode is activated in the strong coupling region to ensure a constant output power and constant transfer efficiency against the variation of the coupling coefficient. In the weak coupling region, the pulse width modulation (PWM) mode with a fixed frequency and a variable duty cycle is adopted, and a control strategy based on transmitter-only parameters is presented to obtain stable output power.

4.1 System Structure and Characteristic Analysis

The circuit diagram of the PT-symmetric MC-WPT system is shown in Fig. 4.1, where the inverter powered by a DC source U_{dc} produces an AC square voltage u_{ac}, and by using a combination of self-oscillating method and pulse width modulation (SO-PWM), the phase difference between the fundamental voltage of the inverter and its output current is always maintained at 180° [1]. Thereby, the inverter can be equivalent to a negative resistor $-R_N$, which provides gain to the transmitter

B. Zhang et al., *Parity-Time Symmetric Wireless Power Transfer*, CPSS Power Electronics Series, https://doi.org/10.1007/978-981-96-3149-0_4

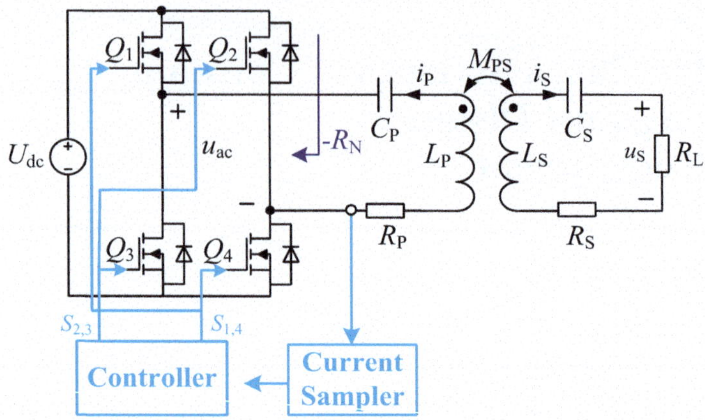

Fig. 4.1 Circuit diagram of the PT-symmetric MC-WPT system

resonator. The transmitter and receiver are coupled together by inductor L_P and L_S with mutual inductance M_{PS}. C_P and C_S are the capacitances of the tuning capacitors. R_P and R_S are the internal resistances of the transmitter and receiver, respectively. The currents of transmitting and receiving coils are denoted as i_P and i_S, respectively. u_S is the output voltage. R_L is the resistance of the heating element.

4.1.1 Operating Frequency

The equivalent circuit of the system is described in Fig. 4.2, where u_P is the fundamental component of u_{ac}. For simplification, only the fundamental component of voltage and current is considered. Therefore, based on Kirchhoff's voltage law (KVL), the system can be described by the following phasor equation:

$$\begin{bmatrix} \frac{-R_N+R_P}{L_P}+j\left(\omega-\frac{\omega_0^2}{\omega}\right) & j\omega k_{PS}\sqrt{\frac{L_S}{L_P}} \\ j\omega k_{PS}\sqrt{\frac{L_P}{L_S}} & \frac{R_L+R_S}{L_S}+j\left(\omega-\frac{\omega_0^2}{\omega}\right) \end{bmatrix}\begin{bmatrix} \dot{I}_P \\ \dot{I}_S \end{bmatrix}=0, \tag{4.1}$$

Fig. 4.2 Equivalent circuit of the PT-symmetric MC-WPT system

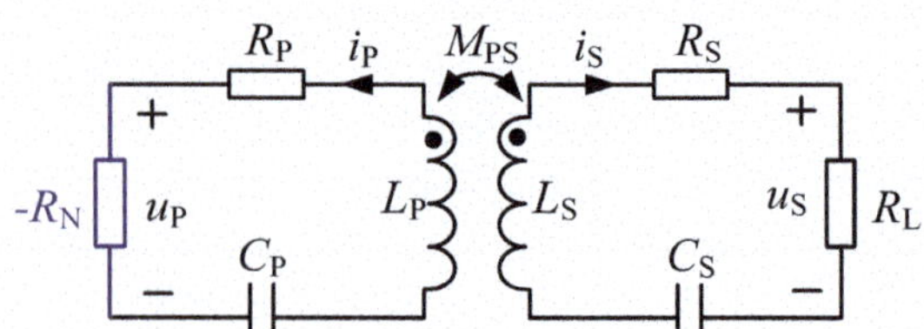

where ω is the operating frequency of the inverter, and $\omega_0 = 1/\sqrt{L_P C_P} = 1/\sqrt{L_S C_S}$ is the resonant frequency of the transmitter and receiver. $k_{PS} = M_{PS}/\sqrt{L_P L_S}$ is the coupling coefficient between the transmitting and receiving coils.

To find a steady solution with real ω, the real and imaginary parts of the determinant of the coefficient matrix of (4.1) are separated as

$$\begin{cases} \left(\omega - \dfrac{\omega_0^2}{\omega}\right)\left(\dfrac{R_L + R_S}{L_S} + \dfrac{-R_N + R_P}{L_P}\right) = 0 \\ \dfrac{-R_N + R_P}{L_P}\dfrac{R_L + R_S}{L_S} - \left(\omega - \dfrac{\omega_0^2}{\omega}\right)^2 + \omega^2 k_{PS}^2 = 0 \end{cases}. \tag{4.2}$$

Then, the steady operating angular frequency ω can be derived as

$$\omega = \begin{cases} \omega_0, \ k_{PS} < k_C \\ \omega_{H,L} = \omega_0 \sqrt{\dfrac{2 - \gamma^2 \pm \sqrt{(2-\gamma^2)^2 + 4(k_{PS}^2 - 1)}}{2(1 - k_{PS}^2)}}, \ k_C \le k_{PS} \le 1 \end{cases}, \tag{4.3}$$

where $\gamma = (R_L + R_S)/(\omega_0 L_S)$, and $k_C = \sqrt{\gamma^2 - \gamma^4/4}$ is the critical coupling coefficient.

According to (4.3), the steady operation region can be divided into the exact PT-symmetric region (i.e., the strong coupling region $k_{PS} \ge k_C$) and the broken PT-symmetric region (i.e., the weak coupling region $k_{PS} < k_C$). In the exact PT-symmetric region, the system supports two steady states with operating angular frequencies $\omega = \omega_{H,L}$. Here, $\omega = \omega_0$ is also a solution of (4.1) in the exact PT-symmetric region, but this mode has been proven to be unstable [2]. In the broken PT-symmetric region, only one operating angular frequency is located at $\omega = \omega_0$.

4.1.2 *Self-oscillating Operating Mode*

In the exact PT-symmetric region, the inverter is controlled by a self-oscillating (SO) control strategy, where the output current of the inverter is sensed and then converted into two complementary square waves by a zero-crossing comparator and a dead-time generation circuit to drive the inverter [3]. Consequently, the phase difference between u_P and i_P will automatically remain at 180°, as shown in Fig. 4.3. The output of the full-bridge inverter can always be equivalent as a negative resistance, thereby ensuring that the system oscillates at $\omega = \omega_{H,L}$ [4]. Ignoring the dead time, the fundamental component of the inverter output voltage can be expressed as $U_P = 2\sqrt{2}\,U_{dc}/\pi$, because the duty cycle is fixed at 0.5.

In the exact PT-symmetric region, due to $\omega = \omega_{H,L} \neq \omega_0$ and (4.2), the current ratio can be derived as

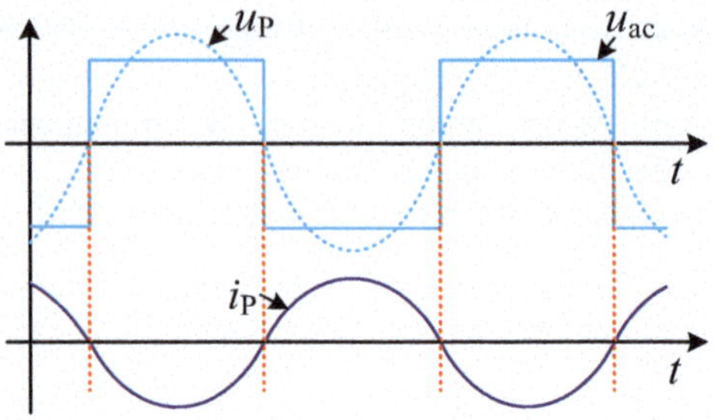

Fig. 4.3 Steady-state voltage and current of $-R_N$ in the SO operating mode

$$\frac{I_P}{I_S} = \left|\frac{\dot{I}_P}{\dot{I}_S}\right| = \sqrt{\frac{L_S}{L_P}}. \tag{4.4}$$

Meanwhile, due to $U_P = R_N I_P$, and $U_S = R_L I_S$, the voltage ratio can be deduced as

$$\frac{U_P}{U_S} = \left|\frac{\dot{U}_P}{\dot{U}_S}\right| = \sqrt{\frac{L_P}{L_S}}\frac{R_L + R_S}{R_L} + \sqrt{\frac{L_S}{L_P}}\frac{R_P}{R_L}. \tag{4.5}$$

Thus, the transfer efficiency and output power can be obtained as

$$\eta_{T\_SO} = \frac{I_S^2 R_L}{I_P^2 R_P + I_S^2 R_S + I_S^2 R_L} = \frac{R_L}{\frac{L_S}{L_P} R_P + R_S + R_L}, \tag{4.6}$$

$$P_{o\_SO} = \frac{U_S^2}{R_L} = \frac{8U_{dc}^2 R_L}{\pi^2\left[\frac{L_P}{L_S}(R_L + R_S)^2 + 2R_P(R_L + R_S) + \frac{L_S}{L_P}R_P^2\right]}. \tag{4.7}$$

It can be found that the constant output power with constant transfer efficiency against the variation of coupling coefficient k_{PS} can be achieved, which guarantees the robustness of the MC-WPT system regardless of transfer distance and misalignment in the strong coupling region.

4.1.3 PWM Operating Mode

In the broken PT-symmetric region, the PWM operating mode with a fixed frequency (i.e., $\omega = \omega_0$) and a variable duty cycle D_1 $(0 \le D_1 \le 0.5)$ is adopted. In this case, since both transmitter and receiver are fully tuned, the phases of u_P and i_P are opposite, as shown in Fig. 4.4. In other words, the negative resistor can be also achieved, and the fundamental component of the inverter output voltage can be expressed as $U_P = 2\sqrt{2}\sin(D_1\pi)U_{dc}/\pi$ by Fourier analysis of Fig. 4.4.

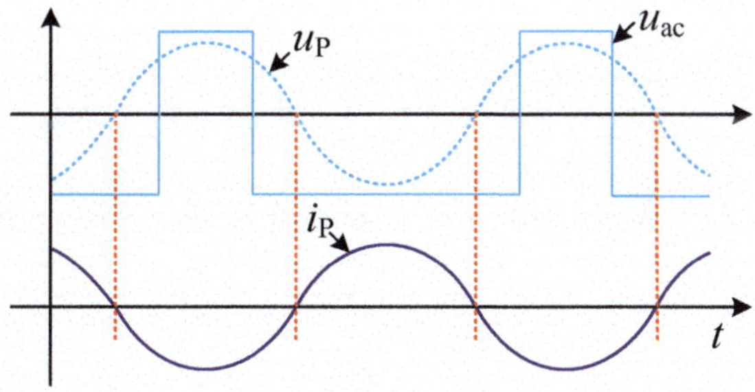

Fig. 4.4 Steady-state voltage and current of $-R_N$ in the PWM operating mode

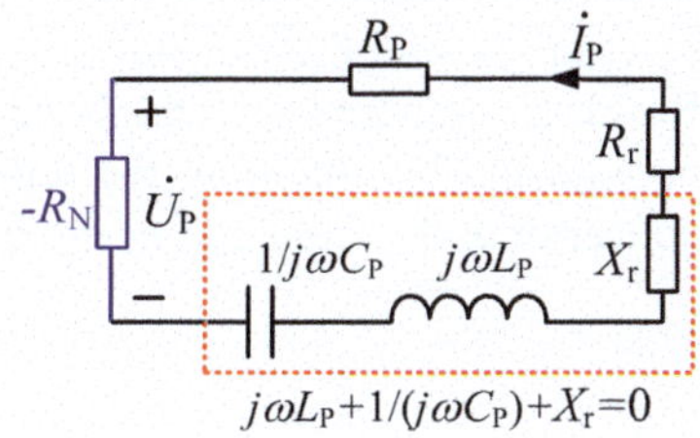

Fig. 4.5 Equivalent circuit on the transmitting side

Figure 4.5 shows the equivalent circuit on the transmitting side, where X_r and R_r represent the reflected reactance and the reflected resistance from the receiving side, respectively.

According to Kirchhoff's law, the steady equation of the circuit shown in Fig. 4.5 can be written as

$$(-R_N + R_P + R_r)\dot{I}_P = 0 \tag{4.8}$$

Due to $R_N = U_P/I_P$ and $U_P = 2\sqrt{2}\sin(D_1\pi)U_{dc}/\pi$, the reflected resistance R_r can be expressed as

$$R_r = \frac{2\sqrt{2}\sin(d_1\pi)U_{dc}}{\pi I_P} - R_P. \tag{4.9}$$

Then the power transferred from the transmitter to the receiver P_S can be calculated as

$$P_T = I_P^2 R_r. \tag{4.10}$$

Combining (4.9) and (4.10), the transfer efficiency of the PWM operating mode can be described as

$$P_{\mathrm{o\_PWM}} = P_{\mathrm{T}}\frac{R_{\mathrm{L}}}{R_{\mathrm{L}}+R_{\mathrm{S}}} = \frac{\left[\frac{2\sqrt{2}\sin(d_1\pi)U_{\mathrm{dc}}}{\pi}\right]^2\left[\frac{2\sqrt{2}\sin(d_1\pi)U_{\mathrm{dc}}}{\pi I_{\mathrm{P}}}-R_{\mathrm{P}}\right]}{\left[R_{\mathrm{P}}+\left(\frac{2\sqrt{2}\sin(d_1\pi)U_{\mathrm{dc}}}{\pi I_{\mathrm{P}}}-R_{\mathrm{P}}\right)\right]^2}\frac{R_{\mathrm{L}}}{R_{\mathrm{L}}+R_{\mathrm{S}}}. \tag{4.11}$$

By submitting $\omega = \omega_0$ into (4.1), the transfer efficiency can be obtained as

$$\eta_{\mathrm{T\_PWM}} = \frac{\left(\omega k_{\mathrm{PS}}\sqrt{L_{\mathrm{P}}L_{\mathrm{S}}}\right)^2 R_{\mathrm{L}}}{R_{\mathrm{P}}(R_{\mathrm{L}}+R_{\mathrm{S}})^2+(R_{\mathrm{L}}+R_{\mathrm{S}})\left(\omega k_{\mathrm{PS}}\sqrt{L_{\mathrm{P}}L_{\mathrm{S}}}\right)^2}. \tag{4.12}$$

It can be found that the output power can be regulated by the output voltage of the inverter U_{P} and the transmitting current I_{P}, regardless of the coupling coefficient. Besides, compared with the SO control mode, the output voltage of the inverter can be adjusted by changing the duty cycle D_1 without adding an additional DC-DC converter at the front end of the inverter.

4.2 Control and Implementation

4.2.1 Mode Switching Condition

Based on the analysis above, the SO operating mode is selected in the strong coupling region, while the PWM operating mode is chosen in the weak coupling region. The coupling strength can be indirectly judged by the value of the reflected resistance. Therefore, the reflection resistance at k_{C} is defined as the critical reflection resistance R_{C}, which can be given by submitting $\omega = \omega_0$ and $k_{\mathrm{PS}} = k_{\mathrm{C}}$ into (4.1), there is

$$R_{\mathrm{C}} = \frac{\omega_0^2 k_{\mathrm{C}}^2 L_{\mathrm{P}}L_{\mathrm{S}}}{R_{\mathrm{L}}+R_{\mathrm{S}}} = \frac{\omega_0^2 L_{\mathrm{P}}L_{\mathrm{S}}\left(\gamma^2-\gamma^4/4\right)}{R_{\mathrm{L}}+R_{\mathrm{S}}}. \tag{4.13}$$

It can be seen that the critical reflection resistance R_{C} is only related to the inherent parameters of the system, and has nothing to do with the coupling coefficient. When the actual value of reflected resistance is larger than the critical reflected resistance (i.e., $R_{\mathrm{r}} \geq R_{\mathrm{C}}$), it means that the system is in a strong coupling region. Otherwise, it is in a weak coupling region. It should be noted that since the phase difference between u_{P} and i_{P} always maintains 180°, the actual value of the reflected resistance in the whole coupling region can be calculated by (4.9).

As can be seen from (4.9), when estimating the reflected resistance R_{r}, only the RMS value I_{P} of the transmitting coil current needs to be sampled, so there is no need for wireless communication and real-time coupling coefficient detection. However, a preset-value of R_{P} is required in the calculation. Under normal circuit operating conditions, R_{P} is generally considered to be constant, but due to the proximity effect,

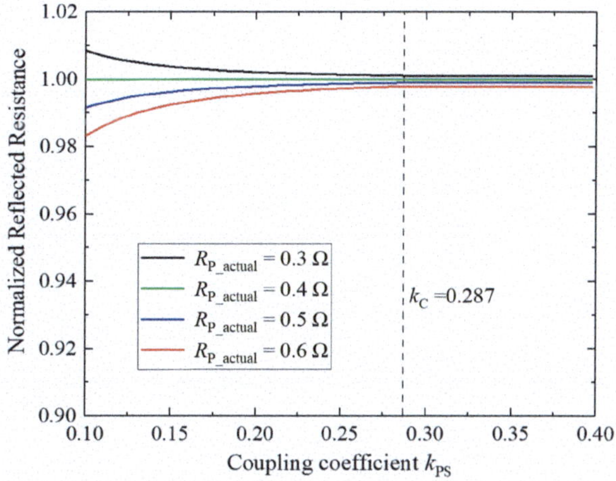

Fig. 4.6 Normalized reflected resistance under the different internal resistances and coupling coefficients

the variable operating frequency will cause a slight change in the internal resistance of the coil. In this design, the internal resistance of the transmitter R_P was measured offline using a precision impedance analyzer at the resonant frequency, and the maximum change of the internal resistance of the coil is 10%.

Here, the ratio of the actual value of the reflected resistance R_{r_actual} to the estimated value $R_{r\_estimated}$ is defined as the normalized reflected resistance, R_{r_actual} and $R_{r\_estimated}$ can be calculated from $R_{r\_actual} = U_P/I_P\text{-}R_{P\_actual}$ and $R_{r\_estimated} = U_P/I_P\text{-}R_{P\_preset}$, respectively, where R_{P_actual} and R_{P_preset} are the actual-vale and preset-value of R_P, respectively. Assuming the preset-value of R_P is 0.4 Ω (i.e., $R_{P\_preset} = 0.4$ Ω), the normalized reflected resistance under different internal resistances and coupling coefficients are shown in Fig. 4.6. It can be found that even if the actual internal resistance R_P changes by 25 and 50% relative to the preset-value, the relative error in reflected resistance is within 0.9 and 1.69%. Moreover, with the increase of the coupling coefficients, the error gradually decreases. The errors at the critical coupling coefficient $k_C = 0.287$ are only 0.1 and 0.42%, respectively. Therefore, the influence of R_P changes on the estimation of the reflected resistance can be negligible.

4.2.2 *Control Scheme*

Figure 4.7 shows the steady-state operating waveform and driving signal of the inverter based on SO-PWM control, where i'_P is the square wave signal with the same frequency and phase as the output current i_P of the inverter, PWM signal is the drive control signal generated by the digital controller DSP, S_m is the mode control signal, S_{1_4} and S_{2_3} are the two-way drive signals of the full-bridge inverter, which control the opening and closing of MOSFET tubes Q_1 (Q_4) and Q_2 (Q_3), respectively. As can be seen from Fig. 4.7, the fundamental component u_P of the output voltage of

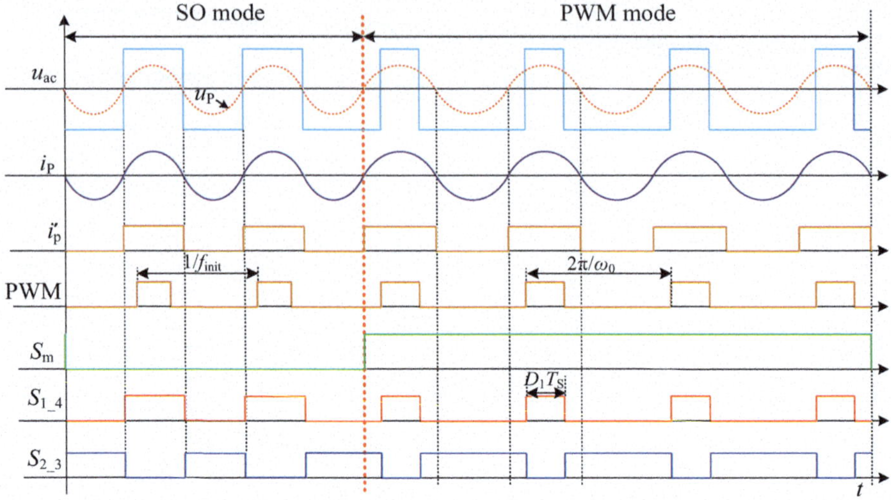

Fig. 4.7 The steady-state operating waveforms and driving signals of the inverter based on SO-PWM control

the inverter will be in phase with its output current i_P, which indicates that the output characteristic of the inverter is equivalent to a negative resistor. Besides, when the system operates in the strong coupling region, the SO control mode is activated. In this case, the mode control signal S_m is low level, the PWM signal is shielded, and the drive signals S_{1_4} and S_{2_3} are generated according to the square wave signal i'_P. When the system works in the weak coupled region, it switches to PWM control mode. In this case, the square wave signal i'_P is shielded, and the driver signals S_{1_4} and S_{2_3} are generated according to the PWM signal. The frequencies of S_{1_4} and S_{2_3} should always track the natural frequencies of the system, and the duty cycle D_1 of S_{1_4} is dynamically adjusted by (4.11). The coupling region of the system is judged according to the actual reflection resistance.

Figure 4.8 gives the control flowchart of the system, the inverter operation is initiated at frequency ω_0, the mode control signal S_m is initiated to high level. Then the real reflected resistance R_r is estimated by (4.11). According to (4.3), two periodic steady-states ω_1 and ω_2 can coexist under the exact PT-symmetric region. Since the periodic waveforms in the time domain correspond to limit cycles in the state space, the system can evolve into one state or another depending on the initial conditions, and the final state will tend to evolve into the same state as the starting state [5, 6], so a smaller one is selected to reduce the switching losses of the inverter. From this, when R_r is larger than the critical reflected resistance R_C, the frequency of PWM wave is first set to be lower than ω_0, so that the system has a low-frequency initial state. Then, the mode control signal S_m is set to a low level, and the SO operating mode is activated. In this case, the constant output power can be achieved automatically. Otherwise, when R_r becomes smaller than the critical reflected resistance R_C, the control mode change to the PWM mode. In this mode, the output power is regulated

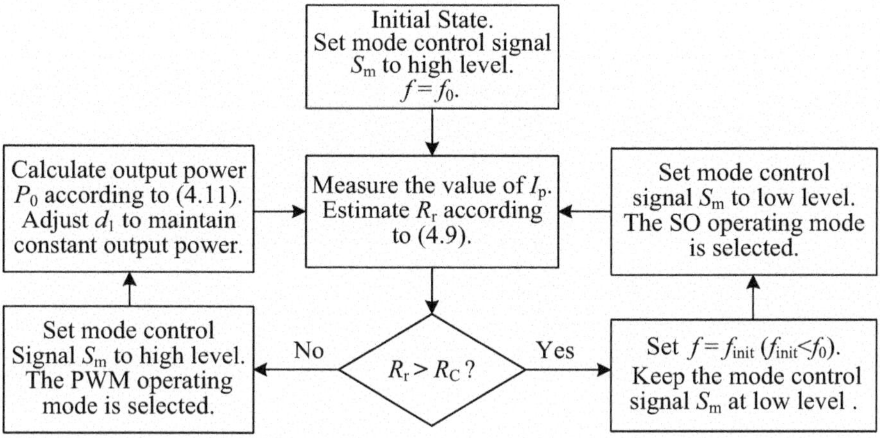

Fig. 4.8 Flowchart of the closed-loop control

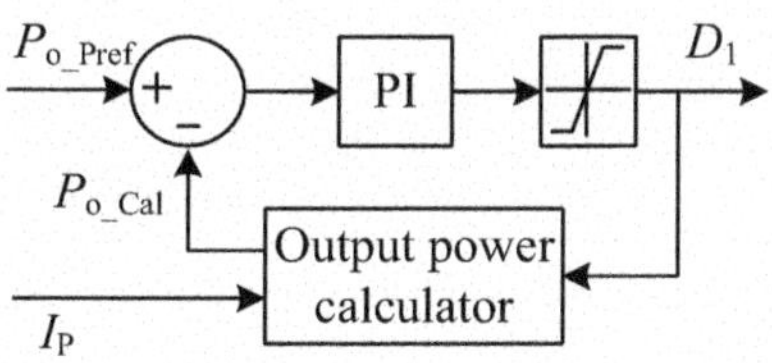

Fig. 4.9 Output power control block diagram in the PWM operating mode

by adjusting the duty cycle D_1. The output power control block diagram in the PWM mode is shown in Fig. 4.9, where the sampled value of I_P and the duty cycle D_1 are fed into the output power calculator to evaluate output power P_{o_Cal} in real-time according to (4.11), and then compare it with the reference output power P_{o_Ref}. the difference $\Delta P = P_{o\_Ref}$-P_{o_Cal} is nullified with the help of a proportional-integral (PI) controller. Particularly, the output of PI controller is the value of the required duty D_1, which is sent to the microcontroller pulse PWM peripheral to drive the inverter switches Q_1 ~ Q_4.

4.3 System Design and Verification

Figure 4.10 shows the structure block diagram of the PT-symmetric MC-WPT system based on the SO-PWM control. The overall architecture of the system is divided into two parts: the main circuit and the control circuit. The main circuit includes a full-bridge inverter, series compensation networks, and load, where the full-bridge inverter is composed of four MOSFETs (Q_1 ~ Q_4), its output characteristics are equivalent to a negative resistor $-R_N$ based on the SO-PWM control.

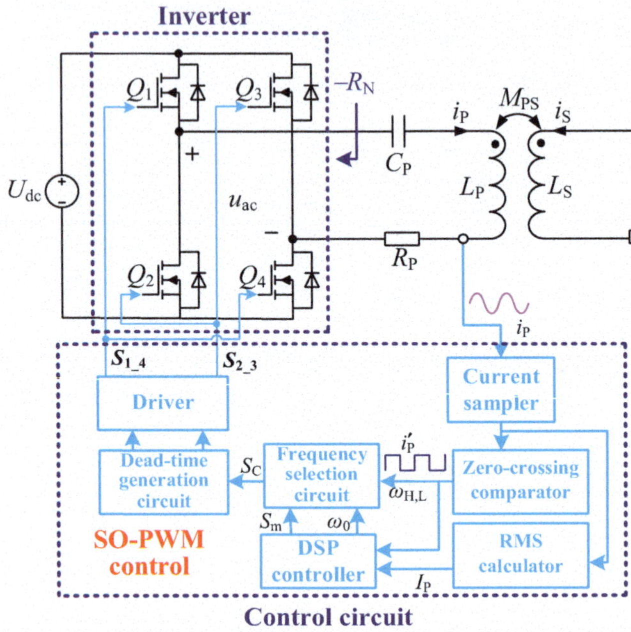

Fig. 4.10 PT-symmetric MC-WPT system based on the SO-PWM control

Figure 4.11 gives the scheme of the controller and current sampler, where a current transformer (CT) CU8965 and a differential amplifier circuit are first used to convert the current i_P into a voltage signal, and then the output signal of the differential amplifier circuit passes through a zero-crossing comparator (TL3016) and an RMS calculator (AD637), respectively. A digital-signal-processor (TMS320F28377D) is employed to control the operating mode according to the reflected resistance R_r. In the strong coupling region, the mode control signal S_m is set to a low level, then the driving control signal S_C of the inverter is generated by a zero-crossing comparator, so that the inverter automatically oscillates at $\omega = \omega_{H,L}$ in response to the change of coupling conditions. In the weak coupling region, the mode control signal S_m becomes a high level, the processor processes the output power control algorithm and then outputs a PWM wave with a fixed frequency ω_0 and a required duty cycle d_1 to generate the driving control signal S_C. Finally, a dead-time generation circuit is used to convert the driving control signal S_C into two complementary square waves with dead time to drive the inverter through drive circuits (Si8271) [2].

Figure 4.12 shows the experimental prototype, and the detailed circuit parameters are listed in Table 4.1. In this design, the rated output power P_o delivered to the load resistor is set as 500 W, and according to (4.7), the required DC input voltage can be obtained as $U_{dc} = 248$ V. The coils are fabricated using the Litz wire with 0.05 mm × 800 strands and 24 turns, and the inner and outer diameter of the coil are 16 and 26 cm, respectively (Fig. 4.12).

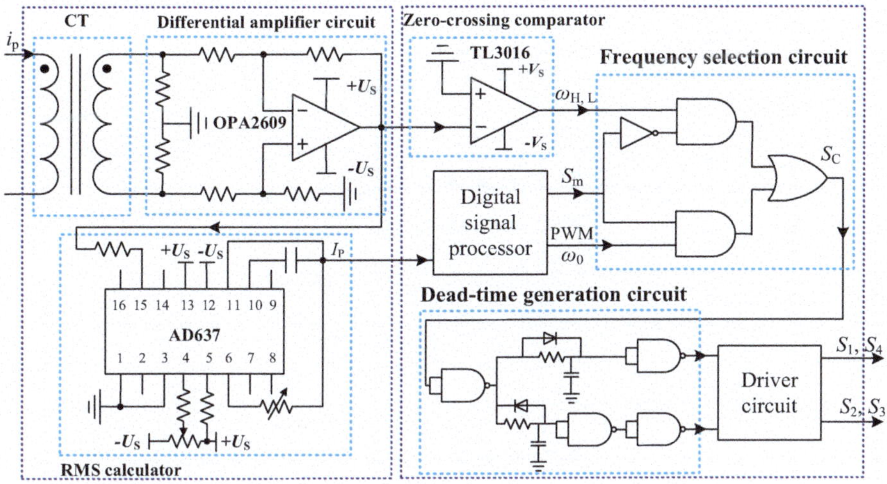

Fig. 4.11 Scheme of the controller and current sampler

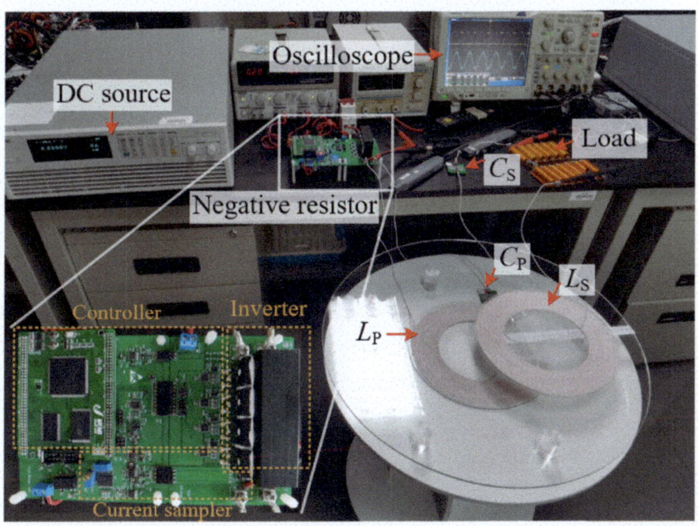

Fig. 4.12 Experiment setup of the PT-symmetric MC-WPT system

gives the output power and transfer efficiency under different lateral misalignments, it can be seen that when the lateral misalignment variations from 0 to 15 cm occurs, the system can achieve a constant output power of around 500 W with maximum fluctuations within 3.6% while maintaining a transfer efficiency of more than 91.6%. In addition, the transfer efficiency is maintained near 99% in the exact

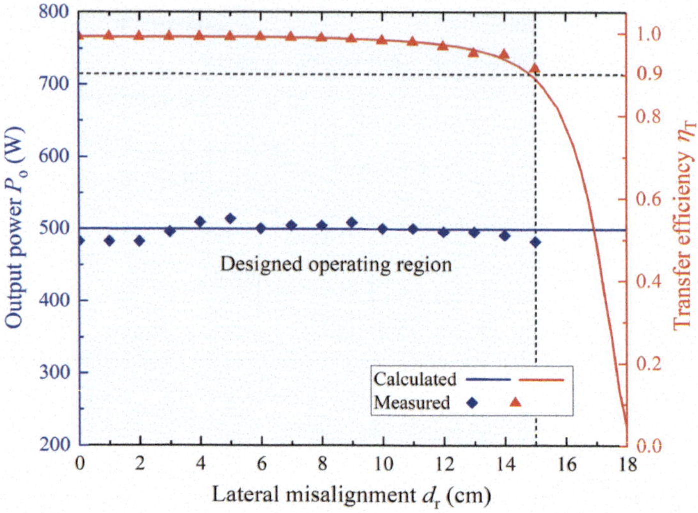

Fig. 4.13 Theoretical and experimental results of output power and transfer efficiency as a function of the lateral misalignment

Table 4.1 The circuit parameters of the experimental prototype

Description	Symbol	Values
Inductance of transmitting coil	L_P	180.3 μH
Inductance of receiving coil	L_S	180 μH
Compensation capacitance of transmitter	C_P	1.56 nF
Compensation capacitance of receiver	C_S	1.56 nF
Internal resistance of transmitter	R_P	0.41 Ω
Internal resistance of receiver	R_S	0.38 Ω
DC voltage source	U_{dc}	248 V
Resonant frequency	f_0	299.6 kHz
Load resistance	R_L	98 Ω
Output power	P_o	500 W

PT-symmetric region ($d_r < 5.5$ cm) and even at the worst misalignment conditions ($d_r = 15$ cm), the transfer efficiency can maintain about 91.6%, which is acceptable.

Figure 4.14 shows the comparison results between the system and the standard PT-symmetric MC-WPT system, it can be seen that in the broken PT-symmetric region, the output power of the standard PT-symmetric MC-WPT system no longer remains constant, and the huge power fluctuation will directly damage power components. Fortunately, the PT-symmetric MC-WPT system based on SO-PWM control can maintain a constant output power in the whole coupling region, while maintaining the same transfer efficiency as the standard PT-symmetric MC-WPT system. As can be seen from Fig. 4.14, when the transfer efficiency is greater than 91.6%, the misalignment tolerance can still be increased by 172%.

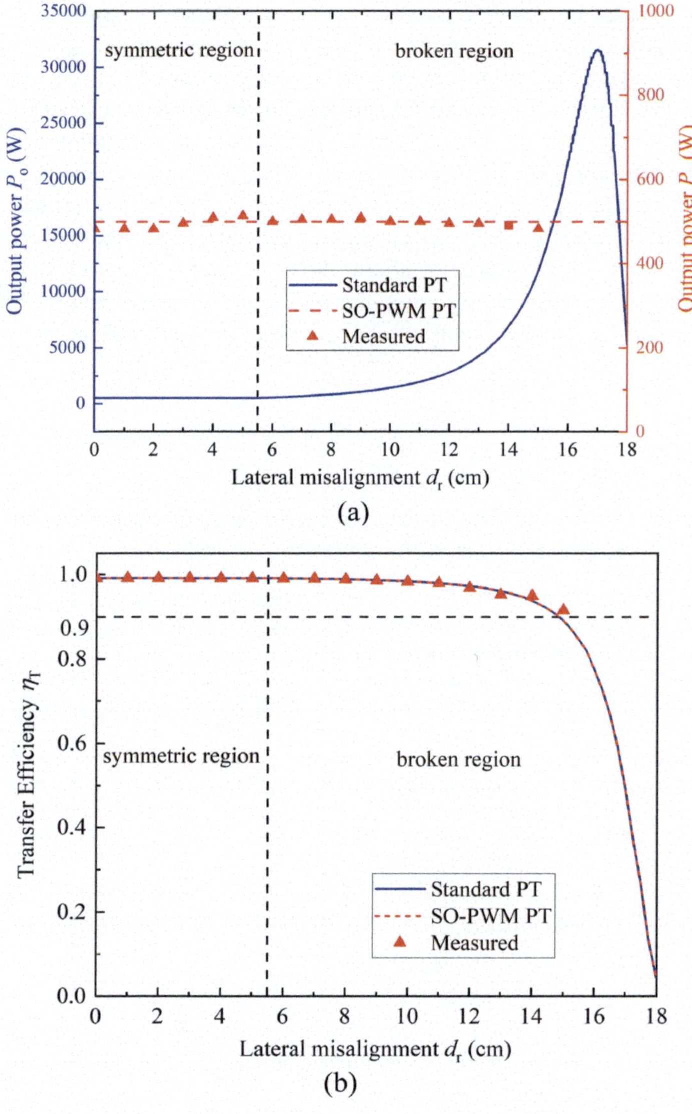

Fig. 4.14 Comparison with the standard PT-symmetric MC-WPT system of **a** output power and **b** transfer efficiency

4.4 Summary

This chapter mainly introduces a PT-symmetric MC-WPT system with stable output power in the whole coupling range based on the SO-PWM control method. Firstly, the system structure and circuit model are given, and the operating frequency characteristic is analyzed. Then, the SO and PWM operating modes and the corresponding system characteristics are discussed. Meanwhile, the influence of the parasitic inductance of the heating element is analyzed. In addition, the mode switching condition is derived, and the control scheme is given. Finally, a 500 W experimental prototype is implemented. The experimental results demonstrate that with a transfer distance of 5 cm, the PT-symmetric MC-WPT system using SO-PWM control method can achieve a constant output power of about 500 W within a range of 15 cm lateral misalignment, while maintaining a transfer efficiency of more than 91.6%, which proves the feasibility and reliability of application in kitchen appliances.

References

1. Wu L, Zhang B, Jiang Y, Zhou J (2022) A robust parity-time-symmetric WPT system with extended constant-power range for cordless kitchen appliances. IEEE Trans Ind Appl 58(1):1179–1189
2. Zhou J, Zhang B, Xiao W, Qiu D, Chen Y (2019) Nonlinear parity-time-symmetric model for constant efficiency wireless power transfer: application to a drone-in-flight wireless charging platform. IEEE Trans Ind Electron 66(5):4097–4107
3. Wu L, Zhang B, Zhou J (2020) Efficiency improvement of the parity-time-symmetric wireless power transfer system for electric vehicle charging. IEEE Trans Power Electron 35(11):12497–12508
4. Zhu H, Zhang B, Wu L (2020) Output power stabilization for wireless power transfer system employing primary-side-only control. IEEE Access 8:63735–63747
5. Aroudi AE, Huang J, Al-Numay MS, Li Z (2020) On the coexistence of multiple limit cycles in H-bridge wireless power transfer systems with zero current switching control. IEEE Trans Circ Syst I Regul Pap 67(5):1729–1739
6. Tang C-S, Sun Y, Dai X, Wang Z-H, Su Y-G, Hu A (2013) Transition control of bifurcated frequencies in inductive power transfer systems through time delay perturbation. Acta Phys Sin 62(15):158401–158401

Chapter 5
Transfer Characteristics of Constant Controllable Power Against Distance and Misalignment

The PT-symmetric theory opens up a new approach for the design and control of MC-WPT systems with misalignment tolerance (that is, designing and controlling MC-WPT systems from the perspective of realizing the PT-symmetric condition). Based on the PT-symmetric principle, the parameter design and control for achieving system anti-interference characteristics can be quickly obtained. However, for the existing PT-based MC-WPT system with constant output power, load changes are not considered in output power control [1, 2]. In practical applications, the high controllability of output power is required when the load resistance changes. Therefore, the main objective of this chapter is to propose a novel output power control strategy on the primary side for solving the issue of the output power fluctuation caused by the variation of the coupling coefficient and load in the magnetic coupling-based MC-WPT system. The proposed control strategy is realized by a primary-side controller and an inverter with front-end DC-DC converter, without the need for the information of mutual inductance, wireless communication of feeding back the load conditions to the primary side, and secondary-side control circuit, thereby compressing the volume of the secondary side, simplifying the control algorithm and improving the robustness of the system. Besides, the output power of the system can automatically keep stable at desired values under reasonable variation in load and coupling conditions by the proposed control strategy [3].

5.1 System Modeling and Characteristic Analysis

The circuit model of a series-series compensated MC-WPT system based on PT symmetry is shown in Fig. 5.1, where L_i, C_i, R_i, and i_i ($i = 1, 2$) denote the coil inductance, tuning capacitance, intrinsic loss resistance, and current of the primary-side and secondary-side resonators, respectively. M is mutual inductance between the two coils. The negative resistance $-R_{\text{n}}$ is the only gain element in the system that

B. Zhang et al., *Parity-Time Symmetric Wireless Power Transfer*, CPSS Power Electronics Series, https://doi.org/10.1007/978-981-96-3149-0_5

provides power to the primary LC loop. Through the magnetic coupling between the primary-side coil and secondary-side coil, the power transfers to the secondary-side loss elements, i.e., R_2 and load R_L. The reference positive direction of i_1 and i_2 are shown by the arrows in Fig. 5.1. The AC voltage u_1 of negative resistance is in phase with i_1, and u_2 is the output voltage, i.e., the voltage of load R_L.

Considering the fundamental harmonic of voltage and current, u_1, u_2, i_1, and i_2 can be easily expressed in terms of phasors $\dot{U}_1$, $\dot{U}_2$, $\dot{I}_1$, and $\dot{I}_2$, respectively. For the circuit model in Fig. 5.1, according to Kirchhoff's voltage law, the voltage equations of the primary-side and secondary-side circuits can be derived as

$$\begin{cases} (-R_n + R_1)\dot{I}_1 + \left(j\omega L_1 - j\dfrac{1}{\omega C_1}\right)\dot{I}_1 - j\omega M\dot{I}_2 = 0 \\ -j\omega M\dot{I}_1 + (R_L + R_2)\dot{I}_2 + \left(j\omega L_2 - j\dfrac{1}{\omega C_2}\right)\dot{I}_2 = 0 \end{cases}, \tag{5.1}$$

where j is the imaginary unit, and ω is the operating angular frequency of the system. k denotes the coupling coefficient and $k = M/\sqrt{L_1 L_2}$. Then, we rewrite (5.1) into matrix form as

$$\begin{pmatrix} \frac{1}{\omega L_1 C_1} - \omega + j\frac{-R_n + R_1}{L_1} & \omega k\sqrt{\frac{L_2}{L_1}} \\ \omega k\sqrt{\frac{L_1}{L_2}} & \frac{1}{\omega L_2 C_2} - \omega + j\frac{R_L + R_2}{L_2} \end{pmatrix} \begin{pmatrix} \dot{I}_1 \\ \dot{I}_2 \end{pmatrix} = 0. \tag{5.2}$$

Let the natural angular frequency of the primary-side and secondary-side circuits be $\omega_1 = \omega_2 = \omega_0$, i.e., $\omega_0 = 1/\sqrt{L_1 C_1} = 1/\sqrt{L_2 C_2}$, which is one of the PT-symmetric conditions. Therefore, (5.2) can be rewritten as

$$\begin{pmatrix} \frac{\omega_0}{\omega} - \frac{\omega}{\omega_0} + j\frac{-R_n + R_1}{\omega_0 L_1} & \frac{\omega}{\omega_0} k\sqrt{\frac{L_2}{L_1}} \\ \frac{\omega}{\omega_0} k\sqrt{\frac{L_1}{L_2}} & \frac{\omega_0}{\omega} - \frac{\omega}{\omega_0} + j\frac{R_L + R_2}{\omega_0 L_2} \end{pmatrix} \begin{pmatrix} \dot{I}_1 \\ \dot{I}_2 \end{pmatrix} = 0, \tag{5.3}$$

which has nontrivial solutions only if the coefficient matrix has a zero determinant

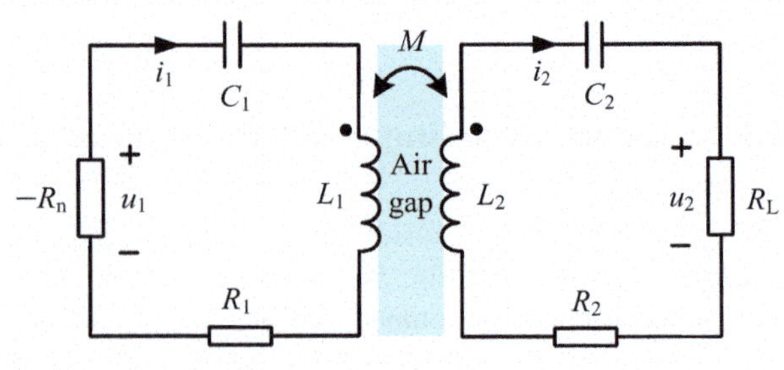

Fig. 5.1 Circuit model of series-series compensated MC-WPT system based on PT symmetry

$$\left(\frac{\omega_0}{\omega}-\frac{\omega}{\omega_0}+j\frac{-R_n+R_1}{\omega_0 L_1}\right)\left(\frac{\omega_0}{\omega}-\frac{\omega}{\omega_0}+j\frac{R_L+R_2}{\omega_0 L_2}\right)-\left(\frac{\omega k}{\omega_0}\right)^2=0. \tag{5.4}$$

The real and imaginary parts of (5.4) can be separated respectively as

$$\left(\frac{\omega_0}{\omega}-\frac{\omega}{\omega_0}\right)^2-\left(\frac{\omega k}{\omega_0}\right)^2-\frac{(-R_n+R_1)(R_L+R_2)}{\omega_0^2 L_1 L_2}=0, \tag{5.5}$$

$$\left(\frac{\omega_0}{\omega}-\frac{\omega}{\omega_0}\right)\left(\frac{-R_n+R_1}{\omega_0 L_1}+\frac{R_L+R_2}{\omega_0 L_2}\right)=0. \tag{5.6}$$

From (5.5) and (5.6), two cases are discussed as follows:

(1) $\omega \neq \omega_0$

When $\omega \neq \omega_0$, the R_n will automatically meet

$$\frac{-R_n+R_1}{L_1}+\frac{R_L+R_2}{L_2}=0, \tag{5.7}$$

and ω satisfies

$$\omega_{\pm}=\omega_0\sqrt{\frac{A\pm\sqrt{A^2+4\left(k^2-1\right)}}{2\left(1-k^2\right)}}, \tag{5.8}$$

in which $A=2-[(R_L+R_2)/(\omega_0 L_2)]^2$.

So as to obtain a steady-state solution of (5.3) with a real ω while allowing R_L or k to vary, the system ought to satisfy follow conditions

$$\begin{cases} A^2+4\left(k^2-1\right)\geq 0 \\ A\pm\sqrt{A^2+4\left(k^2-1\right)}\geq 0 \end{cases}, \tag{5.9}$$

where $0 \leq k \leq 1$.

Thus, the condition can be derived as

$$R_L \leq R_C=\omega_0 L_2\sqrt{2\left(1-\sqrt{1-k^2}\right)}-R_2, \tag{5.10}$$

where R_C is the critical load resistance. In this condition, the system is PT-symmetric.

(2) $\omega=\omega_0$

When $\omega=\omega_0$, from (5.5) and (5.6) the R_n can be derived as

$$-R_n=-\frac{L_1 L_2 \omega_0^2 k^2}{(R_L+R_2)}-R_1. \tag{5.11}$$

In this case, the PT symmetry of the system is broken.

From (5.10), it can be observed that the load can be divided into two parts: $R_L \leq R_C$ and $R_L > R_C$. From the viewpoint of physical meaning, the two parts correspond to the exact PT-symmetric region and the broken PT-symmetric region, respectively. We note that $\omega = \omega_0$ is also a solution of (5.3), but it is unstable in the exact PT-symmetric region [4]. Therefore, the operating frequency f of the system can be derived as

$$f = \begin{cases} \sqrt{2}\pi f_0 \sqrt{\dfrac{B \pm \sqrt{B^2 + 4(k^2 - 1)}}{(1 - k^2)}} & (R_L \leq R_C) \\ f_0 & (R_L > R_C) \end{cases}, \tag{5.12}$$

where the frequency $f_0 = \omega_0/(2\pi)$, $f = \omega/(2\pi)$, and $B = 2 - \left[(R_L + R_2)/(2\pi f_0 L_2)\right]^2$

As can be seen from (5.12), in the exact PT-symmetric region, a bifurcation of the system's operating frequency will occur and two frequency branches exist, i.e., the high-frequency branch and the low-frequency branch, respectively. In the broken PT-symmetric region, the system works at a resonant state, where the operating frequency is f_0.

Substituting (5.5) and (5.7) into the Eq. (5.3) in the exact PT-symmetric region yields the current ratio as

$$\frac{I_1}{I_2} = \sqrt{\frac{L_2}{L_1}}, \tag{5.13}$$

where I_1 and I_2 are the root-mean-square (RMS) value of fundamental components of i_1 and i_2, respectively. Meanwhile, the ratio of phasor $\dot{I}_1$ and $\dot{I}_2$ is deduced as

$$\frac{\dot{I}_1}{\dot{I}_2} = \sqrt{\frac{L_2}{L_1}}\left(\frac{1 - f_0^2/f^2}{k} - j\frac{R_L + R_2}{2\pi f L_2 k}\right). \tag{5.14}$$

Thus, the phase difference between i_1 and i_2 is derived as

$$\varphi_1 - \varphi_2 = \begin{cases} \arctan \dfrac{f(R_L + R_2)}{2\pi\left(f_0^2 - f^2\right)L_2}, & f > f_0 \\ -\pi + \arctan \dfrac{f(R_L + R_2)}{2\pi\left(f_0^2 - f^2\right)L_2}, & f < f_0 \end{cases}, \tag{5.15}$$

where the φ_1 and φ_2 are the phase angles of $\dot{I}_1$ and $\dot{I}_2$, respectively. The phase difference is related to the load and operating frequency.

As for the broken PT-symmetric region, $f = f_0$ and (5.11) are substituted into Eq. (5.3) to obtain the current ratio as

$$\frac{I_1}{I_2} = \frac{R_L + R_2}{2\pi f_0 k\sqrt{L_1 L_2}}, \tag{5.16}$$

and the phase difference $\varphi_1 - \varphi_2 = -\pi/2$. In general, the transfer efficiency of a MC-WPT system is expressed as [5]

$$\eta = \frac{I_2^2 R_L}{I_1^2 R_1 + I_2^2 R_L + I_2^2 R_2}. \tag{5.17}$$

The expression of η in the exact PT-symmetric region and the broken PT-symmetric region can be obtained for the PT-based MC-WPT system by substituting (5.13) and (5.16) into (5.17), respectively

$$\eta = \begin{cases} \dfrac{R_L}{\frac{L_2}{L_1} R_1 + R_2 + R_L}, & (R_L \le R_C) \\ \dfrac{4\pi^2 R_L f_0^2 k^2 L_1 L_2}{r_1 (R_L + R_1)^2 + (R_2 + R_L) 4\pi^2 f_0^2 k^2 L_1 L_2}, & (R_L > R_C) \end{cases}. \tag{5.18}$$

The negative resistance $-R_n$ satisfies

$$I_1 R_n = U_1, \tag{5.19}$$

where the U_1 is the RMS value of fundamental components u_{1_FHA} of u_1. The output power P_L can be normally given by

$$P_L = I_2^2 R_L. \tag{5.20}$$

Furthermore, the output power P_L of the PT-symmetric MC-WPT system in the exact PT-symmetric region can be deduced by substituting (5.7) and (5.13) into (5.19) and (5.20)

$$P_L = \frac{R_L U_1^2}{\frac{L_1}{L_2}(R_L + R_2)^2 + 2R_1 (R_L + R_2) + \frac{L_2}{L_1} R_1^2}, R_L \le R_C. \tag{5.21}$$

The output power P_L of the PT-symmetric MC-WPT system in the broken PT-symmetric region can be derived by substituting (5.11) and (5.16) into (5.19) and (5.20)

$$P_L = \frac{4\pi^2 f_0^2 k^2 L_1 L_2 R_L U_1^2}{\left[4\pi^2 f_0^2 k^2 L_1 L_2 + R_1 (R_L + R_2)\right]^2}, R_L > R_C. \tag{5.22}$$

From (5.13), (5.18), and (5.21), the equivalent circuit of the exact PT-symmetric region is obtained when the phase relation of $\dot{I}_1$ and $\dot{I}_2$ is neglected. As shown in Fig. 5.2, the equivalent circuit of the exact PT-symmetric region comprises an

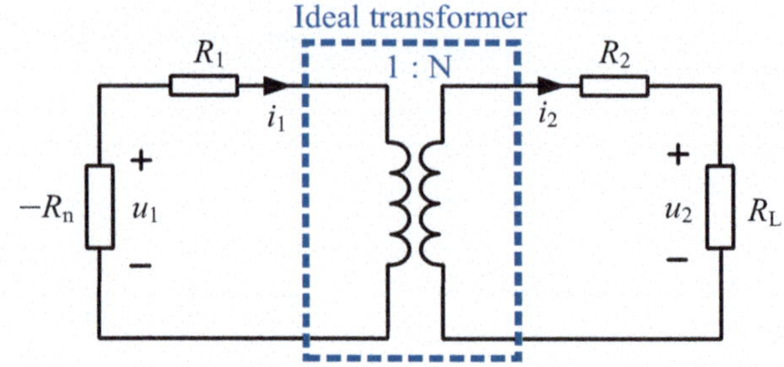

Fig. 5.2 Equivalent circuit of the exact PT-symmetric region (the phase relation of i_1 and i_2 is neglected)

ideal transformer whose ratio is 1:N, where $N = \sqrt{L_2/L_1}$. It can be seen that the current ratio, the output power P_L and the transfer efficiency η are independent on the coupling coefficient k in the exact PT-symmetric region, which is superior compared with transfer characteristics of conventional MC-WPT system.

5.2 System Control and Implementation

5.2.1 System Structure

The proposed MC-WPT system consists of two insulated parts named as primary side and secondary side, and series-series compensations are used on both primary side and secondary side. The proposed system structure is shown in Fig. 5.3, while the detailed circuit is depicted in Fig. 5.4.

The front-end DC-DC converter in Fig. 5.3, which could be a buck, boost, or buck-boost converter, is utilized to regulate the input voltage U_d of the full-bridge inverter at the primary side for output power stabilization control. The full-bridge inverter, which is a voltage-source-type inverter (VSI) and widely used in MC-WPT systems due to its high voltage utilization and simplicity, is adopted to realize the ac negative resistance in this paper. Then, the output power stabilization control based on primary-side information is realized only by the proposed primary-side controller, which consists of the negative resistance control, load identification, and output power closed-loop control. As can be seen from Fig. 5.3, the system does not have dual-side wireless communication and secondary-side control circuits, which compress the volume of the secondary side.

On the secondary side, through the full-bridge rectifier with a capacitive output filter, the obtained AC voltage u_2 is converted to DC voltage U_o for household applications, battery charging, and other power converters et al. The main purpose of this paper is to propose and verify a MC-WPT system and its output power stabilization control strategy so that the DC resistance is used as the load in analysis.

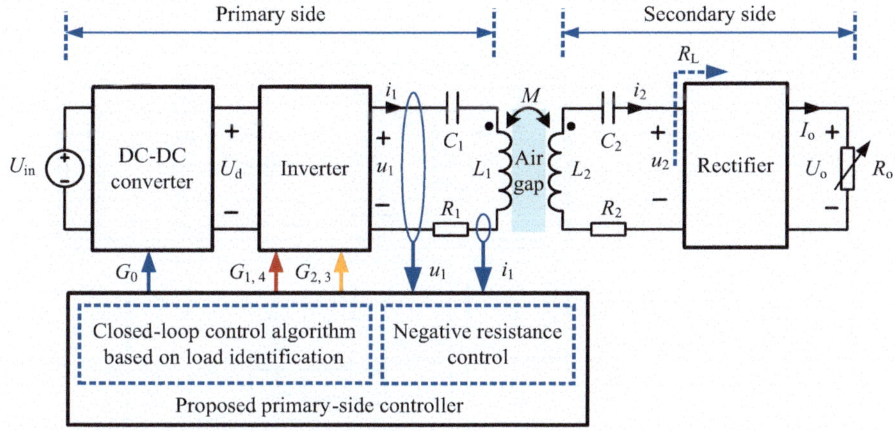

Fig. 5.3 System structure of the proposed MC-WPT system

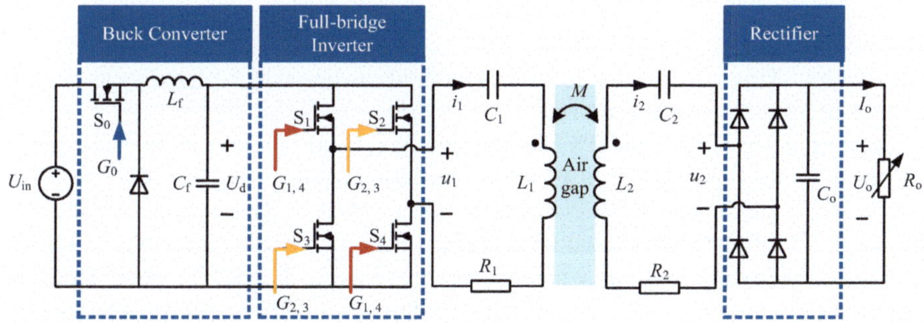

Fig. 5.4 Main circuit of the proposed MC-WPT system

5.2.2 Negative Resistance Control

Based on the property of negative resistance, it is reasonable to control the output voltage in phase with the output current of the inverter by using the current in the primary-side resonant tank as a feedback signal. Therefore, the output of this self-oscillating controlled inverter can be equivalent as a negative resistance [6], and the control circuit thereof is depicted in Fig. 5.5. The primary-side current i_1, i.e., output current of the inverter is captured via the current sense transformer and converted into a voltage signal. Then the voltage signal is amplified by the differential amplifier and sent to the zero-crossing comparator via a time-leading compensator to detect the zero crossing points of i_1. Finally, the output signal of the comparator is sent to the dead time generation circuit to generate pulse width modulation (PWM) signals, i.e., $PWM_{G1,4}$ and $PWM_{G2,3}$, which drive the corresponding switch S_1–S_4 of the full-bridge inverter as shown in Fig. 5.6. The time-leading compensator after the amplifier is used for compensating the time delay from current sense transformer

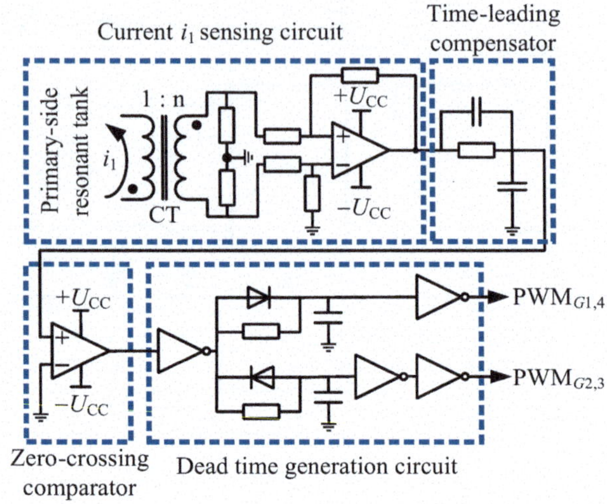

Fig. 5.5 Self-oscillating control circuit diagram of the full-bridge inverter for negative resistor

[7], zero-crossing comparator, dead time circuit and drivers, to ensure that u_1 is in phase with i_1. This self-oscillating control method can be easily realized by an analog circuit and no processor is required.

The output waveforms and PWM signals of the full-bridge inverter by the negative resistance control are shown in Fig. 5.6. The duty width of the PWM signals for switch S_1–S_4 are 50% without considering the dead time. The output current i_1 of the inverter is almost a sinusoidal wave due to the high intrinsic quality factor (Q) of the primary-side resonant tank, while the output voltage u_1 is a square wave. Through the control circuit, the output current i_1 is in phase with both u_1 and its fundamental harmonic u_{1_FHA}, which is consistent with the property of negative resistance. Based on FHA, the RMS value U_1 of u_{1_FHA} is given by

$$U_1 = \frac{2\sqrt{2}}{\pi} U_d. \tag{5.23}$$

5.2.3 *Online Load Identification Approach*

In practical applications, the equivalent resistive load R_o that calculated by U_o/I_o may be uncertain or variable. In order to control the output power only by the primary-side controller, it is necessary to identify the load R_o indirectly in real-time. The property of PT-symmetric circuit indicates that R_o has a relationship with the negative resistance $-R_n$, so it is possible to estimate R_o employing primary-side information. Therefore, a novel online load identification approach by PT symmetry is proposed.

From (5.7), the R_L is derived as

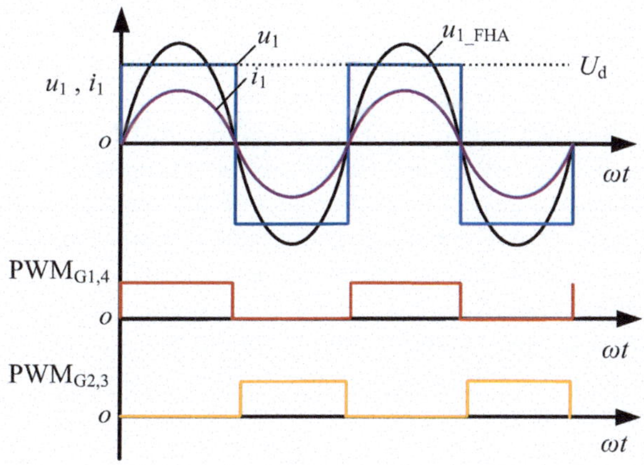

Fig. 5.6 PWM signals and output waveforms of the full-bridge inverter

$$R_L = \frac{L_2}{L_1}(R_n - R_1) - R_2. \tag{5.24}$$

Then, the load R_o can be deduced by substituting $R_L = 8R_o/\pi^2$ into (5.24)

$$R_o = \frac{\pi^2}{8}\left[\frac{L_2}{L_1}(R_n - R_1) - R_2\right]. \tag{5.25}$$

Obviously, combining (5.19) and (5.25), the estimated value $R_{o,\,est}$ of the load R_o is calculated by

$$R_{o,\,est} = \frac{\pi^2}{8}\left[\frac{L_2}{L_1}\left(\frac{U_1}{I_1} - R_1\right) - R_2\right]. \tag{5.26}$$

Once the system is fabricated, the circuit parameters, i.e. the values of L_i and R_i ($I = 1, 2$) are acquired by previous measurements and assumed to be constant approximately. Therefore, through acquiring and calculating the RMS values of U_1 and I_1 in real time without the need for complicated measuring of the phase or active power, the estimated value $R_{o,\,est}$ can be quickly obtained according to (5.26) in the exact PT-symmetric region, which is easy to implement in the digital processor.

5.2.4 Output Power Control

As transmission characteristics are shown in Sect. 5.1, the output power is coupling-independent and only related to the voltage U_1 and the load R_L in the exact PT-symmetric region. With (5.21) and (5.23), the output power P_L is deduced as

$$P_{\mathrm{L}} = \frac{\frac{8}{\pi^2} U_{\mathrm{in}}^2 D^2 R_{\mathrm{L}}}{\frac{L_1}{L_2}(R_{\mathrm{L}} + R_2)^2 + 2R_1(R_{\mathrm{L}} + R_2) + \frac{L_2}{L_1} R_1^2}. \tag{5.27}$$

It indicates that when the actual load resistance is estimated, the duty cycle D of the buck converter can be adjusted accordingly, and the desired output power P_{L} is achieved consequently, regardless of the coupling coefficient. Therefore, the closed-loop control of duty cycle D is essential to achieve accurate and robust output power control when the load varies. Assuming that the rectifier power losses are ignored, the output power $P_{\mathrm{o}} = I_{\mathrm{o}} U_{\mathrm{o}} = P_{\mathrm{L}}$. Since there is no wireless communication desired in the system, the output power P_{o} should be estimated based on the load identification in Sect. 5.2.3. From (5.13) and (5.20), the estimated value $P_{\mathrm{o,\,est}}$ of output power is calculated by

$$P_{\mathrm{o,\,est}} = \frac{8L_1}{\pi^2 L_2} I_1^2 R_{\mathrm{o,\,est}}. \tag{5.28}$$

Finally, a closed-loop PI control method is proposed for output power regulation in the exact PT-symmetric region. As described in Fig. 5.7, the closed-loop control block diagram consists of the online load identification, the output power estimation, and the PWM signal generation of the PI-controlled buck converter. While the load varies, through acquiring the output current i_1 and the output voltage u_1 in real time and calculating RMS value of fundamental harmonic respectively, the DC load resistance and the output power can be estimated using (5.26) and (5.28). The estimated values $P_{\mathrm{o,\,est}}$ is compared with a preset reference $P_{\mathrm{o,\,ref}}$, which represents the desired output power P_{o}. The error $\Delta P_{\mathrm{o}} = P_{\mathrm{o,\,ref}} - P_{\mathrm{o,\,est}}$ is fed into the PI controller that generates the duty cycle D, and the switch driver generates the PWM signal that drives the buck converter. Eventually, the voltage U_1 of negative resistance is regulated to maintain the constant P_{o}. Obviously, the proposed control method in the exact PT-symmetric region is coupling-independent, which simplifies the control algorithm and improves the robustness of the system.

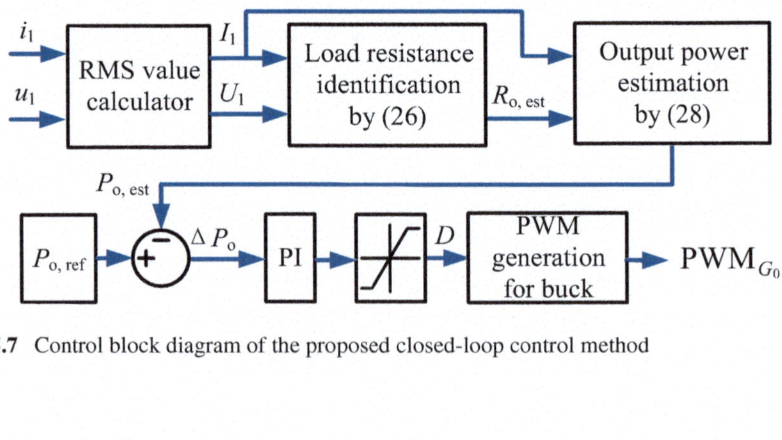

Fig. 5.7 Control block diagram of the proposed closed-loop control method

With the proposed output power stabilization control strategy, the normalized output power and corresponding transfer efficiency as the function of coupling coefficient k and load resistance R_o are theoretically calculated and plotted in Fig. 5.8. The circuit parameters used in the theoretical calculation are listed in Table 5.1. The critical line in Fig. 5.8a is the boundary of the exact PT-symmetric region constrained by (5.10). The theoretical output results indicate that the output power can stabilize at specified levels against reasonable variations of load and coupling coefficient via the proposed control strategy, meanwhile, the transfer efficiency is also coupling-independent in the exact PT-symmetric region and higher as the dc load resistance R_o larger, which is shown in Fig. 5.8b.

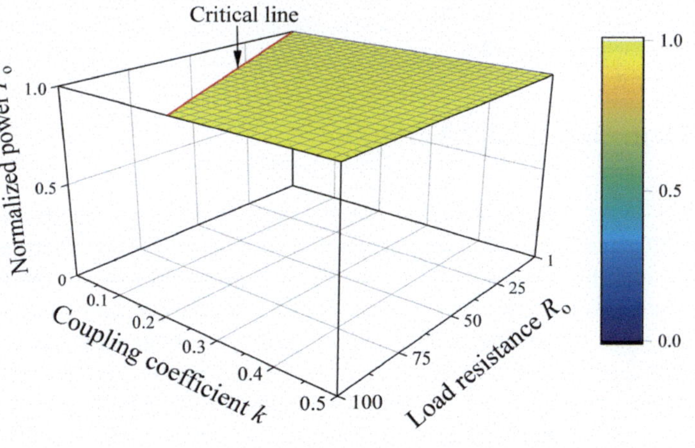

(a)

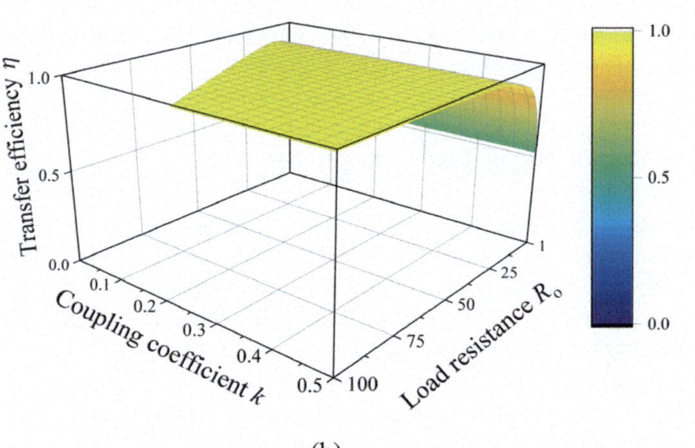

(b)

Fig. 5.8 Theoretical output results under various condition by the proposed output power stabilization control strategy. **a** Normalized output power. **b** Transfer efficiency

Table 5.1 The circuit parameters of the experimental prototype

Description	Symbol	Values
Inductance in primary-side resonator	L_1	192.83 μH
Inductance in secondary-side resonator	L_2	193.20 μH
Series capacitance in primary-side resonator	C_1	1.519 nF
Series capacitance in secondary-side resonator	C_2	1.518 nF
AC loss resistance in primary-side resonator	r_1	0.361 Ω
AC loss resistance in secondary-side resonator	r_2	0.366 Ω
Resonant frequency of resonator	f_0	294 kHz
DC input voltage of system	U_{in}	180 V
Inductance of front-end buck converter	L_f	1.8 mH
Output filter capacitance of rectifier	C_o	220 μF/250 V

5.3 System Design and Validation

Figure 5.9 shows the experimental prototype. the power flows through the front-end buck converter, the full-bridge inverter, a pair of magnetically coupled resonators, the rectifier, and the capacitive filter. Finally, it is provided to load resistors. In the primary-side controller, the current of the inverter is converted into a voltage signal by a current sense transformer CU8965, and then the voltage signal passes through a differential amplifier circuit consisting of the high-speed operational amplifier LM6172, a zero-crossing comparator TL3016 and the dead time generation circuit containing quad-2 input NAND gate (Fairchild 74VHC00MTCX). PWM signals drive the switch S_0–S_4 through isolated gate drivers (Si8271). A floating-point digital-signal processor TMS320F28377d is employed to sample the RMS value via an RMS calculator AD637 and carry out the output power closed-loop control. For the coreless coils of resonators, the helix shape coil made by Litz wire (ϕ 0.05 mm $\times$ 1000 strands, outer ϕ 2.3 mm) is chosen considering the skin effect at several hundred kilohertz. The diameter of each coil is 30 cm and the axial width is about 6cm with single layer 20-turn (axial arrangement) structure. For the series capacitors of resonators, the high voltage doorknob ceramic capacitors are chosen for their low losses and high current-bearing capability at high frequency. The circuit parameters of resonators measured by a precise impedance analyzer (Wayne Kerr 6500B) are listed in Table 5.1. The frequency sweeping measurement results indicate that the loss resistances of resonators vary little within 250–350 kHz, so they are considered constant in the analysis, which are equal to the values shown in Table 5.2

Figure 5.10 shows experimental results of $P_{o,ref} = 400$ W and $R_o = 15\ \Omega$ under different transfer distance *d*. As can be seen form Fig. 5.10a, the output power nearly stabilizes at the reference level with the power error Δ_P ($\Delta_P = \left|P_o - P_{o,ref}\right|/P_{o,ref}$) less than 2.85% against the coupling coefficient varying from 0.35 to 0.036. The experimental results in Fig. 5.10b indicate that the transfer efficiency (around 95%) as

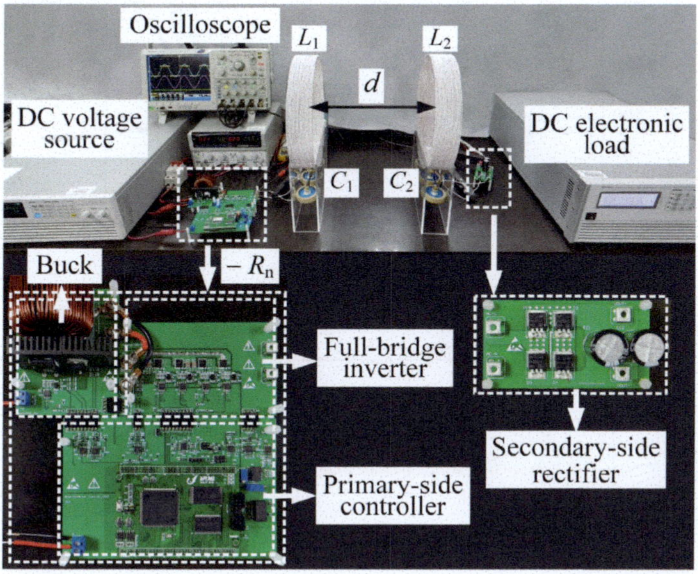

Fig. 5.9 Photo of the designed experimental prototype

well as the current ratio remains almost constant within a range of 34 cm ($k = 0.036$–0.35). The operating frequency f and the phase difference φ_1–φ_2 of currents measured by the oscilloscope are displayed in Fig. 5.10c, which show that the experimental prototype operates at the low frequency branch, and φ_1–φ_2 gradually increases and approaches $-\pi/2$ as d increases. The experimental results coincide with calculated values in the exact PT-symmetric region.

Table 5.2 shows experimental results of $R_{\text{o, est}}$ and the corresponding error Δ_R ($\Delta_R = \left|R_{\text{o, est}} - R_\text{o}\right|/R_\text{o}$) when $P_{\text{o, ref}} = 200$ W. The error Δ_R is less than 4.3% under $k = 0.15$ and 0.096, which decreases with the increase of R_o. The average value of $R_{\text{o, est}}$ can be used to reduce the errors.

Experimental results and waveforms of $P_{\text{o, ref}} = 200$ W are shown in Figs. 5.11 and 5.12, respectively. It indicates that the output power P_o can be stable at desired value against various loads under $k = 0.15$ or $k = 0.096$. The maximum error Δ_P is 4.10% corresponding to output power 191.8 W at $R_\text{o} = 10\ \Omega$, while the minimum error Δ_P is 1.75% corresponding to output power 196.5 W at $R_\text{o} = 55\ \Omega$.

Similarly, the output power can also be controlled at $P_{\text{o, ref}} = 400$ W under different loads when $k = 0.15$ or $k = 0.096$, experimental results and waveforms of which are shown in Figs. 5.13 and 5.14. Experimental results show that the maximum error Δ_P is 3.40% corresponding to output power 386.4 W at $R_\text{o} = 10\ \Omega$, while the minimum error Δ_P is 1.68% corresponding to output power 393.3 W at $R_\text{o} = 55\ \Omega$.

As can be seen form the experimental waveforms, the working frequency is self-regulated at different operating points by the negative resistance control circuit, but there are slight phase differences between the current i_1 and voltage u_1 at certain

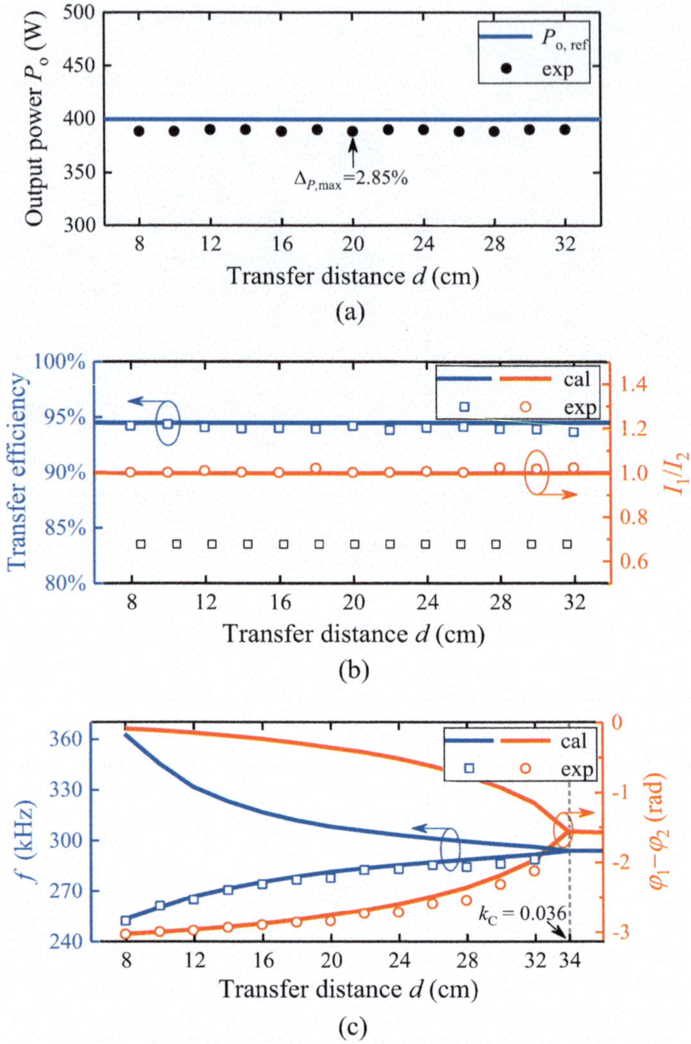

Fig. 5.10 Experimental results of $P_{o, ref} = 400$ W and $R_o = 15\ \Omega$ under different transfer distance d. **a** Output power. **b** Efficiency. **c** Operating frequency and phase differences of i_1, i_2

operating points, which are mainly caused by the minor variation in leading time of the compensator at different operating frequency. Whereas the inverter can be still considered to a negative resistance since the errors are small at most of operating points and experimental results nearly correspond to the theory. The voltage u_1 will automatically change to sustain stable output power at each operating point. Note that the output power error Δ_P tends to decrease, i.e., P_o are closer to $P_{o, ref}$, as the load resistance R_o increases. The experimental results validate the correctness of proposed output power stabilization control strategy.

Table 5.2 Experimental results of online load identification

	$d = 15$ cm ($k = 0.15$)		$d = 20$ cm ($k = 0.096$)	
R_o (Ω)	$R_{o,\,est}$ (Ω)	Δ_R (%)	$R_{o,\,est}$ (Ω)	Δ_R (%)
10	10.43	4.25	10.40	4.01
15	15.61	4.03	15.52	3.45
20	20.55	2.73	20.51	2.55
25	26.07	4.26	25.79	3.14
30	31.00	3.33	30.85	2.85
35	36.22	3.47	36.10	3.14
40	40.95	2.37	41.00	2.50

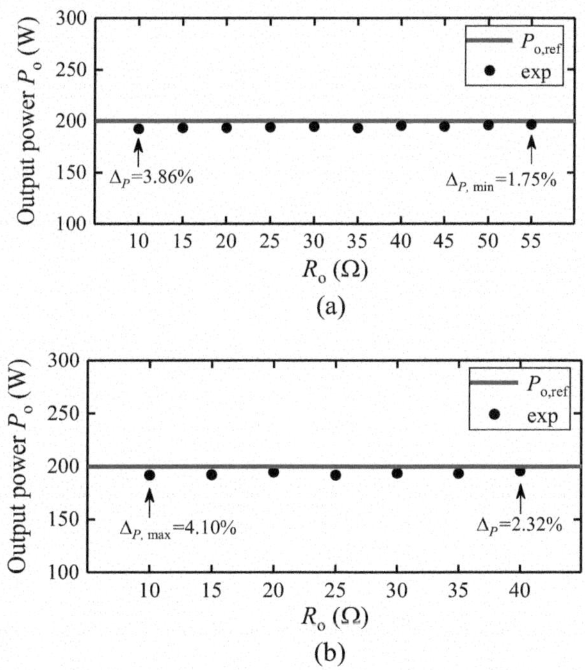

Fig. 5.11 Experimental results of output power and errors when $P_{o,\,ref} = 200$ W under different load R_o. **a** $d = 15$ cm. **b** $d = 20$ cm

The overall efficiency is calculated by P_{in}/P_o, where P_{in} is the system input power read from DC power supply directly. As shown in Fig. 5.15a, the maximum overall efficiency is 91.9% at $R_o = 55\ \Omega$ when $P_{o,\,ref} = 400$ W. Generally, system losses mainly consist of buck converter losses P_{buck} (switching and conduction losses), inverter losses $P_{inverter}$ (switching and conduction losses), resonator losses $P_{resonator}$, rectifier losses $P_{rectifier}$ (switching and conduction losses) and other losses P_{others}. The system losses are measured at the maximum overall efficiency point, results of which are plotted in Fig. 5.15b. It shows that the losses P_{buck} and $P_{inverter}$ account for

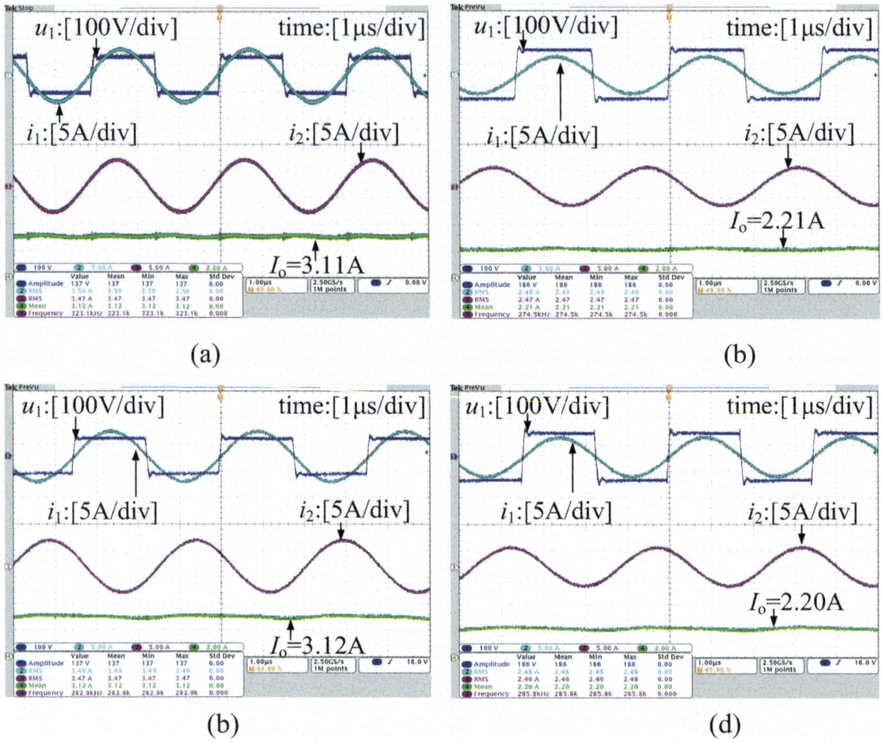

Fig. 5.12 Experimental waveforms when $P_{o,\,ref} = 200$ W under different conditions. **a** $R_o = 20$ Ω, $d = 15$ cm. **b** $R_o = 40$ Ω, $d = 15$ cm. **c** $R_o = 20$ Ω, $d = 20$ cm. **d** $R_o = 40$ Ω, $d = 20$ cm

a large proportion of system losses. Hence, the design of inverter and buck converter are vital to improve the system performance.

5.4 Summary

In this chapter, a novel primary-side-only control strategy that realizes stable output power under reasonable variation in load and coupling conditions is proposed for a series-series compensated WPT system. It consists of the negative resistance control of PT-symmetric circuit, the online load identification approach by PT symmetry, and the closed-loop control method. It has the advantage of eliminating the need for coupling information, dual-side wireless communication as well as secondary-side control circuits, which simplifies the control algorithm and makes the receiver compact and lightweight. Through the proposed strategy, the output power is coupling-independent in the exact PT-symmetric region, and able to stabilize at specified values according to load estimation results. Experimental results

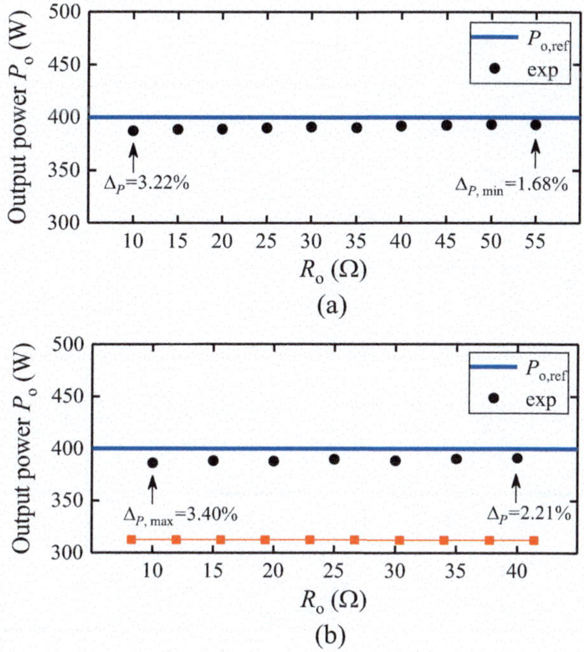

Fig. 5.13 Experimental results of output power and errors when $P_{o,\,ref} = 400$ W under different load R_o. **a** $d = 15$ cm. **b** $d = 20$ cm

validate the performance of the proposed control strategy and system with enough accuracy and high overall efficiency. It provides a simple, but effective solution to restrain output power fluctuation, and is suitable for applications such as constant power charging, dynamic powering, and household heating appliances where stable and flexible output power is needed under reasonable variation of load and some degree of free positioning.

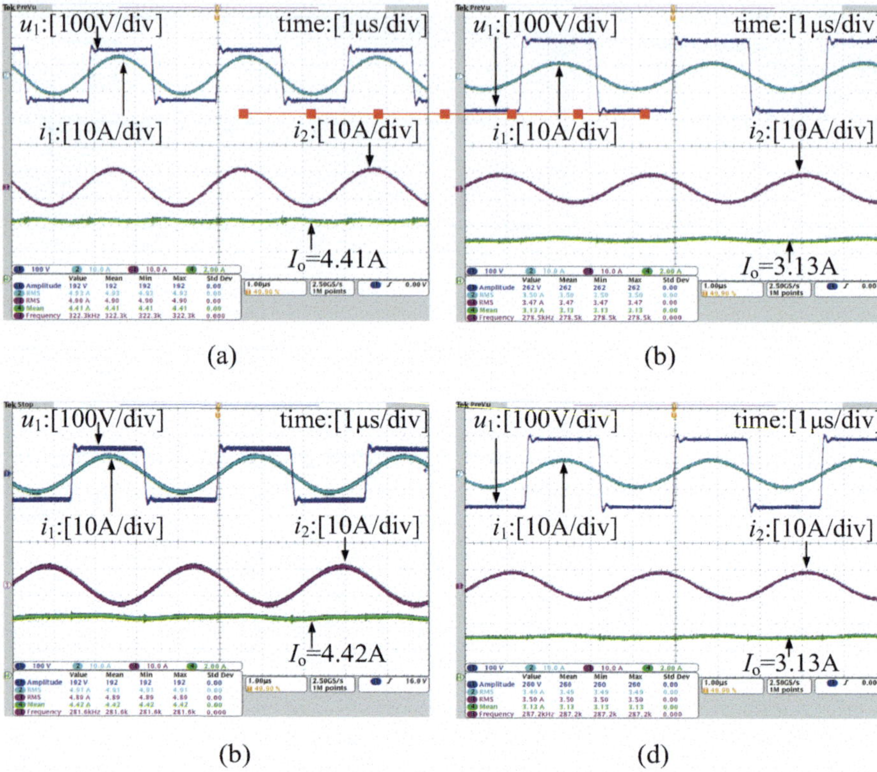

Fig. 5.14 Experimental waveforms when $P_{o,\,ref} = 400$ W under different conditions. **a** $R_{on} = 20$ Ω, $d = 15$ cm. **b** $R_o = 40$ Ω, $d = 15$ cm. **c** $R_o = 20$ Ω, $d = 20$ cm. **d** $R_o = 40$ Ω, $d = 20$ cm

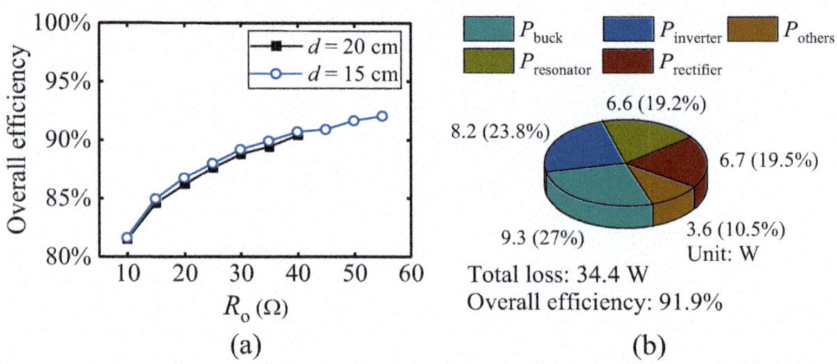

Fig. 5.15 Experimental results of overall efficiency and losses when $P_{o,\,ref} = 400$ W. **a** Overall efficiency. **b** Loss breakdown of R_o =55 Ω, $d = 15$ cm

References

1. Wu L, Zhang B, Zhou J (2020) Efficiency improvement of the parity-time-symmetric wireless power transfer system for electric vehicle charging. IEEE Trans Power Electron 35(11):12497–12508
2. Wu L, Zhang B, Jiang Y, Zhou J (2022) A robust parity-time-symmetric WPT system with extended constant-power range for cordless kitchen appliances. IEEE Trans Ind Appl 58(1):1179–1189
3. Zhu H, Zhang B, Wu L (2020) Output power stabilization for wireless power transfer system employing primary-side-only control. IEEE Access 8:63735–63747
4. Zhou J, Zhang B, Xiao W et al (2019) Nonlinear parity-time-symmetric model for constant efficiency wireless power transfer: application to a drone-in-flight wireless charging platform. IEEE Trans Ind Electron 66(5):4097–4107
5. Zhong W, Hui SYR (2018) Maximum energy efficiency operation of series-series resonant wireless power transfer systems using on-off keying modulation. IEEE Trans Power Electron 33(4):3595–3603. https://doi.org/10.1109/tpel.2017.2709341
6. Assawaworrarit S, Fan S (2020) Robust and efficient wireless power transfer using a switch-mode implementation of a nonlinear parity-time symmetric circuit. Nat Electron 3(5):273–279
7. Yan K, Chen Q, Hou J, Ren X, Ruan X (2014) Self-oscillating contactless resonant converter with phase detection contactless current transformer. IEEE Trans Power Electron 29(8):4438–4449

Chapter 6
Transfer Characteristics of Constant Voltage/Constant Current

Chapters 4 and 5 mainly introduce the transfer characteristics of constant output power and transfer efficiency of PT-symmetric MC-WPT system, and solve the problem that the system characteristics are changed by the coupling coefficient. However, in practical applications, such as electric vehicles (EVs), its battery resistance changes during charging, and the coupling coefficient also changes during dynamic charging. Therefore, in this chapter, the variations of load resistance and coupling coefficient will be taken into account simultaneously, and the constant voltage (CV) and constant current (CC) output characteristics of the PT-symmetric MC-WPT system will be analyzed. Firstly, the system structure is introduced. Then, the CC charging mode and the CV charging mode of the system are analyzed. Meanwhile, the influences of the internal resistance and self-inductance variation are discussed. Finally, the implementation of a negative resistor and the CC and CV control strategy are given.

6.1 System Structure

The circuit diagram of the PT-symmetric MC-WPT system with CC and CV output characteristics is shown in Fig. 6.1 [1]. Here, a full-bridge inverter, which consists of four power switches (Q_1~Q_4), converts a DC voltage U_{dc} to an AC square voltage u_{ac}. The transmitter and receiver are coupled by two coupled coils L_P and L_S with mutual inductance M_{PS}. C_P and C_S are the tuning capacitors in the transmitting and receiving sides, respectively. The internal resistances of the transmitter and receiver are represented by R_P and R_S. D_{r1}~D_{r4} are the diodes of the rectifier, and C_f is the capacitance of the filter capacitor. S_1 is a short-circuit switch, which is only turned on briefly before charging starts. The currents of transmitting and receiving coils are denoted as i_P and i_S, respectively. u_S is the input voltage of the rectifier. R_L is the

B. Zhang et al., *Parity-Time Symmetric Wireless Power Transfer*, CPSS Power Electronics Series, https://doi.org/10.1007/978-981-96-3149-0_6

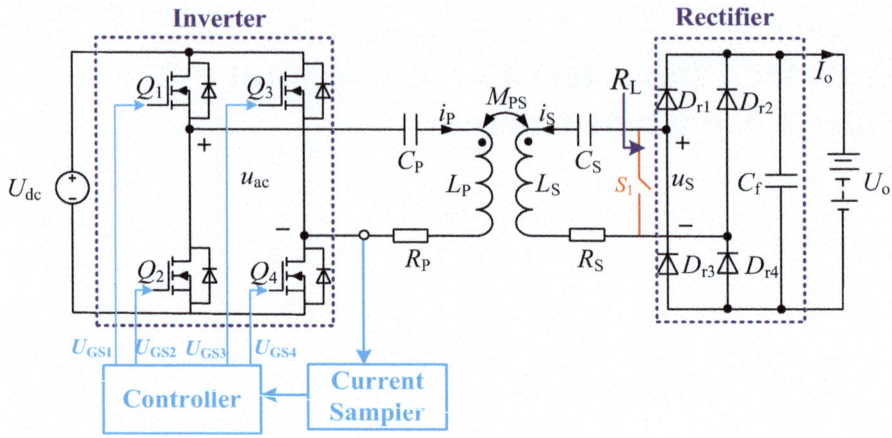

Fig. 6.1 The full circuit diagram of the PT-symmetric MC-WPT system with CC and CV output

equivalent resistance of the rectifier. Finally, I_0 and U_0 are DC output current and DC output voltage.

Considering the fundamental component of voltage and current, u_{ac}, u_S, i_P and i_S, can be easily expressed in terms of phasors $\dot{U}_P$, $\dot{U}_S$, $\dot{I}_P$, and $\dot{I}_S$, respectively. Then the relationship between the input and output of the inverter and the rectifier can be written as follows [2, 3]:

$$\begin{cases} U_P = 2\sqrt{2}U_{dc}\sin(0.5D_S\pi)/\pi \\ I_S = \pi\sqrt{2}I_o/4 \\ U_S = 2\sqrt{2}U_o/\pi \\ R_L = 8R_B/\pi^2 \end{cases}, \tag{6.1}$$

where U_P, U_S, and I_S are the RMS values of $\dot{U}_P$, $\dot{U}_S$, and $\dot{I}_S$, respectively. D_S is the duty cycle of the inverter, and $R_B = U_0/I_0$ is the equivalent resistance of the battery.

For the convenience of analysis, the typical charging curve of lithium-ion batteries is given, as shown in Fig. 6.2 [4, 5]. It can be seen that the charging process of the battery is generally divided into two modes: constant current (CC) charging mode and constant voltage (CV) charging mode. In CC mode, the charging current I_0 is equal to the rated current I_{BR}, which is unchanged, and when the battery voltage reaches the rated voltage U_{BR}, the charger enters the CV charging mode. In CV mode, the charging current gradually decreases, and the charging process ends when the charging current is lower than the cut-off current I_{Bmin}. During the charging process, the equivalent resistance R_B of the battery will gradually increase.

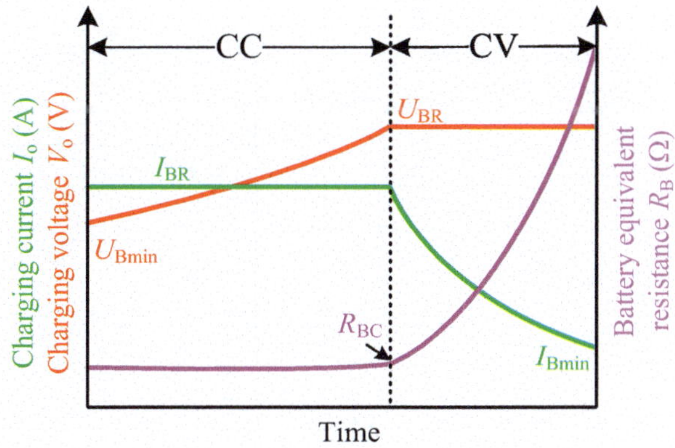

Fig. 6.2 Typical charging curve for lithium-ion batteries

6.2 Characteristic Analysis

6.2.1 CC Mode

When the PT-symmetric MC-WPT system works in CC charging mode, its equivalent circuit can be described as shown in Fig. 6.3. In CC mode, the inverter is controlled based on the phase synchronization method to implement the negative resistor $-R_N$, whereby the transmitter and receiver form a PT-symmetric MC-WPT system. The voltage and current waveform of the inverter is shown in Fig. 6.4, wherein u_{ac} and u_P represent the inverter output voltage and its fundamental wave component, respectively, and T_{iP} represents the cycle of the inverter output current i_P [1].

Using the circuit theory, the system can be described as

$$\begin{bmatrix} (-R_N + R_P) + j\left(\omega L_P - \frac{1}{\omega C_P}\right) & j\omega M_{PS} \\ j\omega M_{PS} & (R_L + R_S) + j\left(\omega L_S - \frac{1}{\omega C_S}\right) \end{bmatrix} \begin{bmatrix} \dot{I}_P \\ \dot{I}_S \end{bmatrix} = 0, \quad (6.2)$$

where ω is the operating frequency of the inverter, $\omega_P = 1/\sqrt{L_P C_P}$ and $\omega_S = 1/\sqrt{L_S C_S}$ are the resonant frequencies of the transmitter and receiver, respectively.

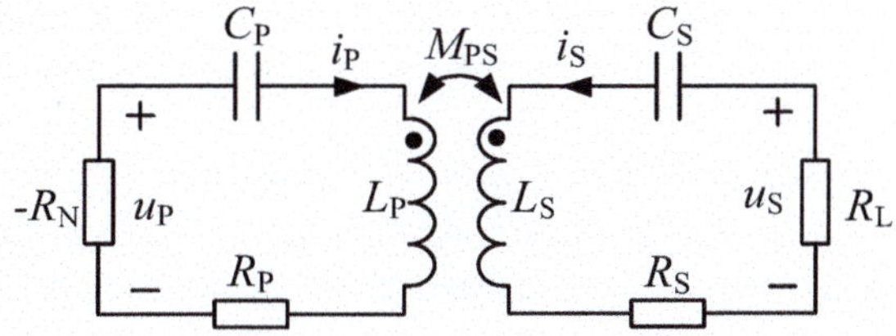

Fig. 6.3 Equivalent circuit of the CC charging mode

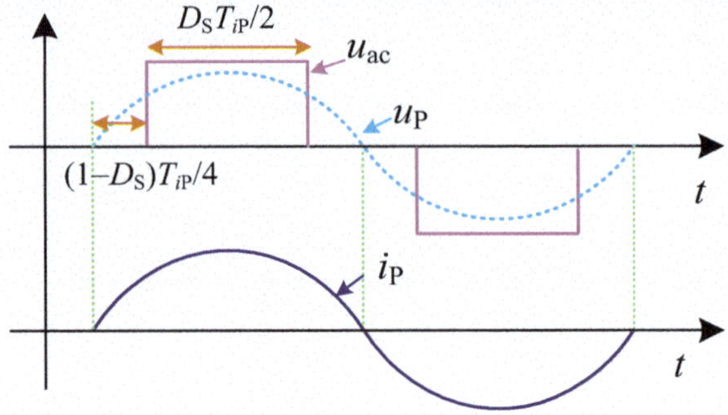

Fig. 6.4 The voltage and current waveform of the inverter in CC mode

According to the PT-symmetric conditions [6], the matched resonance frequency $\omega_P = \omega_S$ (i.e., $L_P C_P = L_S C_S$) is needed, then the characteristic equation can be obtained by the coefficient determinant of (6.2) being zero, that is

$$\left[\frac{-R_N + R_P}{L_P} + j\left(\omega - \frac{\omega_0^2}{\omega}\right)\right]\left[\frac{R_L + R_S}{L_S} + j\left(\omega - \frac{\omega_0^2}{\omega}\right)\right] + \omega^2 k_{PS}^2 = 0. \quad (6.3)$$

Since the equivalent resistance R_B of the battery will gradually increase during the charging process, the working interval of the system is divided here according to the AC equivalent load R_L. The steady-state solutions of (6.3) with real ω can be derived as follows:

$$\omega = \begin{cases} \omega_0, \ R_L > R_C \\ \omega_{H\_CC} = \omega_0 \sqrt{\dfrac{2 - \gamma^2 \pm \sqrt{\left[2 - \gamma^2\right]^2 - 4\left(1 - k_{PS}^2\right)}}{2\left(1 - k_{PS}^2\right)}}, \ R_L \le R_C \text{ or} \\ \omega_{L\_CC} = \omega_0 \sqrt{\dfrac{2 - \gamma^2 \pm \sqrt{\left[2 - \gamma^2\right]^2 - 4\left(1 - k_{PS}^2\right)}}{2\left(1 - k_{PS}^2\right)}}, \ R_L \le R_C \end{cases}, \quad (6.4)$$

where $\gamma = (R_L + R_S)/(\omega_0 L_S)$, ω_{H_CC} and ω_{L_CC} represent the high-frequency branch and the low-frequency branch corresponding to the operating angular frequency in CC mode, and R_C represents the critical equivalent AC load, which is expressed as

$$R_C = \omega_0 L_S \sqrt{2 - 2\sqrt{1 - k_{PS}^2}} - R_S. \quad (6.5)$$

It is noted that $\omega = \omega_0$ is also a solution of (6.2) in the exact PT-symmetric region, but this mode has been proven to require higher saturated gain [7]. Thus, according to the value of load resistance, the steady-state operation region can be divided into the exact PT-symmetric region (i.e., $R_L \le R_C$) and the broken PT-symmetric region (i.e., $R_L > R_C$). When $R_L \le R_C$, the system works in the exact PT-symmetric region, and supports two steady states with operating frequency $\omega = \omega_{H\_CC,\,L\_CC}$, thereby the system oscillates at $\omega = \omega_{H\_CC,\,L\_CC}$ automatically in the exact PT-symmetric region without knowing the values of coupling coefficient and load resistance. When $R_L > R_C$, the system enters into a broken PT-symmetric region, and works in the natural resonant frequency $\omega = \omega_0$.

In CC mode, we let the system work in the exact PT-symmetric region.

Assuming that the system always operates in the exact PT-symmetric region in CC mode, the equivalent resistance of the loaded battery should be satisfied according to the Eq. (6.5) and $R_L = 8R_B/\pi^2$, the equivalent resistance of the loaded battery must be met $R_{BC} \leq \pi^2 R_C/8$, where $R_{BC} = U_{BR}/I_{BR}$ is defined as the ratio of the rated charging voltage to the rated charging current. Then, according to (6.4), the critical coupling coefficient k_C can be calculated to ensure that the system works in the exact PT-symmetric region during the CC charging stage, that is, the coupling coefficient between coils needs to meet the following requirements:

$$k_{PS} \geq k_C = \sqrt{\left[\frac{8R_{BC}/\pi^2 + R_S}{\omega_0 L_S}\right]^2 - \frac{1}{4}\left[\frac{8R_{BC}/\pi^2 + R_S}{\omega_0 L_S}\right]^4}. \tag{6.6}$$

In this case, substituting (6.3) into (6.2), the current ratio can be derived as

$$\frac{I_P}{I_S} = \left|\frac{\dot{I}_P}{\dot{I}_S}\right| = \sqrt{\frac{L_S}{L_P}}. \tag{6.7}$$

By substituting (6.1) into (6.7), the DC output current I_o can be written as

$$I_o = \frac{2\sqrt{2}I_P}{\pi}\sqrt{\frac{L_P}{L_S}}. \tag{6.8}$$

From (6.8), it can be seen that the DC output current I_0 depends on the ratio of L_P and L_S and the current of transmitting coil I_P. Once these parameters are fixed, I_o can be considered as a constant current source. Substituting $I_o = I_{BR}$ into (6.8), the required transmitting coil current reference value I_{Pref} can be obtained as

$$I_{Pref} = \frac{\sqrt{2}\pi}{4}\sqrt{\frac{L_S}{L_P}}I_{BR}. \tag{6.9}$$

From (6.9), it can be seen that the reference current of the transmitting coil I_{Pref} is determined by the rated charging current I_{BR}, and has nothing to do with the coupling coefficient k_{PS} and the load resistance R_B. Therefore, by adjusting $I_P = I_{Pref}$, a constant DC output current independent of the load resistance and coupling coefficient can be obtained.

In addition, substituting (6.3) and (6.7) into $U_P = R_N I_P$ and $U_S = R_L I_S$, the voltage ration can be obtained as

$$\frac{U_{\mathrm{P}}}{U_{\mathrm{S}}}=\sqrt{\frac{L_{\mathrm{P}}}{L_{\mathrm{S}}}}\frac{R_{\mathrm{L}}+R_{\mathrm{S}}}{R_{\mathrm{L}}}+\sqrt{\frac{L_{\mathrm{S}}}{L_{\mathrm{P}}}}\frac{R_{\mathrm{P}}}{R_{\mathrm{L}}}. \tag{6.10}$$

Therefore, the transfer efficiency $\eta_{\mathrm{T\_CC}}$ and power $P_{\mathrm{T\_CC}}$ in the exact PT-symmetric region can be derived as follows:

$$\begin{aligned}\eta_{\mathrm{T\_CC}}&=\frac{I_{\mathrm{S}}^2R_{\mathrm{L}}}{I_{\mathrm{P}}^2R_{\mathrm{P}}+I_{\mathrm{S}}^2R_{\mathrm{S}}+I_{\mathrm{S}}^2R_{\mathrm{L}}}=\frac{R_{\mathrm{L}}}{\frac{L_{\mathrm{S}}}{L_{\mathrm{P}}}R_{\mathrm{P}}+R_{\mathrm{S}}+R_{\mathrm{L}}}\\&=\frac{8R_{\mathrm{B}}/\pi^2}{\frac{L_{\mathrm{S}}}{L_{\mathrm{P}}}R_{\mathrm{P}}+R_{\mathrm{S}}+8R_{\mathrm{B}}/\pi^2},\end{aligned} \tag{6.11}$$

$$\begin{aligned}P_{\mathrm{T\_CC}}&=\frac{U_{\mathrm{S}}^2}{R_{\mathrm{L}}}=\frac{R_{\mathrm{L}}U_{\mathrm{P}}^2}{\frac{L_{\mathrm{P}}}{L_{\mathrm{S}}}(R_{\mathrm{L}}+R_{\mathrm{S}})^2+2R_{\mathrm{P}}(R_{\mathrm{L}}+R_{\mathrm{S}})+\frac{L_{\mathrm{S}}}{L_{\mathrm{P}}}R_{\mathrm{P}}^2}\\&=\frac{8R_{\mathrm{B}}}{\pi^2}I_{\mathrm{o}}^2.\end{aligned} \tag{6.12}$$

From (6.11) and (6.12), it can be found that the transfer efficiency and transfer power are independent of the coupling coefficient k_{PS} in the exact PT-symmetric region. When the coupling coefficient changes, the value of the R_{P} and R_{S} may be slightly different due to the magnetic material. However, the variation is not serious, and the internal resistance of coils is generally considered constant [8, 9]. Besides, the equivalent load resistance R_{L} is generally much larger than the internal resistance, so the influence of the internal resistance change caused by the coupling coefficient on the transfer efficiency and output power can be ignored.

6.2.2 CV Mode

When the battery voltage reaches the rated voltage V_{BR}, the MC-WPT system goes into CV charging mode. Figure 6.5 gives the equivalent circuit of the system in CV mode.

According to Fig. 6.5 and Kirchhoff's law, the steady-state model of the system in CV mode can be expressed as

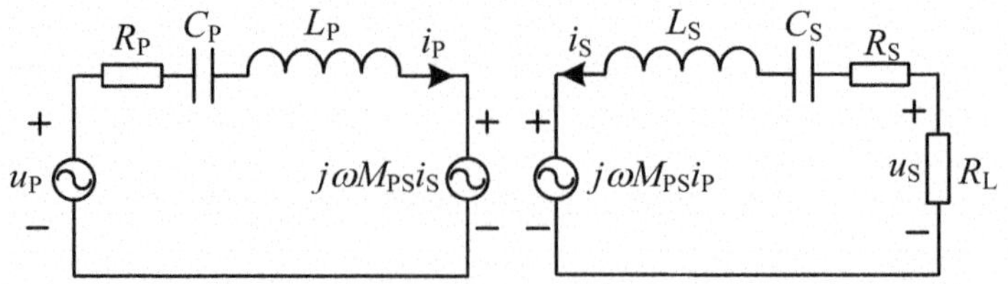

Fig. 6.5 Equivalent circuit of the CV charging mode

$$\begin{cases} \dot{U}_S = j\omega M_{PS}\dot{I}_P \dfrac{R_L}{R_L + Z_S} \\ \dot{I}_P = \dfrac{\dot{U}_P - j\omega M_{PS}\dot{I}_S}{Z_P} \end{cases}, \tag{6.13}$$

where $Z_P = R_P + j\omega L_P + 1/(j\omega C_P)$ and $Z_S = R_S + j\omega L_S + 1/(j\omega C_S)$ represent the impedance of the transmitting and receiving sides, respectively.

Then the voltage ratio can be derived as

$$\begin{aligned} \frac{\dot{U}_P}{\dot{U}_S} &= \frac{(R_L + Z_S)Z_P + (\omega M_{PS})^2}{j\omega M_{PS}R_L} \\ &= \frac{(R_L + R_S)\left(\omega L_P - \frac{1}{\omega C_P}\right) + R_P\left(\omega L_S - \frac{1}{\omega C_S}\right)}{\omega M_{PS}R_L} + j\frac{\Delta}{\omega M_{PS}R_L}, \end{aligned} \tag{6.14}$$

where $\Delta = [\omega L_P - 1/(\omega C_p)][\omega L_S - 1/(\omega C_S)] - (R_L + R_S)R_P - (\omega M_{ps})^2$.

If R_P and R_S are considered negligible, i.e., $R_P \approx 0$ and $R_S \approx 0$, (6.14) can be simplified as

$$\frac{\dot{U}_P}{\dot{U}_S} = \frac{\left(\omega L_P - \frac{1}{\omega C_P}\right)}{\omega M_{PS}} + j\frac{\Delta_1}{\omega M_{PS}R_L}, \tag{6.15}$$

where $\Delta_1 = [\omega L_P - 1/(\omega C_p)][\omega L_S - 1/(\omega C_S)]$-$(\omega M_{ps})^2$.

From (6.15), it can be found that when $\Delta_1 = 0$, the voltage ratio $\dot{U}_P/\dot{U}_S$ will be independent of the load resistance R_L. Let $\Delta_1 = 0$ and $L_P C_P = L_S C_S$, the operating frequency of the system can be derived as

$$\omega = \begin{cases} \omega_{H\_CV} = \dfrac{\omega_0}{\sqrt{1 - k_{PS}}} \\ \omega_{L\_CV} = \dfrac{\omega_0}{\sqrt{1 + k_{PS}}} \end{cases}, \tag{6.16}$$

where ω_{H_CV} and ω_{L_CV} represent the high-frequency branch and low-frequency branch corresponding to the operating angular frequency in CV mode, respectively.

Substituting (6.16) into (6.15), the input voltage of the rectifier U_S in CV mode can be deduced as

$$U_S = \left| \dot{U}_P \frac{\frac{\omega_0}{\sqrt{1 \pm k_{PS}}} k_{PS}\sqrt{L_P L_S}}{\frac{\omega_0}{\sqrt{1 \pm k_{PS}}} L_P - \frac{L_P \omega_0^2 \sqrt{1 \pm k_{PS}}}{\omega_0}} \right| = U_P\sqrt{\frac{L_S}{L_P}}. \tag{6.17}$$

Then the DC output voltage U_o can be obtained by the input/output relationship of the rectifier as

$$U_{\text{o}} = \sin(0.5D_{\text{S}}\pi)U_{\text{dc}}\sqrt{\frac{L_{\text{S}}}{L_{\text{P}}}}. \tag{6.18}$$

It can be seen that when the operating angular frequency of the system satisfies (6.16), a constant DC output voltage independent of load and coupling coefficient can be achieved. Comparing (6.4) and (6.16), if the internal resistances are ignored, and the load resistance is set to 0, it can be deduced that $\omega_{\text{H\_CC}} = \omega_{\text{H\_CV}}$ and $\omega_{\text{L\_CC}} = \omega_{\text{L\_CV}}$. $\omega_{\text{H\_CC}} = \omega_{\text{H\_CV}}$ and $\omega_{\text{L\_CC}} = \omega_{\text{L\_CV}}$ can be automatically obtained by controlling the output characteristics of the inverter to negative resistance, there is no need to know the actual value of the coupling coefficient in advance, so before starting the charge, it is only necessary to short-circuit the load and control the inverter to run in the negative resistance state, then the operating frequency of the inverter can be sampled as the operating frequencies of CV mode without knowing the coupling coefficient k_{PS} in advance. Furthermore, for stationary wireless charging of EVs, once the car is parked, its position (i.e., the value of k_{PS}) will not change. Based on (6.16), the operating frequency during the CV mode is fixed after the car is parked.

In CV mode, the transfer efficiency and power can be further deduced as follows [10]:

$$\begin{aligned} \eta_{\text{T\_CV}} &= \frac{(\omega M_{\text{PS}})^2 R_{\text{L}}}{R_{\text{P}}\left[(R_{\text{L}}+R_{\text{S}})^2 + X_{\text{S}}^2\right] + (R_{\text{L}}+R_{\text{S}})(\omega M_{\text{PS}})^2} \\ &= \frac{(\omega M_{\text{PS}})^2 8R_{\text{B}}/\pi^2}{R_{\text{P}}\left[\left(8R_{\text{B}}/\pi^2 + R_{\text{S}}\right)^2 + X_{\text{S}}^2\right] + \left(8R_{\text{B}}/\pi^2 + R_{\text{S}}\right)(\omega M_{\text{PS}})^2}, \end{aligned} \tag{6.19}$$

$$\begin{aligned} P_{\text{o\_CV}} &= \frac{U_{\text{P}}^2 R_{\text{L}} (\omega M_{\text{PS}})^2}{\left[R_{\text{P}}(R_{\text{L}}+R_{\text{S}}) - X_{\text{P}}X_{\text{S}} + (\omega M_{\text{PS}})^2\right]^2 + \left[R_{\text{P}}X_{\text{S}} + (R_{\text{L}}+R_{\text{S}})X_{\text{P}}\right]^2} \\ &= \frac{U_0^2}{8R_{\text{B}}/\pi^2}, \end{aligned} \tag{6.20}$$

where $X_{\text{P}} = \omega L_{\text{P}} - 1/(\omega C_{\text{P}})$ and $X_{\text{S}} = \omega L_{\text{S}} - 1/(\omega C_{\text{S}})$.

6.2.3 *Influence of the Internal Resistance Variation*

According to the derivation of (6.2)–(6.8), it can be seen that the circuit mode of CC mode takes the internal resistance (i.e., R_{P} and R_{S}) into consideration, so the accurate constant output current can be achieved in CC mode, as shown in Fig. 6.6a. Meanwhile, in the CV mode, the influence of the internal resistance on the output voltage under the different load resistances and coupling coefficients is shown in Fig. 6.6b, wherein the normalized output voltage is defined as $U_{\text{S\_actual}}/U_{\text{S}}$, $U_{\text{S\_actual}}$ can be derived by submitting $\omega_{\text{H\_CC}|R_{\text{L}}=0}$ or $\omega_{\text{L\_CC}|R_{\text{L}}=0}$ into (6.14), U_{S} can be calculated

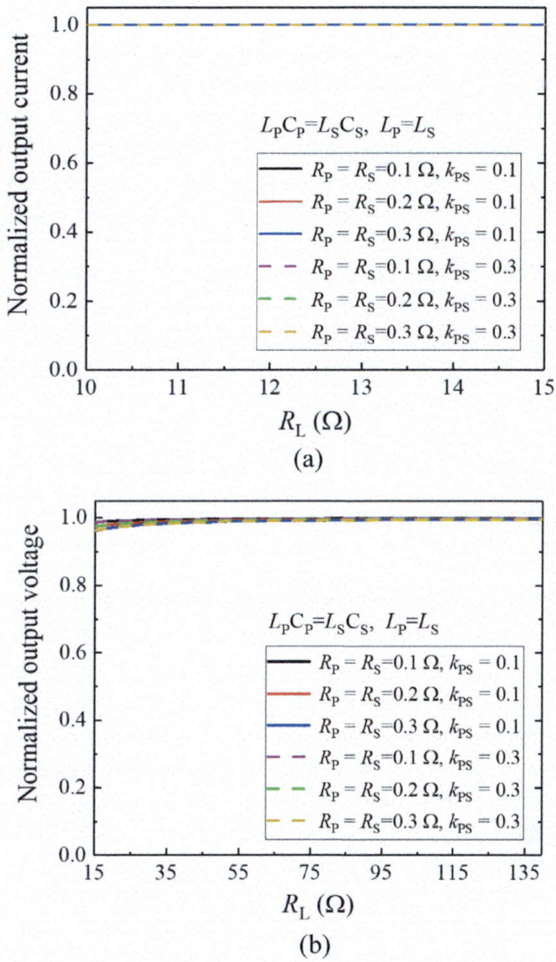

Fig. 6.6 Influence of the internal resistance on **a** output current and **b** output voltage

by (6.17). The calculation results are obtained under the conditions of $L_PC_P = L_SC_S$ and $L_P = L_S$.

It can be seen from Fig. 6.6b that the internal resistance of 0.1 Ω and 0.3 Ω only leads to no more than 1.3% and 3.8% relative error in output voltage, respectively. And the error gradually decreases with the increase of the load resistance.

6.2.4 Influence of the Self-Inductance Variation

In practical applications such as wireless charging of EV, the magnetic material such as ferrite is used for shielding in the wireless EV charging system. Due to the influence of the magnetic material on the magnetic flux of the coils, the self-inductance of the coils varies with the relative position of the transmitting and receiving coils. Meanwhile, the resonant frequency of the transmitter and receiver is also changed due to the self-inductance variation.

According to (6.8) and (6.18), it can be found that the output current and voltage are directly related to the ratio of L_P and L_S, but have nothing to do with the value of L_P and L_S. Furthermore, as long as $\omega_P = \omega_S$, (6.8) and (6.18) are valid. In other words, there is no need to consider the specific value of the resonant frequency. Although the coil self-inductances L_P and L_S vary with coil-misalignment due to the influence of the magnetic material, the changing trend of their variation is the same. Therefore, a constant self-inductance ratio and $\omega_P = \omega_S$ can be basically satisfied when misalignment occurs. Especially when symmetrical transmitting and receiving coils are used, the influence of self-inductance changes on output current and voltage can be eliminated.

6.3 Control and Implementation

6.3.1 Implementation of Negative Resistor

To achieve CC/CV charging using only a single-stage inverter, the negative resistor is implemented by the phase synchronization method. Compared with the previous method, including the use of a nonlinear operational amplifier [7] and a self-oscillating controlled inverter [6], the advantage is that the negative resistor can be reliably controlled to work in high-frequency mode or low-frequency mode. More importantly, the power provided by the negative resistor can be regulated without an additional DC-DC converter.

Figure 6.7 gives the voltage and current waveforms of the inverter and its derive signals, wherein u_{ac} is the output voltage of the inverter, u_P is the fundamental component of u_{ac}, i_P is the output current of the inverter (i.e., the current flowing through the transmitting coil), U_{GS1} ~ U_{GS4} represent the drive signals of the MOS tubes Q_1 ~ Q_4, respectively. $\varphi_P = D_S\pi$ is the phase shift angle, T_{iP} is the period of the current i_P. It can be found that as long as the phase relationship between the drive signals U_{GS1} ~ U_{GS4} and the output current i_P of the inverter always maintains the relationship shown in Fig. 6.7, the fundamental wave component of the output voltage of the inverter will be in phase with its output current (note that the output voltage and current of the current shown in Fig. 6.1 are non-associated reference direction), so the inverter can be equivalent to a negative resistance.

The detailed principle to generate the required drive signal are given as follows:

The drive signal is generated by a digital signal processor (DSP) controller, and the main modules used include a pulse capture module (CAP) and different pulse width modulation modules (PWM1 and PWM2). Figure 6.8 gives the generation process of driver pulses based on the phase synchronization method. As shown in Fig. 6.8, the complementary drive signals U_{GS1} and U_{GS2} are generated by the PWM1

Fig. 6.7 Operating waveforms of inverter

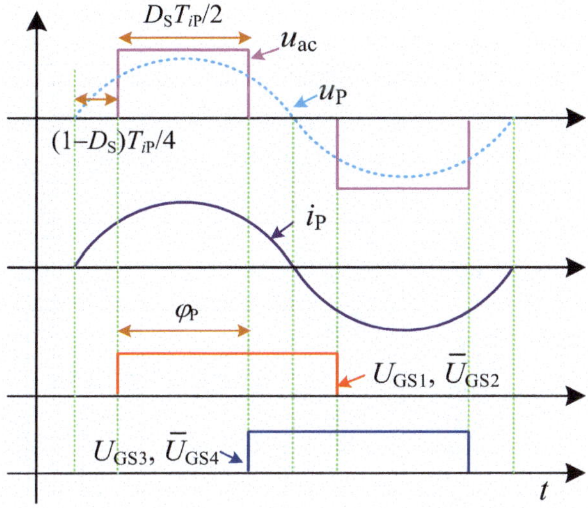

module, and the complementary drive signals U_{GS3} and U_{GS4} are generated by the PWM2 module. The generation process is described as follows: first, the output current i_P of the inverter is sampled, and the square wave signal i'_P with the same frequency and phase is generated. Then, the rising edge of the square wave signal i'_P is captured through the CAP module to obtain the phase and frequency of the inverter output current in real-time. When the CAP module captures the rising edge of the square wave signal i'_P, the synchronization signal S_{n1} of PWM1 is generated, and the value of PWM1's counter CTR1 is immediately updated to the initial value Pha1. In addition, when the value of CTR1 is reduced to 0, the synchronization signal S_{n2} of PWM2 is generated, and the value of the PWM2's counter CTR2 is immediately updated to the initial value Pha2. Here, the counters CTR1 and CTR2 of PWM1 and PWM2 work in the down-count mode. When CTR1 is equal to 0, U_{GS1} is forced to output a high level, and U_{GS2} is forced to output a low level. When the counter CTR1 is equal to half of the period (PRD/2), U_{GS1} is forced to output a low level, and U_{GS2} is forced to a high level. Similarly, the driver signal U_{GS3} and U_{GS4} can be generated based on CTR2. The difference between them is the initial phase of the counter (Pha1 and Pha2). The values of Pha1 and Pha2 determine the phase and amplitude of the inverter output voltage, respectively. In order to maintain the output current and output voltage of the inverter in phase, the values of Pha1 and Pha2 should be satisfied

$$\begin{cases} \text{Pha1} = \dfrac{(1-D_S)}{4}\text{PRD} \\ \text{Pha2} = \dfrac{D_S}{2}\text{PRD} \end{cases}, \tag{6.21}$$

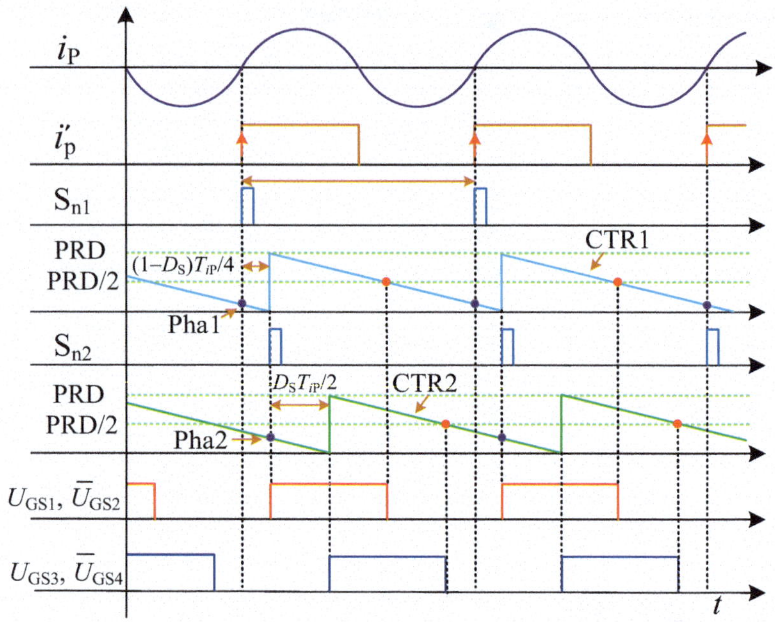

Fig. 6.8 The generation process of driver signals based on the phase synchronization method

where PRD $= T_S/T_{CLK}$, T_S is the switching period captured in real-time by CAP module ($T_S = 2\pi/\omega$), and T_{CLK} represents the time required for the counter to decrease by 1. Besides, by limiting the value range of PRD, the stable state of the operating angular frequency of the system can be controlled. For instance, if the system is to operate on the high-frequency branch ω_{H_CC} in the exact PT-symmetric region, the value of PRD can be limited to a range less than $\text{PRD}_0 = 2\pi/(\omega_0 T_{CLK})$ by DSP. On the contrary, if the system is to oscillate at low-frequency branch ω_{L_CC}, the value of PRD should be set greater than PRD_0.

Therefore, based on the phase synchronization method, the fundamental component of the output voltage of the inverter and its current can be automatically maintained in the same phase. Such a characteristic ensures that the system oscillates automatically at $\omega = \omega_{H\_CC,\ L\_CC}$ in the exact PT-symmetric region (or $\omega = \omega_0$ in the broken PT-symmetric region) against the variations of coupling coefficient and load resistance. Furthermore, the power injected into the system can be easily adjusted by changing the value of the Pha2, and the advantages of the phase synchronization method can be summarized into two points. One is that the negative resistor with the characteristics of continuous output power adjustment is realized by the single-stage inverter. The other is that the control is flexible and reliable based on digital control.

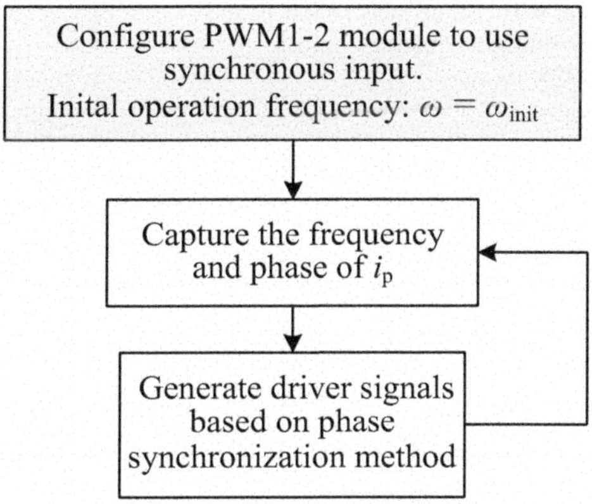

Fig. 6.9 Flowchart of the operating frequency control in CC mode

6.3.2 Operating Frequency Control in CC Mode

Figure 6.9 shows the control implementation of operating frequency in CC mode. Firstly, the inverter operation is initiated at $\omega = \omega_{\text{init}}$, where ω_{init} is selected equal to 100 kHz, and PRD is set to be less than PRD_0, so that the system works at the high-frequency branch $\omega_{\text{H\_CC}}$ in the exact PT-symmetric region. Then, the frequency and phase of transmitting coil current i_P are captured, and the new driver signals of the inverter is generated based on the phase synchronization method shown in Fig. 6.8. The driver signals are used to drive the inverter switches to update the operating frequency of the system. In addition, according to the latest frequency and phase of i_P, the drive signals are regenerated, and the operating frequency is constantly updated. The switching frequency becomes steady when the phase difference between u_P and i_P has been zeroed.

6.3.3 Operating Frequency Control in CV Mode

Figure 6.10 gives the control implementation of operating frequency in CV mode. Based on the theoretical analysis in Sect. 6.2.2, the operating frequency of CV (ω_{CV}) can be obtained by turning on the switch S_1. Hence, the inverter is first controlled based on the phase synchronization method shown in Fig. 6.9a, and the duty cycle of the inverter is constant at $D_S = D_{\text{init}} = 0.01$ to prevent overcurrent when S_1 is turned on. Then the transmitter has to judge whether S_1 is turned on. Due to the influence of magnetic materials, the resonance frequency ω_0 will vary between ω_{0_min} and ω_{0_max} (i.e., $\omega_{0\_\text{min}} \leq \omega_0 \leq \omega_{0\_\text{max}}$) under different misalignment conditions. In this design, the high-frequency branch $\omega_{\text{H\_CC}}$ is chosen by setting PRD to be less than PRD_0 in the exact PT-symmetric region.

According to (6.4), if $\omega \leq \omega_{0\_\text{max}}$, it means that the system goes into the broken PT-symmetric region (in this case, it can be concluded that the switch S_1 is not turned on

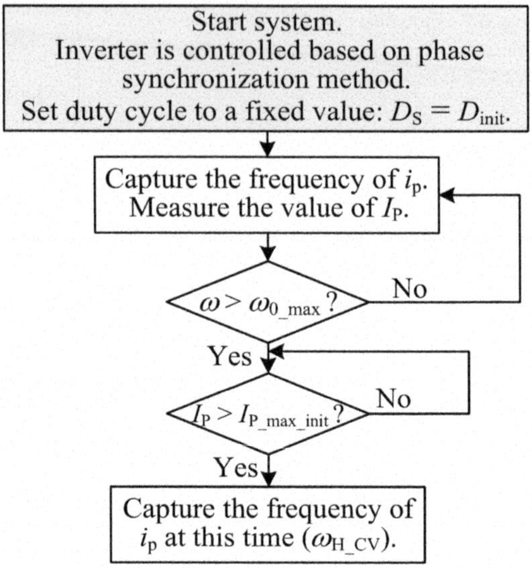

Fig. 6.10 Flowchart of the operating frequency control in CV mode

or that there is no receiver). Otherwise, the system works in the exact PT-symmetric region (in this case, there are two situations, one is that the equivalent load resistance R_L is less than R_C and the switch S_1 is turned off, and the other is that the switch S_1 is turned on). Assuming that the minimum value of the load resistance is R_{L_min}, based on (6.7) and (6.10), the maximum value of I_P in the exact PT-symmetric region when S_1 is turned off (i.e., $I_{P\_max\_init}$) can be calculated by

$$I_{P\_max\_init} = \frac{2\sqrt{2}U_{dc}\sin(0.5D_{init}\pi)L_S}{\pi\left[L_P\left(R_{L\_min}+R_S\right)+L_SR_P\right]}. \tag{6.22}$$

It can be seen that $\omega > \omega_{0\_max}$ and $I_P > I_{P\_max\_init}$ will be satisfied at the same time when the load is short-circuited (i.e., the switch S_1 has been turned on), so the transmitter can determine when the switch S_1 is turned on by measuring the operating frequency and the RMS value of i_P without communication. This allows the receiver to turn on the switch S_1 at any time after the EV is parked (the S_1 is automatically turned off after T_d seconds, where T_d is preset). Finally, the frequency of i_P can be sampled as the operating frequency ω_{H_CV} of CV mode when $\omega > \omega_{0\_max}$ and $I_P > I_{P\_max\_init}$.

6.3.4 Charging Process Control

Figure 6.11 presents the control flowchart of the whole charging process. As shown in Fig. 6.11, after starting the system, it is first necessary to wait for the short-circuit switch S_1 to close, and then obtain the operating angular frequency ω_{H_CV} in CV mode based on the method shown in Fig. 6.10. Because the system needs work in the exact PT-symmetric region during the CC charging mode, $k_{PS} \geq k_C$ should be satisfied. By submitting $k_{PS} = k_C$ and $R_L = 0$ into (6.4), the operating frequency at the critical coupling coefficient k_C when S_1 is turned on (i.e., ω_C) can be obtained. In addition, the system is controlled to oscillate at the high-frequency branch ω_{H_CC} in the exact PT-symmetric region. According to (6.4), it can be found that the high-frequency branch ω_{H_CC} decreases with the decrease of the coupling coefficient. Therefore, it can be judged whether the coupling coefficient meets $k_{PS} \geq k_C$ by comparing the working angular frequency of real-time sampling with the size of ω_C. If $\omega_{H\_CV} \geq \omega_C$, it indicates that the system is working within the strong coupling region (i.e., $k_{PS} \geq k_C$). Then, the charging process will be started after waiting for T_d seconds (i.e., after the S_1 is turned off again). Because the battery may be in a high state of charge (SOC) when the charging process is just started, the charger should directly enter the CV mode in this case. If $\omega \leq \omega_{0\_max}$, it means that the system works in the broken PT-symmetric region (i.e., $R_L > R_C \geq 8R_{BC}/\pi^2$). Then, the CV mode is activated directly. Otherwise, the CC charging mode is selected. During the CC charging process, it can be seen from (6.8) that the DC output current can be maintained constant by adjusting I_P equal to the reference current of the transmitting coil I_{Pref}. The difference between I_P and I_{Pref} is nullified with the help of a proportional-integral (PI) controller. Meanwhile, the real-time value of U_P is calculated based on (6.1). By submitting $R_L = 8R_{BC}/\pi^2$ and $U_S = 2\sqrt{2}\,U_{BR}/\pi$ into (6.10), the output voltage of the inverter when the battery voltage reaches the rated voltage (i.e., U_{PC}) can be calculated as

$$U_{PC} = \frac{2\sqrt{2}U_{BR}\pi\left[\left(8R_{BC}/\pi^2 + R_S\right)L_P + R_P L_S\right]}{8R_{BC}\sqrt{L_P L_S}}. \quad (6.23)$$

From (6.23), it can be seen that U_{PC} is only related to rated parameters. After system parameters are determined, U_{PC} can be directly calculated according to (6.23). As shown in Fig. 6.11, when $U_P \geq U_{PC}$, the CC charging mode will be changed to CV charging mode. In CV mode, the operating angular frequency and the duty cycle are set as $\omega = \omega_{H\_CV}$ and $D_S = D_{CV}$, respectively, where D_{CV} can be calculated from (6.18) as

$$D_{CV} = \frac{2\arcsin\left(U_{BR}\sqrt{L_P}/U_{dc}\sqrt{L_S}\right)}{\pi}. \quad (6.24)$$

Then, the PWM modules generates the required drive singles according to ω_{H_CV} and D_{CV}. After that, the constant DC output voltage can be achieved.

From the above analysis, it can be seen that the whole control process does not require any communication between the transmitting side and the receiving side, and does not need to add additional dc-dc converters, and the CC mode can be automatically switched to the CV mode without changing the system topology.

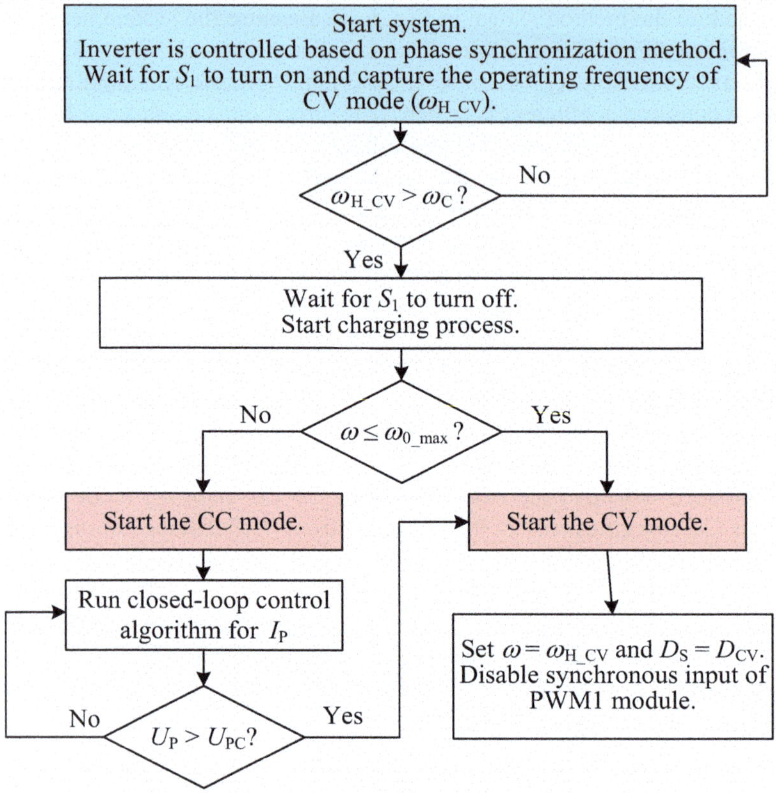

Fig. 6.11 Flowchart of the closed-loop control for charging process

6.4 System Design and Verification

Figure 6.12 shows the structure block diagram of the PT-symmetric MC-WPT system based on the phase synchronization method. The main circuit includes a full-bridge inverter, series compensation network, short-circuit switch S_1, high-frequency rectifier and load battery. The full-bridge inverter is composed of four MOSFETs (Q_1 ~ Q_4), and four switching tubes Q_1 ~ Q_4 are respectively controlled by four driving signals U_{GS1} ~ U_{GS4}, which inverts the DC voltage U_{dc} to a high-frequency quasi-square wave voltage u_{ac}. The rectifier consists of four Schottky diodes (D_{r1} ~ D_{r4}), which rectifies the received high-frequency energy into a DC output. Switch S_1 is connected in parallel to the input of the rectifier and is closed only briefly before starting charging. R_L represents the AC equivalent load as viewed from the rectifier input. The control circuit consists of five modules, and the detailed scheme is shown in Fig. 6.13. Here, a current transformer (CT) CU8965 and a differential amplifier circuit consisting of a high-speed operational amplifier LM6172 are first used to convert the current into a voltage signal. Then, an RMS calculator AD637 is

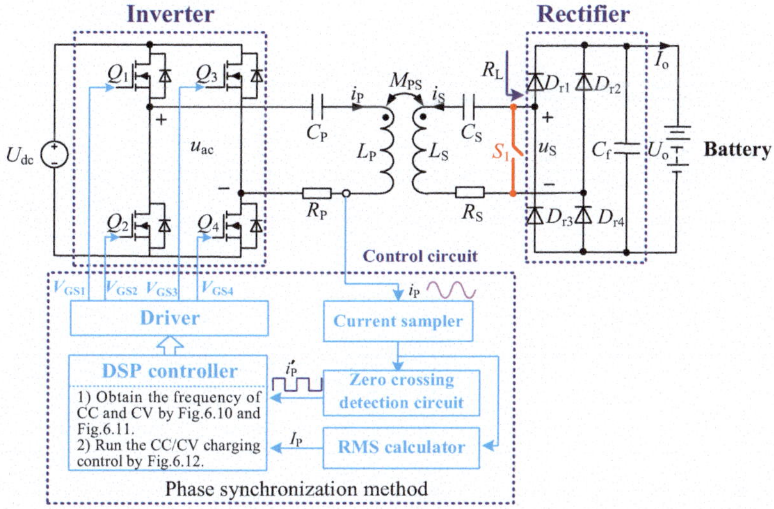

Fig. 6.12 PT-symmetric MC-WPT system based on the phase synchronization method

employed to measure the value of I_P, and a zero-crossing comparator comprising a high-speed comparator TL3016 is used to capture the frequency and phase of i_p. A floating-point DSP (TMS320F28377D) runs the control algorithm and outputs four PWM waves sent to a drive circuit to generate the driving signal. The drive circuit consists of four isolated gate drivers (Si8271).

Figure 6.14 gives the experimental prototype, where the DD coils are selected as the coupling coils due to their excellent anti-offset performance. To ensure that the coupling coefficient k_{PS} is larger than the critical value k_C at the worst misalignment case, the coils are optimized with the help of the finite-element analysis software Maxwell. To reduce the high-frequency internal resistance of the coils, the coils are fabricated using the Litz wire with 0.05 mm × 1000 strands and 11 turns, and the dimension is 550 mm × 360 mm × 5 mm. In addition, a ferrite layer is added for coupling enhancement and flux guidance. All other parameters are listed in Table 6.1.

For the convenience of testing, the electronic load replaces the battery as the load. Depending on the battery charging profile, the equivalent resistance of the battery can range from 10 to 140 Ω (CC mode: 10–15 Ω, CV mode: 15–140 Ω), and then the critical coupling coefficient can be calculated as $k_C = 0.076$. Besides, for EV charging applications, the range of transfer distance d is between 100–250 mm (depending on the vehicle type) is required.

Figure 6.15 shows the measured results of the self-inductance of the transmitting and receiving coils and the coupling coefficient between them under different transfer distances. As can be seen from Fig. 6.15a, although the self-inductance of coils changes with the transfer distance, the self-inductance ratio of the transmitting coil L_P and the receiving coil L_S remains constant. Besides, it can be seen from Fig. 6.15b that $k_{PS} > k_C$ ($k_C = 0.076$) can always be satisfied.

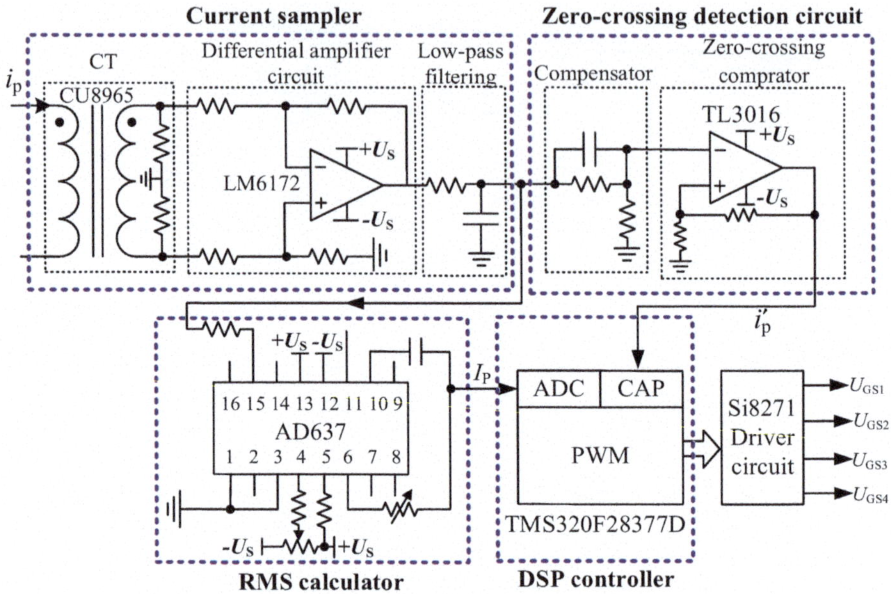

Fig. 6.13 The detailed schematic diagram of the control circuit

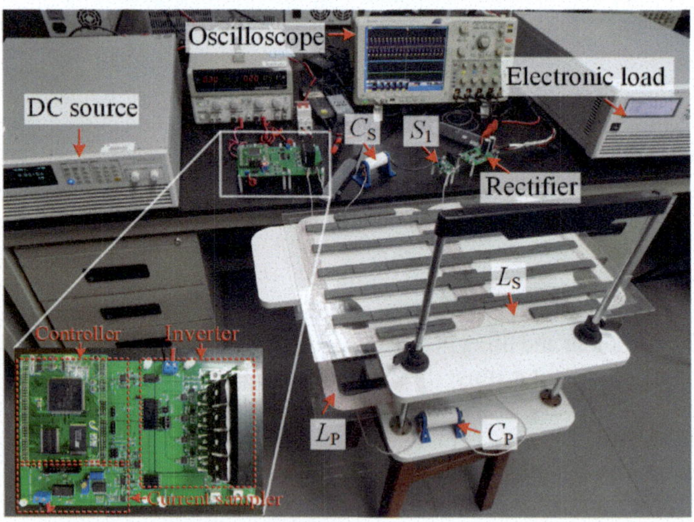

Fig. 6.14 The experimental prototype system

Table 6.1 The circuit parameters of the experimental prototype

Description	Symbol	Values
Inductance of transmitting coil	L_P	277–304 μH
Inductance of receiving coil	L_S	278–305 μH
Compensation capacitance of transmitter	C_P	10.34 nF
Compensation capacitance of receiver	C_S	10.33 nF
Internal resistance of transmitter	R_P	0.37–0.4 Ω
Internal resistance of receiver	R_S	0.35–0.37 Ω
DC voltage source	U_{dc}	150 V
Rated charging current	I_{BR}	8 A
Rated charging voltage	U_{BR}	120 V
Transfer distance	d	100–250 mm
Equivalent load resistance of battery	R_B	CC mode: 10–15 Ω CV mode: 15–140 Ω

Figure 6.16 shows the charging profile with the charging current and charging voltage, where the transfer distance d is set as 100 mm, 150 mm, 200 mm, and 250 mm. It can be seen that the charging current and charging voltage are in good agreement with the theoretical charging profile. In CC mode, the charging current only changes within 1%, and in CV mode, the charging voltage only changes within 4%, which are small and acceptable.

Figure 6.17 gives the measured experimental results of the transfer efficiency of the system at different transfer distances, it can be seen that in the CC mode, the transfer efficiency increases with the increase of the load resistance, while in the CV mode, the transfer efficiency first rises and then drops with the increase of the load resistance. Meanwhile, in the CC mode, when the load resistance R_B is the same, the transfer efficiency remains constant under different transfer distances, which is consistent with the theoretical analysis. Furthermore, the maximum transfer efficiency can reach 98.5% at a 10 cm air gap ($k_{PS} = 0.429$), and the transfer efficiency gradually decreases as the transfer distance increases. But when the transfer distance is increased to 25 cm ($k_{PS} = 0.099$), the minimum transfer efficiency is still close to 90%.

6.5 Summary

This chapter introduces a PT-symmetric MC-WPT system with load- and position-independent constant output current and voltage for EVs charging. Firstly, the system structure of the PT-symmetric MC-WPT system for EVs charging and the typical charging process with CC and CV modes of the battery are introduced. Then, based on the circuit theory, the operating frequency, output current and voltage, transfer

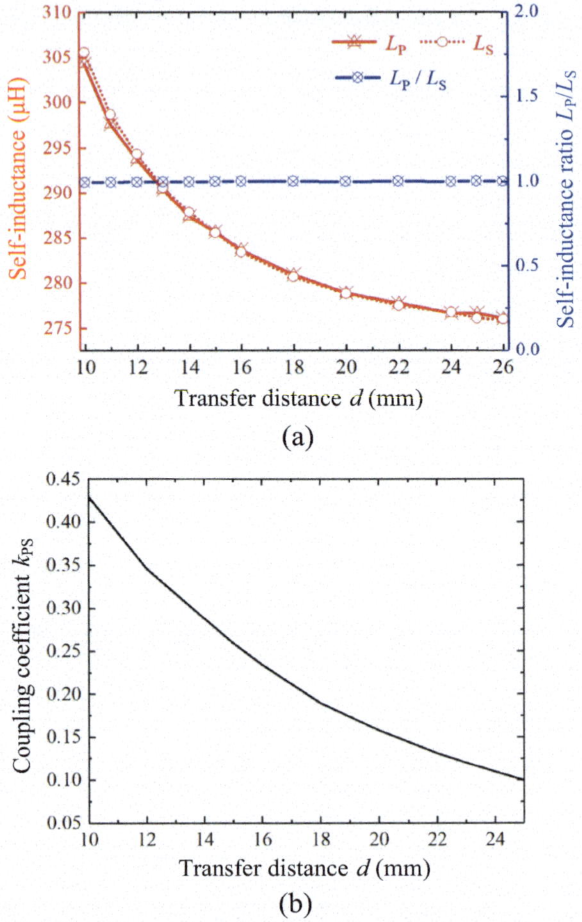

Fig. 6.15 Measured self-inductance and coupling coefficient as a function of the distance between the two coils. **a** Self-inductance of L_P and L_S and their ratio. **b** Coupling coefficient k_{PS}

efficiency, and output power of the system are derived. Meanwhile, the CC and CV charging modes and control strategies are established. In addition, the implementation of the negative resistor and the whole system are given. Finally, an experimental prototype with an 8 A charging current and a 120 V charging voltage is built. The analytical and experimental results demonstrate that within the transfer distance of 10–25 cm, the CC/CV outputs can be maintained during the charging process.

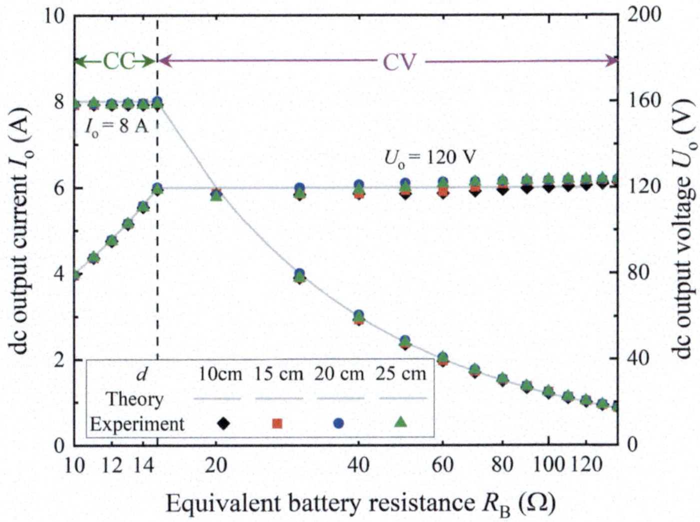

Fig. 6.16 Theoretical and experimental results of charging current and voltage under different transfer distance

Fig. 6.17 Theoretical and experimental results of transfer efficiency of the system

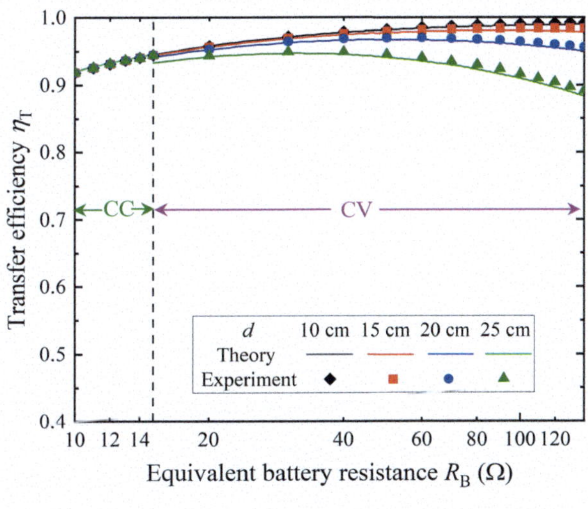

References

1. Wu L, Zhang B, Jiang Y (2022) Position-independent constant current or constant voltage wireless electric vehicles charging system without dual-side communication and DC–DC converter. IEEE Trans Industr Electron 69(8):7930–7939
2. Li Y, Hu J, Liu M et al (2019) Reconfigurable intermediate resonant circuit based WPT system with load-independent constant output current and voltage for charging battery. IEEE Trans Power Electron 34(3):1988–1992

3. Li Y, Hu J, Chen F et al (2018) Dual-phase-shift control scheme with current-stress and efficiency optimization for wireless power transfer systems. IEEE Trans Circuits Syst I Regul Pap 65(9):3110–3121
4. Song K, Li Z, Jiang J, Zhu C (2018) Constant current/voltage charging operation for series-series and series-parallel compensated wireless power transfer systems employing primary-side controller. IEEE Trans Power Electron 33(9):8065–8080
5. Yenil V, Cetin S (2022) Load independent constant current and constant voltage control of LCC-series compensated wireless EV charger. IEEE Trans Power Electron 37(7):8701–8712
6. Zhou J, Zhang B, Xiao W et al (2019) Nonlinear parity-time-symmetric model for constant efficiency wireless power transfer: application to a drone-in-flight wireless charging platform. IEEE Trans Industr Electron 66(5):4097–4107
7. Assawaworrarit S, Yu X, Fan S (2017) Robust wireless power transfer using a nonlinear parity-time-symmetric circuit. Nature 546(7658):387–390
8. Su YG, Chen L, Wu XY et al (2018) Load and mutual inductance identification from the primary side of inductive power transfer system with parallel-tuned secondary power pickup. IEEE Trans Power Electron 33(11):9952–9962
9. Huang Z, Lam CS, Mak PI et al (2020) A single-stage inductive-power-transfer converter for constant-power and maximum-efficiency battery charging. IEEE Trans Power Electron 35(9):8973–8984
10. Liu F, Yang Y, Jiang D et al (2017) Modeling and optimization of magnetically coupled resonant wireless power transfer system with varying spatial scales. IEEE Trans Power Electron 32(4):3240–3250

Chapter 7
PT-Symmetric Multi-load Magnetic-Field Coupled Wireless Power Transfer System

With the development of MC-WPT technology and the increasing number of electronic products, multi-load MC-WPT system has gradually become more and more popular in recent years. However, the output stability of the existing multi-load MC-WPT system is easily affected by the coupling coefficient. When the coupling coefficient changes, it is difficult to maintain constant output power and constant transfer efficiency, resulting in anti-offset ability and low spatial degree of freedom. In addition, how to achieve the power distribution among different loads according to the actual needs is also the main challenge of a multi-load MC-WPT system. In order to solve the above problems, inspired by the PT-symmetric MC-WPT system, a novel multi-load MC-WPT system that combines the PT-symmetric principle and time-sharing control strategy is proposed, which achieves constant output power and constant transfer efficiency under the condition of variable coupling coefficient. This chapter first establishes the circuit model of the PT-symmetric multi-load MC-WPT system with a single natural resonant frequency. Then, the operating frequency, output power, and transfer efficiency of the PT-symmetric multi-load MC-WPT system with a single natural resonant frequency are studied. Finally, the PT-symmetric multi-load MC-WPT system with different natural resonant frequencies is further introduced, and the system structure, time-sharing control strategy, and power distribution control are analyzed.

7.1 Single-Frequency Multi-load System

7.1.1 Circuit Modeling

The method of implementing a negative resistor using operational amplifiers is not suitable for medium to large power applications due to its low output power level (only mW level) [1, 2]. Conversely, employing power electronic converters to construct a

B. Zhang et al., *Parity-Time Symmetric Wireless Power Transfer*, CPSS Power Electronics Series, https://doi.org/10.1007/978-981-96-3149-0_7

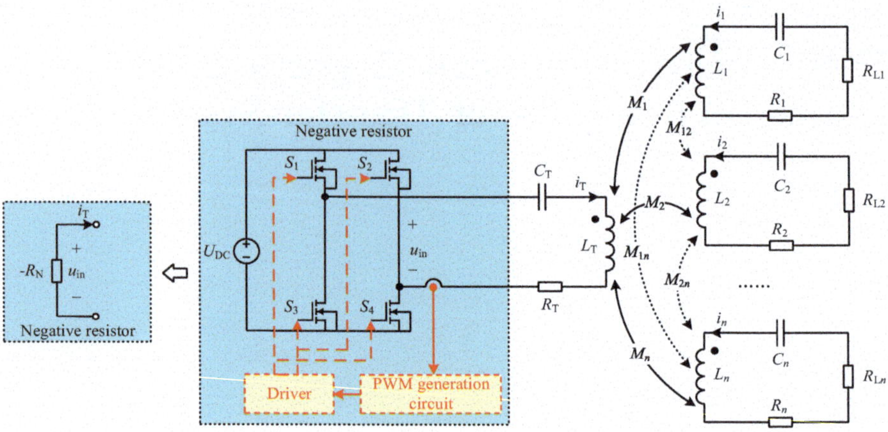

Fig. 7.1 The schematic diagram of PT-symmetric multi-load MC-WPT system using SS compensation topology

negative resistor offers an effective approach with high output power (hundreds of W) and a wide adjustable range [3, 4]. A full-bridge inverter is used to construct a nonlinear saturation gain instead of the operation amplifier to achieve higher output power. The schematic diagram of the PT-symmetric multi-load MC-WPT system with a single natural resonant frequency using SS compensation topology is shown in Fig. 7.1 [5], wherein the negative resistance $-R_N$ is used as a power supply to provide energy for the PT-symmetric multi-load MC-WPT system. L_T, C_T, R_T, and i_T are the coil inductance, resonant capacitor, internal resistance, and coil current of the transmitter, respectively. L_i, C_i, R_i, and i_i are the coil inductance, resonant capacitor, internal resistance and coil current of the i-th receiver, respectively. R_{Li} is the i-th load resistance. M_i is the mutual inductance between the transmitter and the i-th receiver, and M_{ij} is the cross-coupling mutual inductance between the receivers ($i, j = 1, 2, \ldots, n$).

Figure 7.2 shows the waveform of u_{in}, i_T and the gate drive signals of S_1, S_2, S_3 and S_4 (u_{g1}, u_{g2}, u_{g3} and u_{g4}). To simplify the analysis, the delay time and dead time are not considered in modeling. As shown in Fig. 7.2, the full-bridge inverter is controlled by zero crossing points of the current waveform in the transmitting resonator. Thus, u_{in} and its RMS value U_{in} can be defined as

$$u_{in} = \text{sgn}(i_T)U_{DC}, \tag{7.1}$$

$$U_{in} = \frac{2\sqrt{2}}{\pi}U_{DC}, \tag{7.2}$$

where sgn() is the sign function.

Assuming the direction of current i_T and voltage u_{in} shown in Fig. 7.1 is positive, Fig. 7.2 shows that the phase difference between u_{in} and i_T is always zero,

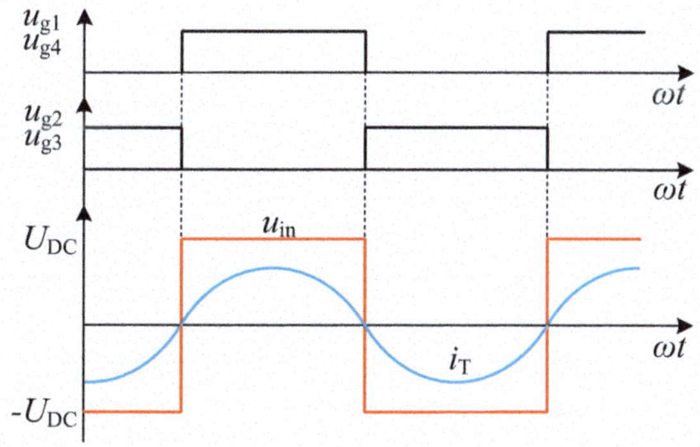

Fig. 7.2 The voltage and current of the nonlinear negative resistance

which means that the negative resistor generates the energy. Therefore, the full-bridge inverter can also be equivalent to a negative resistance $-R_N$, as shown in the sub-picture of Fig. 7.1.

According to the Kirchhoff's voltage law (KVL), the circuit model of the PT-symmetric multi-load MC-WPT system can be deduced as

$$\begin{bmatrix} -R_N + R_T + j\left(\omega L_T - \frac{1}{\omega C_T}\right) & j\omega M_1 & \cdots & j\omega M_n \\ j\omega M_1 & R_{L1} + R_1 + j\left(\omega L_1 - \frac{1}{\omega C_1}\right) & \cdots & j\omega M_{1n} \\ \vdots & \vdots & \cdots & \vdots \\ j\omega M_n & j\omega M_{1n} & \cdots & R_{Ln} + R_n + j\left(\omega L_n - \frac{1}{\omega C_n}\right) \end{bmatrix} \begin{bmatrix} \dot{I}_T \\ \dot{I}_1 \\ \vdots \\ \dot{I}_n \end{bmatrix} = 0 \tag{7.3}$$

where j is the imaginary unit, ω is the operating angle frequency, $\dot{I}_T$ and $\dot{I}_i$ are the current vectors of i_T and i_{Li}, respectively. M_i is the mutual inductance, and M_{ij} is the cross-coupling mutual inductance between the receiving coils ($i, j = 1, 2, \ldots, n$).

To initially explore the feasibility of the PT-symmetric multi-load MC-WPT system and simplify the calculation, the dual-load PT-symmetric MC-WPT system is taken as an example for detailed analysis in this chapter, the equivalent circuit is shown in Fig. 7.3. Assuming that the receiving coils L_1 and L_2 are located on either side of the transmitting coil or are far away from each other, the cross-coupling has little effect on the system and can be ignored approximately, that is, $M_{12} \approx 0$. Thus, the circuit model of the dual-load PT-symmetric MC-WPT system can be presented as

$$\begin{bmatrix} \frac{-R_N+R_T}{L_T} + j\left(\omega - \frac{\omega_T^2}{\omega}\right) & j\omega k_1\sqrt{\frac{L_1}{L_T}} & j\omega k_2\sqrt{\frac{L_2}{L_T}} \\ j\omega k_1\sqrt{\frac{L_T}{L_1}} & \frac{R_{L1}+R_1}{L_1} + j\left(\omega - \frac{\omega_1^2}{\omega}\right) & 0 \\ j\omega k_2\sqrt{\frac{L_T}{L_2}} & 0 & \frac{R_{L2}+R_2}{L_2} + j\left(\omega - \frac{\omega_2^2}{\omega}\right) \end{bmatrix} \begin{bmatrix} \dot{I}_T \\ \dot{I}_1 \\ \dot{I}_2 \end{bmatrix} = 0, \tag{7.4}$$

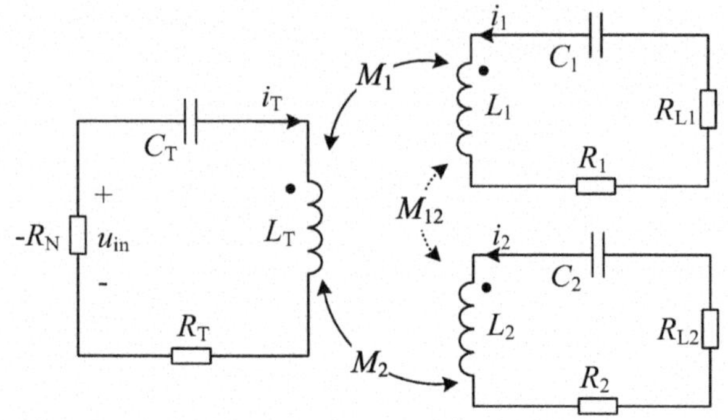

Fig. 7.3 Equivalent circuit of dual-load PT-symmetric MC-WPT system

where $\omega_T = 1/\sqrt{L_T C_T}$ is the natural resonant angular frequency of the transmitter, $\omega_1 = 1/\sqrt{L_1 C_1}$ and $\omega_2 = 1/\sqrt{L_2 C_2}$ are the natural resonant angular frequencies of two receivers, respectively. $k_1 = M_1/\sqrt{L_T L_1}$ and $k_2 = M_2/\sqrt{L_T L_2}$ are the coupling coefficients between the transmitter and two receivers, respectively.

7.1.2 Operating Frequency Characteristic

In the single-load PT-symmetric MC-WPT system, the PT-symmetric conditions require a matched resonance $\omega_T = \omega_R$ (ω_T and ω_R are the natural resonant angular frequencies of the transmitter and receiver), and the parameters and structure of the transmitter and receiver are symmetric and equal [1, 3]. Similarly, the multi-load PT-symmetric MC-WPT system has to satisfy the following PT-symmetric conditions:

$$\omega_T = \omega_1 = \omega_2 = \omega_0, \tag{7.5}$$

$$\frac{R_{L1}}{L_1} = \frac{R_{L2}}{L_2}, \tag{7.6}$$

$$-\frac{-R_N + R_T}{L_T} = \frac{R_1 + R_{L1}}{L_1} = \frac{R_2 + R_{L2}}{L_2}, \tag{7.7}$$

Similarly, in order to have non-zero solution, the determinant of (7.4) must be zero. Furthermore, taking ω to be real, substituting (7.5)–(7.4) and separating the real and imaginary parts, we have

$$(k_1^2 + k_2^2)\omega^4 - (\omega^2 - \omega_0^2)^2 - \left(\frac{R_1 + R_{L1}}{L_1}\right)^2 \omega^2 = 0, \tag{7.8}$$

$$\left[(k_1^2 + k_2^2)\omega^4 - (\omega^2 - \omega_0^2)^2 - \left(\frac{R_1 + R_{L1}}{L_1}\right)^2\right](\omega^2 - \omega_0^2) = 0. \tag{7.9}$$

Then, the frequency solutions can be derived as

$$\omega = \begin{cases} \dfrac{\omega_0\sqrt{2 - Q_{\mathrm{L1}}^{-2} \pm \sqrt{\left(2 - Q_{\mathrm{L1}}^{-2}\right)^2 + 4\left(k_1^2 + k_2^2 - 1\right)}}}{\sqrt{2\left[1 - \left(k_1^2 + k_2^2\right)\right]}}, & k_{\mathrm{C}}^2 \le k_1^2 + k_2^2 < 1 \\ \omega_0,\ 0 < k_1^2 + k_2^2 < k_{\mathrm{C}}^2 \end{cases} \tag{7.10}$$

where $Q_{\mathrm{L1}} = \omega_0 L_1/(R_{\mathrm{L1}} + R_1)$, and $k_{\mathrm{C}}=\sqrt{1 - \frac{1}{4}\left(2 - Q_{\mathrm{L1}}^{-2}\right)^2}$ is defined as the critical coupling coefficient in the multi-load PT-symmetric MC-WPT system.

According to (7.10), the dual-load PT-symmetric MC-WPT system can be divided into two working regions with the critical coupling coefficient k_{C} as the dividing point, as shown in Fig. 7.4a. When the coupling coefficients k_1 and k_2 meet $k_{\mathrm{C}}^2 \le k_1^2 + k_2^2 < 1$, which is called the strong coupling region or PT-symmetric region, the operating frequency of the system is bifurcated, and there are two frequency branches, namely, high-frequency branch f_{H} and low-frequency branch f_{L}, as shown in Fig. 7.4b. At this time, the system supports two modes with two solutions of operating frequency, the gain of the system is fully balanced with all the losses. It is worth noting that the system can only work at one operating frequency at any time, and the system automatically adjusts the operating frequency with the change of k_1 and k_2, due to the nonlinear saturation gain. When the coupling coefficients k_1 and k_2 satisfy $0 < k_1^2 + k_2^2 < k_{\mathrm{C}}^2$, which is called the weak coupling region or broken PT-symmetric region, the operating frequency is always equal to the natural resonant frequency, and only one mode is located at $\omega = \omega_0$. The simulation parameters of the system are as follows: DC input voltage $U_{\mathrm{DC}} = 50$ V, natural resonance frequency $f_{\mathrm{T}} = f_1 = f_2 = 100$ kHz, inductance of transmitting coil and receiving coil $L_{\mathrm{T}} = L_1 = L_2 = 182.7$ uH, internal resistance $R_{\mathrm{T}} = R_1 = R_2 = 0.5\ \Omega$, load $R_{\mathrm{L1}} = R_{\mathrm{L2}} = 10\ \Omega$.

7.1.3 Transfer Characteristics

Substituting (7.5)–(7.10) into, the RMS values of the currents i_{T}, i_1, i_2 can be obtained, then the system's total output power P_{total} and transfer efficiency η_{total} in the strong coupling region and P'_{total} and η'_{total} in the weak coupling region, the i-th load's output power P_{oi} and transfer efficiency η_{i} ($i = 1, 2$) can be expressed, as follows:

(1) In the strong coupling region $\left(k_C^2 \le k_1^2 + k_2^2 < 1\right)$.

The current gains of each load are

$$\frac{I_1}{I_{\mathrm{T}}} = \frac{k_1}{\sqrt{k_1^2 + k_2^2}}\sqrt{\frac{L_{\mathrm{T}}}{L_1}}, \tag{7.11}$$

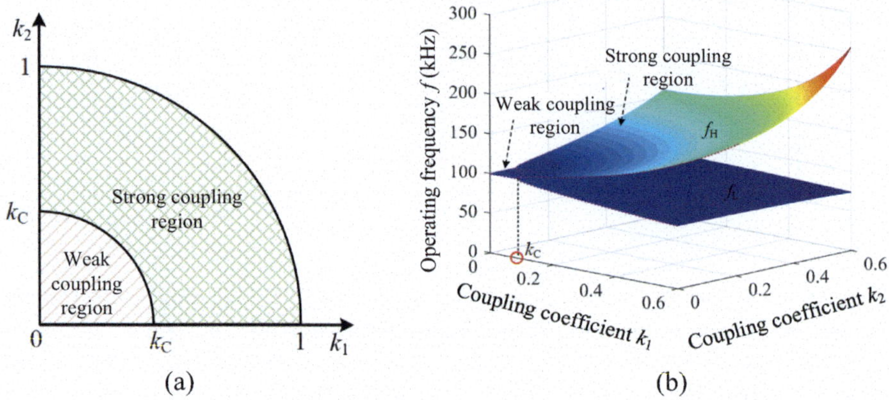

Fig. 7.4 Operating frequency characteristic of the PT-symmetric dual-load MC-WPT system with a single natural resonant frequency. **a** Schematic of the operating region. **b** Operating frequency

$$\frac{I_2}{I_{\mathrm{T}}} = \frac{k_2}{\sqrt{k_1^2 + k_2^2}} \sqrt{\frac{L_{\mathrm{T}}}{L_2}}. \tag{7.12}$$

Due to $U_{\mathrm{in}} = I_{\mathrm{T}} R_{\mathrm{N}}$ (U_{in} is the RMS value of input voltage u_{in}), $U_{\mathrm{o1}} = I_1 R_{\mathrm{L1}}$, and $U_{\mathrm{o2}} = I_2 R_{\mathrm{L2}}$, the voltage gains of each load are

$$\frac{U_{\mathrm{o1}}}{U_{\mathrm{in}}} = \frac{k_1}{\sqrt{k_1^2 + k_2^2}} \frac{1}{\left(1 + \frac{R_1}{R_{\mathrm{L1}}}\right)\sqrt{\frac{L_{\mathrm{T}}}{L_1}} + \frac{R_{\mathrm{T}}}{R_{\mathrm{L1}}}\sqrt{\frac{L_1}{L_{\mathrm{T}}}}}, \tag{7.13}$$

$$\frac{U_{\mathrm{o2}}}{U_{\mathrm{in}}} = \frac{k_2}{\sqrt{k_1^2 + k_2^2}} \frac{1}{\left(1 + \frac{R_2}{R_{\mathrm{L2}}}\right)\sqrt{\frac{L_{\mathrm{T}}}{L_2}} + \frac{R_{\mathrm{T}}}{R_{\mathrm{L2}}}\sqrt{\frac{L_2}{L_{\mathrm{T}}}}}. \tag{7.14}$$

When the internal resistances of the transmitting and receiving coils are ignored, let $R_{\mathrm{T}} = R_1 = R_2 \approx 0$, the voltage gains of each load become

$$\frac{U_{\mathrm{o1}}}{U_{\mathrm{in}}} = \frac{k_1}{\sqrt{k_1^2 + k_2^2}} \sqrt{\frac{L_1}{L_{\mathrm{T}}}}, \tag{7.15}$$

$$\frac{U_{\mathrm{o2}}}{U_{\mathrm{in}}} = \frac{k_2}{\sqrt{k_1^2 + k_2^2}} \sqrt{\frac{L_2}{L_{\mathrm{T}}}}. \tag{7.16}$$

Further analysis shows that when $k_1 = k_2$, current gains, and voltage gains can be simplified to

$$\frac{I_1}{I_T} = \frac{\sqrt{2}}{2}\sqrt{\frac{L_T}{L_1}},\ \frac{I_2}{I_T} = \frac{\sqrt{2}}{2}\sqrt{\frac{L_T}{L_2}}, \tag{7.17}$$

$$\frac{U_{o1}}{U_{in}} = \frac{\sqrt{2}}{2}\sqrt{\frac{L_1}{L_T}},\ \frac{U_{o2}}{U_{in}} = \frac{\sqrt{2}}{2}\sqrt{\frac{L_2}{L_T}}. \tag{7.18}$$

From the above analysis, it can be seen that when the inductance parameters of the coils are fixed, the output current and output voltage of each load can remain constant. However, when $k_1 \neq k_2$, the output current and output voltage of the load are correlated with the coupling coefficient and are not constant.

Then, the output power and transfer efficiency of the whole system and each load can be derived as

$$P_{o1} = \frac{R_{L1}U_{in}^2\frac{k_1^2}{k_1^2+k_2^2}}{\frac{L_T(R_{L1}+R_1)^2}{L_1} + 2(R_{L1}+R_1)R_T + \frac{L_1}{L_T}R_T^2}, \tag{7.19}$$

$$P_{o2} = \frac{R_{L2}U_{in}^2\frac{k_2^2}{k_1^2+k_2^2}}{\frac{L_T(R_{L2}+R_2)^2}{L_2} + 2(R_{L2}+R_2)R_T + \frac{L_2}{L_T}R_T^2}, \tag{7.20}$$

$$P_{total} = P_{o1} + P_{o2} = \frac{R_{L1}U_{in}^2}{\frac{L_T(R_{L1}+R_1)^2}{L_1} + 2(R_{L1}+R_1)R_T + \frac{L_1}{L_T}R_T^2}, \tag{7.21}$$

$$\eta_1 = \frac{k_1^2 R_{L1} L_T}{(k_1^2+k_2^2)[R_T L_1 + L_T(R_{L1}+R_1)]}, \tag{7.22}$$

$$\eta_2 = \frac{k_2^2 R_{L2} L_T}{(k_1^2+k_2^2)[R_T L_2 + L_T(R_{L2}+R_2)]}, \tag{7.23}$$

$$\eta_{total} = \eta_1 + \eta_2 = \frac{R_{L1}L_T}{R_T L_1 + L_T(R_{L1}+R_1)}. \tag{7.24}$$

(2) In the weak coupling region ($0 < k_1^2 + k_2^2 < k_C^2$).

The current gains of each load are

$$\frac{I_1}{I_T} = \frac{\omega_0 k_1 L_1}{R_{L1}+R_1}\sqrt{\frac{L_T}{L_1}}, \tag{7.25}$$

$$\frac{I_2}{I_T} = \frac{\omega_0 k_2 L_2}{R_{L2}+R_2}\sqrt{\frac{L_T}{L_2}}. \tag{7.26}$$

The voltage gains of each load can be derived as

$$\frac{U_{o1}}{U_{in}} = \frac{\omega_0 k_1}{\frac{(R_{L1}+R_1)^2}{R_{L1}L_1}\sqrt{\frac{L_T}{L_1}} + \frac{R_T(R_{L1}+R_1)}{R_{L1}L_1}\sqrt{\frac{L_1}{L_T}}}, \tag{7.27}$$

$$\frac{U_{o2}}{U_{in}} = \frac{\omega_0 k_2}{\frac{(R_{L2}+R_2)^2}{R_{L2}L_2}\sqrt{\frac{L_T}{L_2}} + \frac{R_T(R_{L2}+R_2)}{R_{L2}L_2}\sqrt{\frac{L_2}{L_T}}}. \tag{7.28}$$

When the internal resistances of the transmitting and receiving coils are ignored, the current gains and voltage gains become

$$\frac{I_1}{I_T} = \frac{\omega_0 k_1 L_1}{R_{L1}}\sqrt{\frac{L_T}{L_1}}, \frac{I_2}{I_T} = \frac{\omega_0 k_2 L_2}{R_{L2}}\sqrt{\frac{L_T}{L_2}}, \tag{7.29}$$

$$\frac{U_{o1}}{U_{in}} = \frac{\omega_0 k_1 L_1}{R_{L1}}\sqrt{\frac{L_1}{L_T}}, \frac{U_{o2}}{U_{in}} = \frac{\omega_0 k_2 L_2}{R_{L2}}\sqrt{\frac{L_2}{L_T}}. \tag{7.30}$$

As can be seen from (7.29) and, when the system works in the weak coupling region, regardless of whether k_1 and k_2 are equal, the output current and output voltage of the load are related to the coupling coefficient and cannot be kept constant.

Then, the output power and transfer efficiency of the whole system can be deduced as

$$P'_{total} = \frac{R_{L1}U_{in}^2\left(\frac{R_{L1}+R_1}{L_1}\right)^2}{\omega_0^2\left(k_1^2+k_2^2\right)\left[\frac{L_T}{L_1}(R_{L1}+R_1)^2 + 2(R_{L1}+R_1)R_T + \frac{L_1}{L_T}R_T^2\right]}, \tag{7.31}$$

$$\eta'_{total} = \frac{R_{L1}L_1\left(k_1^2+k_2^2\right)}{\frac{R_T(R_{L1}+R_1)^2}{\omega_0^2 L_T} + L_1(R_{L1}+R_1)\left(k_1^2+k_2^2\right)}. \tag{7.32}$$

From (7.19)–(7.22) and (7.22)–(7.24), it can be seen that the system's total output power and transfer efficiency remain constant in the strong coupling region, while the output power and transfer efficiency of each load vary with coupling coefficients k_1 and k_2. On the contrary, in the weak coupling region, the total output power and transfer efficiency change drastically and are affected by k_1 and k_2 simultaneously, as shown in (7.31) and (7.32).

Figure 7.5 depicts the output power and transfer efficiency curves of the PT-symmetric dual-load MC-WPT system, it can be seen that in the strong coupling region, the total output power and total transfer efficiency of the PT-symmetric dual-load MC-WPT system are only related to the inherent circuit parameters, and are independent of the coupling coefficients k_1 and k_2. Therefore, when the system circuit parameters are determined, the system can always maintain constant output power and constant transfer efficiency. In other words, when load R_{L1} and load R_{L2} move freely in the strongly coupled region, the whole system can always maintain constant output power and constant transfer efficiency. However, for a specific load,

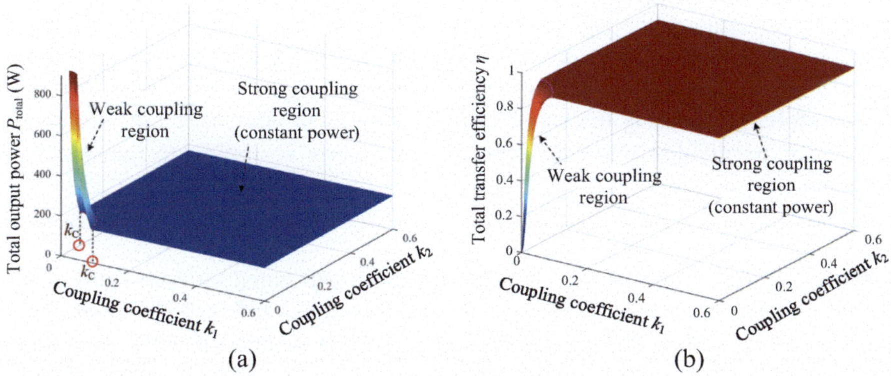

Fig. 7.5 Transfer characteristic of the PT-symmetric dual-load MC-WPT system with a single natural resonant frequency. **a** Output power. **b** Transfer efficiency

its output power is still closely related to the coupling coefficients k_1 and k_2, and cannot be kept constant unless certain conditions are always maintained during the load movement (detailed analysis will be made in Sect. 7.1.4). In the weak coupling region, the total output power and total transfer efficiency of the system are related to k_1 and k_2. When the arbitrary coupling coefficient changes (that is, the load position changes), the total output power and total transfer efficiency of the system will change significantly and cannot be kept constant. The above characteristic analysis can be understood to mean that in a multi-load PT-symmetric MC-WPT system, multiple receivers can be regarded as a whole, which is equivalent to a "single load" for the whole system, so the whole system exhibits the same constant output characteristics as a single-load PT-symmetric MC-WPT system.

7.1.4 Power Distribution

According to (7.19)–(7.20) and (7.22)–(7.23), there is a power distribution and efficiency distribution relationship between loads in the strong coupling region, that is

$$\frac{P_{o1}}{P_{o2}} = \frac{k_1^2}{k_2^2}, \tag{7.33}$$

$$\frac{\eta_1}{\eta_2} = \frac{k_1^2}{k_2^2}. \tag{7.34}$$

According to (7.33) and (7.34), for a dual-load PT-symmetric MC-WPT system with known system parameters, power distribution and transfer efficiency distribution relationship between loads only depends on the coupling coefficients k_1 and k_2. In theory, by adjusting k_1 and k_2 and keeping k_1/k_2 constant, it is possible to achieve constant power output for the system and each load while also achieving arbitrary power distribution between loads.

Therefore, considering the variable load positions in practice, the power distribution can be divided into the following two cases according to the coupling situation.

(1) $k_1 = k_2$

In this case, load 1 and load 2 are symmetrically distributed on both sides of the transmitter, that is, $\beta_1 = \beta_2 = \beta_0$, $d_1 = d_2 = d_0$, as shown in Fig. 7.6a, wherein β is the offset angle between the normal directions, and d is the axial distance. According to (7.33), $P_{o1} = P_{o2} = 50\% P_{total}$ and $\eta_1 = \eta_2 = 50\% \eta_{total}$ can be obtained when $k_1 = k_2$, which means that in the strong coupling region, the total output power of the system is equally distributed to each load. Besides, from the analysis in Sect. 7.1.3, the P_{total} and η_{total} of the dual-load PT-symmetric MC-WPT system are always constant in the strong coupling region, thus the output power and transfer efficiency of each load are also constant at any time and independent of the coupling coefficient, as shown in Fig. 7.6b.

This characteristic can be applied in situations where multiple identical loads require equal amounts of energy, one of the situations is that two loads move simultaneously and cover equal distances and offset angles, and the other situation is that the coil generates a uniform magnetic field, etc. In both cases, the coupling coefficient can be kept constant as the loads move, i.e., $k_1 = k_2$, allowing each load to maintain a constant output and half of the total output power. However, this characteristic also indicates that if the input power of the system is unchanged (that is, the total output power is unchanged), then the output power of each load is fixed and always equals to half of the total output power, which is uncontrollable. Moreover, as can be seen from Fig. 7.6b, it is necessary to pay special attention in practical applications that the coupling coefficients k_1 and k_2 should not be too small ($k_1 = k_2 < k'_C$) so

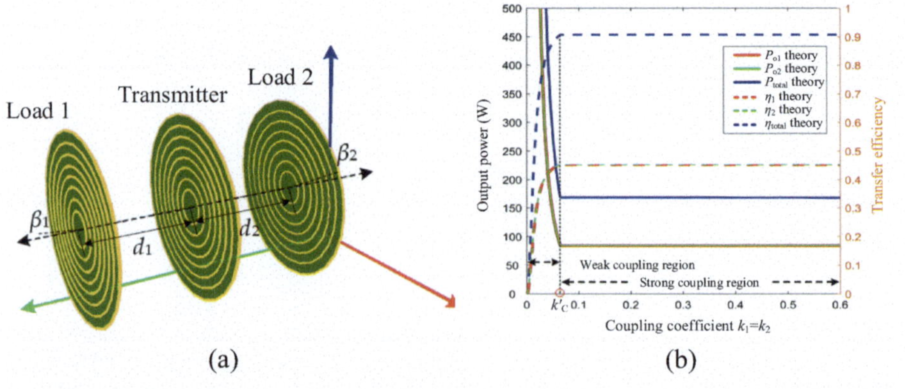

Fig. 7.6 Schematic diagram of power distribution when $k_1 = k_2$. **a** Loads position. **b** Output power and transfer efficiency of the system and each load

that the system works in the weak coupling region. Once the system works in the weak coupling region, the output power will rise rapidly and the transfer efficiency will decrease significantly, resulting in the risk of line burnout, thereby reducing the system's reliability.

(2) $k_1 \neq k_2$

In this case, load 1 and load 2 are located on both sides of the transmitter arbitrarily. Assume that the load 1 is first placed at a certain position (β_1, d_1) and the load 2 is placed at (β_2, d_2), as shown in Fig. 7.7a. Thus, the coupling coefficients k_1 and k_2 can be calculated by [6] or obtained by simulation scanning of HFSS software. Then, according to (7.19)–(7.21), the output power of load 1 and load 2 can be obtained as

$$\begin{cases} P_{o1} = \dfrac{k_1^2}{k_1^2 + k_2^2} P_{\text{total}} \\ P_{o2} = \dfrac{k_2^2}{k_1^2 + k_2^2} P_{\text{total}} \end{cases}. \tag{7.35}$$

It can be found that the output power of each load depends on k_1 and k_2. In this case, the coupling coefficient can be appropriately adjusted according to the power requirements of different loads to achieve the purpose of "on-demand" power distribution. However, in practical applications, the variation range of the coupling coefficient is not arbitrary. Once the difference between the two coupling coefficients is too large, the power distribution will be extremely uneven. The specific performance is as follows: (1) the output power of the load with too large coupling coefficient will be too large, which is easy to damage the load, (2) the output power of the load with too small coupling coefficient is too small, so that it cannot meet the requirements of the normal operation of the load. Therefore, it is necessary to set the output power limit of each load, including the maximum output power and the minimum output power. Assume that the maximum allowable output power of load

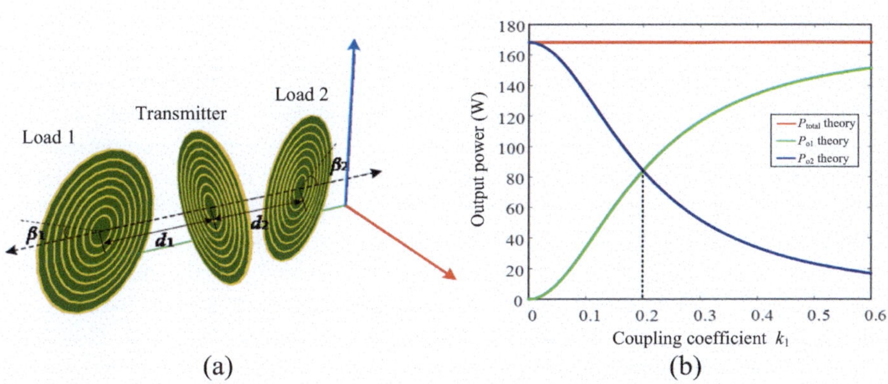

Fig. 7.7 Schematic diagram of power distribution when $k_1 \neq k_2$. **a** Loads position. **b** Power distribution characteristic

1 is P_{o1_max} and the minimum allowable output power is P_{o1_min}, the actual output power of load 1 should satisfy $P_{o1\_max} > P_{o1} > P_{o1\_min}$. Then, the variation range of coupling coefficients k_1 and k_2 can be further deduced as

$$\sqrt{\frac{P_{o1\_max}}{P_{total} - P_{o1\_max}}} \geq \frac{k_1}{k_2} \geq \sqrt{\frac{P_{o1\_min}}{P_{total} - P_{o1\_min}}}. \tag{7.36}$$

In addition, according to the different changes of the load position, there are the following two different situations: (1) one coupling coefficient is fixed, while the other coupling coefficient changes; (2) two coupling coefficients change simultaneously. Suppose that load 2 is fixed at a certain position, that is $k_2 = 0.2$, while the coupling coefficient k_1 of load 1 gradually increases from 0, so that $k_C^2 \leq k_1^2 + k_2^2 < 1$ is always satisfied, then the total output power and the change of output power between load 1 and load 2 are shown in Fig. 7.7b. It can be seen that when k_1 changes, neither load 1 nor load 2 can maintain a constant output, but the total output power remains constant. That is to say, even if load 2 is fixed, its output power is still affected by the position change of load 1. Moreover, the power distribution between the loads is uneven, and the larger the coupling coefficient (that is, the receiving coil is closer to the transmitting coil), the larger the output power.

From the above analysis, there are two methods to achieve the constant output power and transfer efficiency of the dual-load PT-symmetric MC-WPT system and each load simultaneously. One is that $k_1 = k_2$ is always satisfied, and the output power is evenly distributed to each load, that is $P_{o1} = P_{o2} = 50\% P_{total}$ and $\eta_1 = \eta_2 = 50\% \eta_{total}$. The other is to design a control strategy to adjust the coupling coefficients by the power distribution relationship of $k_1 \neq k_2$ and the power demand of different loads, without adding any compensation circuit.

7.1.5 *Comparison with Single-Load PT-Symmetric MC-WPT System*

Through the modeling and transfer characteristics analysis of the single-load PT-symmetric MC-WPT system in Chap. 3 and the single-frequency dual-load PT-symmetric MC-WPT system in this chapter, the comprehensive comparison of the two systems is shown in Table 7.1. It can be found that the dual-load PT-symmetric MC-WPT system needs to meet more stringent PT-symmetric conditions and coupling coefficient conditions. When the system works in the PT-symmetric region, the total output power and total transfer efficiency of the dual-load PT-symmetric MC-WPT system remain constant, and the whole system exhibits the same characteristics as the single-load PT-symmetric MC-WPT system. However, for specific loads, the output power and transfer efficiency of each load are distributed unevenly and vary with the coupling coefficient. Only when the certain coupling

Table 7.1 Comprehensive comparison of single-load PT-symmetric MC-WPT and dual-load PT-symmetric MC-WPT system

	Single-load system	Dual-load system
PT-symmetric condition	$\omega_T = \omega_R = \omega_0, \frac{R_N - R_T}{L_T} = \frac{R_L + R_R}{L_R}$	$\omega_T = \omega_1 = \omega_2 = \omega_0, \frac{R_{L1}}{L_1} = \frac{R_{L2}}{L_2}, \frac{R_N - R_T}{L_T} = \frac{R_{L1} + R_1}{L_1} = \frac{R_{L2} + R_2}{L_2}$
Coupling coefficient condition	$1 - \frac{1}{4}\left[2 - \left(\frac{R_L + R_R}{\omega_0 L_R}\right)^2\right]^2 \le k^2 < 1$	$1 - \frac{1}{4}\left[2 - \left(\frac{R_{L1} + R_1}{\omega_0 L_1}\right)^2\right]^2 \le k_1^2 + k_2^2 < 1$
Total output power	$\frac{R_L U_{in}^2}{\frac{L_T (R_L + R_R)^2}{L_R} + 2(R_L + R_R)R_T + \frac{L_R}{L_T} R_T^2}$	$\frac{R_{L1} U_{in}^2}{\frac{L_T (R_{L1} + R_1)^2}{L_1} + 2(R_{L1} + R_1)R_T + \frac{L_1}{L_T} R_T^2}$
Total transfer efficiency	$\frac{R_L L_T}{R_T L_R + L_T (R_L + R_R)}$	$\frac{R_{L1} L_T}{R_T L_1 + L_T (R_{L1} + R_1)}$
Power distribution	–	$\frac{k_1^2}{k_2^2}$
Voltage gain	$\sqrt{\frac{L_T}{L_R}}$	$\frac{k_i}{\sqrt{k_1^2 + k_2^2}} \sqrt{\frac{L_T}{L_i}}$
Current gain	$\sqrt{\frac{L_R}{L_T}}$	$\frac{k_i}{\sqrt{k_1^2 + k_2^2}} \sqrt{\frac{L_i}{L_T}}$

coefficient condition is satisfied, each load can maintain a constant output power and transfer efficiency.

7.2 Multi-frequency Multi-load System

7.2.1 Directional Energy Transmission

For the single-frequency multi-load PT-symmetric MC-WPT system, the transmitter and all receivers have the same resonant frequency, i.e., $\omega_T = \omega_1 = \omega_2 = \ldots = \omega_n$, and the energy is transmitted to all loads simultaneously, as shown in Fig. 7.8a. However, coupling conditions between loads are generally difficult to be completely consistent, so such systems face great challenges in terms of power distribution uniformity and eliminating cross-load influence [7]. If a multi-load MC-WPT system can achieve the directed energy transfer shown in Fig. 7.8b, then the energy distribution of a multi-load system will become as convenient as a single-load system, and load isolation can be achieved.

To solve the above issues, it is an effective method to introduce multi-frequency into the multi-load MC-WPT system. The following will elaborate on the principle and implementation of the multi-frequency multi-load MC-WPT system to achieve

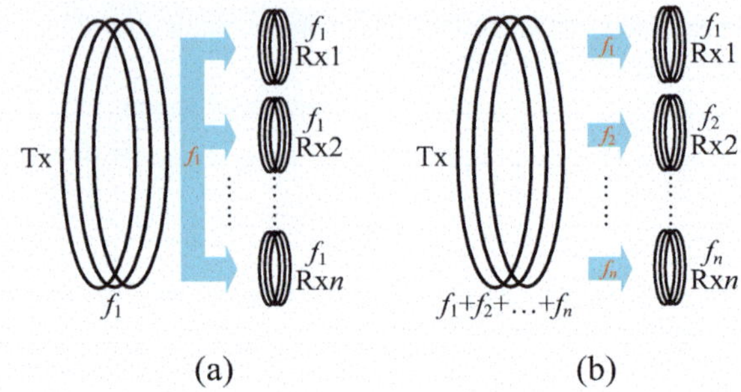

Fig. 7.8 Energy flow diagram of multi-load system. **a** Energy transfer simultaneously. **b** Energy transfer asynchronously

energy directional transmission, which provides a frequency selection basis for the multi-frequency multi-load PT-symmetric MC-WPT system [8].

Without loss of generality, taking the dual-frequency dual-load MC-WPT system as an example, the equivalent circuit is shown in Fig. 7.9, where the frequency of the AC drive power supply U_s is adjustable, and the natural resonant frequencies of the transmitter, receiver 1 and receiver 2 are f_T, f_1, and f_2, respectively, which can be expressed as

$$f_T = \frac{1}{2\pi\sqrt{L_T C_T}}, f_1 = \frac{1}{2\pi\sqrt{L_1 C_1}}, f_2 = \frac{1}{2\pi\sqrt{L_2 C_2}}, \tag{7.37}$$

Assuming that the resonant frequencies of load 1 and load 2 are not equal, that is, $f_1 \neq f_2$, then according to Kirchhoff's voltage law, the equivalent circuit model can be expressed as

$$\begin{bmatrix} Z_T & j\omega M_{T1} & j\omega M_{T2} \\ j\omega M_{T1} & Z_1 & j\omega M_{12} \\ j\omega M_{T2} & j\omega M_{12} & Z_2 \end{bmatrix} \begin{bmatrix} \dot{I}_T \\ \dot{I}_1 \\ \dot{I}_2 \end{bmatrix} = \begin{bmatrix} \dot{U}_s \\ 0 \\ 0 \end{bmatrix}, \tag{7.38}$$

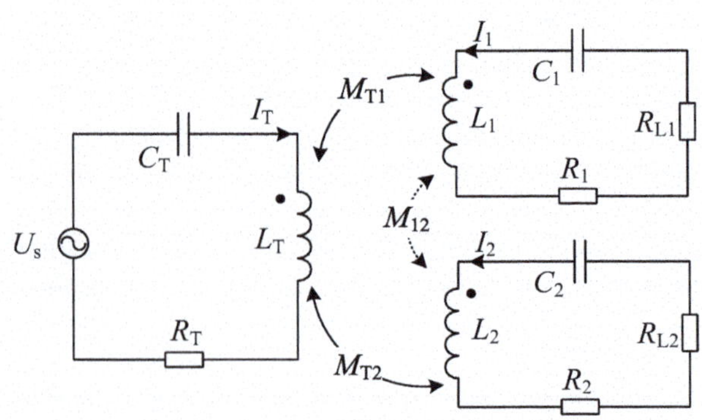

Fig. 7.9 Dual-frequency dual-load MC-WPT system

where $\omega = 2\pi f$ is the operating angular frequency of the system, $Z_{\mathrm{T}} = R_{\mathrm{T}} + j(\omega L_{\mathrm{T}} - \frac{1}{\omega C_{\mathrm{T}}})$, $Z_1 = R_1 + R_{\mathrm{L1}} + j(\omega L_1 - \frac{1}{\omega C_1})$ and $Z_2 = R_2 + R_{\mathrm{L2}} + j(\omega L_2 - \frac{1}{\omega C_2})$ are impedances of the transmitter, receiver 1 and 2, respectively. M_{T1} (M_{T2}) is mutual inductance between the transmitting coil and receiving coil 1 (2). M_{12} is mutual inductance between receiving coils.

According to (7.38), the expressions of the currents flowing through loads 1 and 2 can be derived. Combined with the calculation formula of the output power $P_{\mathrm{o1}}=I_1^2 R_{\mathrm{L1}}$ and $P_{\mathrm{o2}}=I_2^2 R_{\mathrm{L2}}$ and ignored the cross-coupling between loads, the output power P_{o1} and P_{o2} can be obtained as

$$P_{\mathrm{o1}} = \left| -\frac{j\omega M_{\mathrm{T1}} Z_2}{Z_{\mathrm{T}} Z_1 Z_2 + \omega^2 M_{\mathrm{T1}}^2 Z_2 + \omega^2 M_{\mathrm{T2}}^2 Z_1} \right|^2 U_{\mathrm{s}}^2 R_{\mathrm{L1}}, \tag{7.39}$$

$$P_{\mathrm{o2}} = \left| -\frac{j\omega M_{\mathrm{T2}} Z_1}{Z_{\mathrm{T}} Z_1 Z_2 + \omega^2 M_{\mathrm{T1}}^2 Z_2 + \omega^2 M_{\mathrm{T2}}^2 Z_1} \right|^2 U_{\mathrm{s}}^2 R_{\mathrm{L2}}. \tag{7.40}$$

Setting the simulation parameters of the system as $U_{\mathrm{s}} = 10$ V, $L_{\mathrm{T}} = 191.282$ µH, $L_1 = 240.005$ µH, $L_2 = 241.458$ µH, $R_{\mathrm{T}} = R_1 = R_2 = 0.5\ \Omega$, $R_{\mathrm{L1}} = R_{\mathrm{L2}} = 16.67\ \Omega$, the relationship curve between the output power and the operating frequency f can be obtained when matching different resonant frequencies of the transmitter, as shown in Fig. 7.10. It can be seen that when $f_{\mathrm{T}} = f_1$, the output power of load 1 reaches its maximum value near $f = f_1$, while the output power of load 2 is almost 0, that is, all the energy is transferred to load 1. Similarly, when $f_{\mathrm{T}} = f_2$, the output power of load 2 reaches its maximum value near $f = f_2$, while the output power of load 1 is almost 0, that is, all the energy is transferred to load 2. Therefore, by setting the loads to have different resonant frequencies, and adjusting the resonant frequency f_{T} of the transmitter in turn to be equal to the resonant frequency of each load, the directional flow of energy as shown in Fig. 7.8b can be achieved.

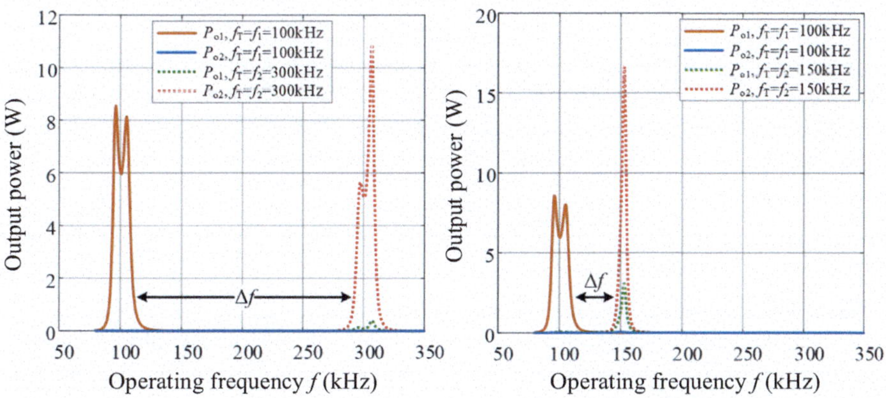

Fig. 7.10 The relationship between the output power of the load and the operating frequency of the system

It is worth noting that for multi-frequency multi-load MC-WPT systems, the resonant frequencies of each load should not be too close, otherwise multiple energy transmission paths will cross and overlap. As can be observed from Fig. 7.10a, when the load resonance frequency difference $\Delta f = |f_1 - f_2|$ is large, the output power of load 1 and load 2 will not overlap, that is, the loads do not affect each other, and the energy is transmitted in a directional manner. However, when the difference value Δf gradually decreases to a certain extent, the output power will inevitably overlap, which means that load 1 and load 2 will obtain energy at the same time, and the energy directional transmission performance is lost.

7.2.2 System Structure

The schematic diagram of the multi-frequency multi-load PT-symmetric MC-WPT system is shown in Fig. 7.11. The system consists of a transmitter and n receivers (loads), wherein the transmitter comprises a negative resistor, which is constructed by a self-oscillating full-bridge inverter, a transmitting coil L_T, and a controllable resonant capacitor array. The controllable resonant capacitor array consists of a time-sharing frequency controller and n parallel controllable switched capacitor branches, each branch is connected in series by a resonant capacitor C_{Ti} and the switch S_i. When the switch S_i is turned on and the others are turned off, the capacitor C_{Ti} is put into operation. It is noted that only one switch is selectively closed at any time. Moreover, each receiver comprises a receiving coil L_i and a resonant capacitor C_i in series, a rectifier, and a load R_{Li}. i_{Li} is the current flowing through the receiving coil L_i, R_{eqi} is the equivalent AC resistance, and meet $R_{eqi} = 8R_{Li}/\pi^2$. M_i is the mutual inductance between the transmitting coil and receiving coil i ($i = 1, 2, \ldots\ldots, n$).

By tuning the capacitor C_i, the natural resonant frequency f_{Ri} of each load is not equal, i.e., $f_{R1} \neq f_{R2} \neq \ldots \neq f_{Ri}$, where $f_{Ri} = 1/(2\pi\sqrt{L_i C_i})$. Meanwhile, the natural resonant frequency f_{Ti} of each capacitor branch composed of capacitor C_{Ti} and switch S_i in series with L_T in the transmitter is not equal, i.e., $f_{T1} \neq f_{T2} \neq \ldots \neq f_{Ti}$, where $f_{Ti} = 1/(2\pi\sqrt{L_T C_{Ti}})$. But the i-th load corresponds to the i-th capacitor branch one by one and has the identical natural resonant frequency, i.e., $f_{Ti} = f_{Ri}$ ($i = 1, 2, \ldots, n$). In addition, the circuit parameters of all receivers and transmitters are set to meet one of the following basic PT-symmetric conditions:

$$-\frac{-R_{Ni} + R_T}{L_T} = \frac{R_{eqi} + R_i}{L_i}. \tag{7.41}$$

Since the negative resistor is constructed by a self-oscillating full-bridge inverter, the resistance value $-R_{Ni}$ automatically follows the load resistance value in the PT-symmetric system, so the parameter specifications of the transmitting coil and the receiving coil should be as consistent as possible.

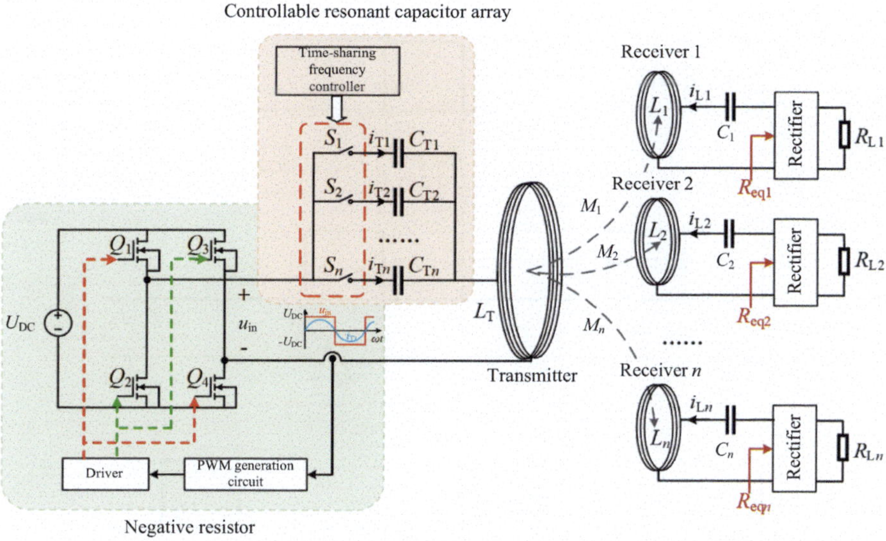

Fig. 7.11 System structure of the multi-frequency multi-load PT-symmetric MC-WPT system

7.2.3 Time-Sharing Control Strategy

Through the turn-on and turn-off of the controllable switch S_i on the transmitter, the dynamic adjustment of the resonant capacitor C_{Ti} can be realized, and then the switching effect of the natural resonant frequency f_{Ti} can be achieved to match the natural resonant frequency of different loads f_{Ri}. Figure 7.12 shows the control sequence diagram of controllable switches $S_1 \sim S_n$ in Fig. 7.11 and currents flowing through each receiver, wherein T is the control period of the time-sharing frequency controller, 0 represents a low level and the switch is off, 1 represents a high level and the switch is on [9]. During the on-time ΔT_i, only the switch S_i is turned on, and its series resonant capacitor C_{Ti} is put into operation, while other switches and capacitors are not working. At this time, the natural resonant frequency of the transmitter is f_{Ti}, only the load i whose natural frequency f_{Ri} is equal to f_{Ti} works, and the current flowing through the receiving coil L_i is approximately sine wave. Otherwise, the switch S_i is turned off and the i-th receiver stops working, resulting in $I_{Li} = 0$. Furthermore, if the delay time and dead time of switches are not considered ($i = 1, 2, \ldots n$), T and ΔT_i can be flexibly adjusted according to the power demand of different loads, as long as $\sum_{\mathrm{i}=1}^{\mathrm{n}} \Delta T_{\mathrm{i}} = T$ is satisfied.

Taking a dual-frequency dual-load PT-symmetric MC-WPT system as an example. When the proper natural resonance frequency of the load is selected, the cross-coupling effect between loads can be ignored. Therefore, the system has two energy transmission paths: $L_T \rightarrow L_1$ and $L_T \rightarrow L_2$, as shown in Fig. 7.13. The time-sharing control period is T, the son-time of switches S_1 and S_2 are ΔT_1 and ΔT_2, respectively, and $\Delta T_1 + \Delta T_2 = T$. In the on-time ΔT_1, the capacitor C_{T1} is put into operation

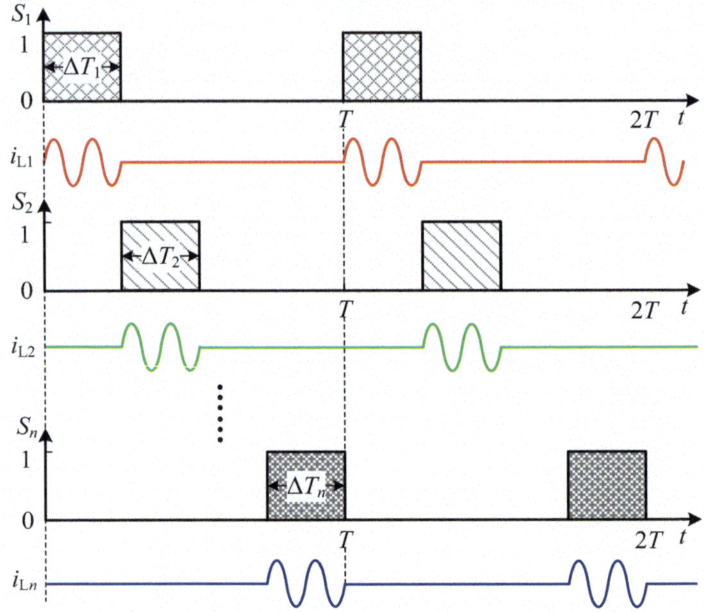

Fig. 7.12 Time-sharing control sequence diagram

and is connected in series with the transmitting coil L_T, then the natural resonant frequency of the transmitter is f_{T1}, which is equal to the natural resonant frequency f_{R1} of load 1, i.e., $f_{T1} = f_{R1}$, but not equal to the natural resonant frequency f_{R2} of load 2. Therefore, the transmitter and the receiver 1 constitute a set of a PT-symmetric MC-WPT system. According to the multi-frequency energy directional flow mechanism described in Sect. 7.2.1, when the difference between the natural resonant frequencies of the receiver 1 and 2 is large enough, the energy will not be transmitted to load 2 through the transmission path, and then load 1 can work normally according to the single-load PT-symmetric MC-WPT system, as shown in mode 1 in Fig. 7.13a. Similarly, in the on-time ΔT_2, the capacitor C_{T2} is put into operation in series with the transmitting coil L_T, and the transmitter and the receiver 2 constitute a set of PT-symmetric MC-WPT system, as shown in mode 2 in Fig. 7.13b.

In fact, the coupling of the transmitting coil with the receiving coil 1 and 2 is an inherent property of the MC-WPT system, that is, the transmission paths $L_T \rightarrow L_1$ and $L_T \rightarrow L_2$ always exist, but only one transmission path is valid at any time, and the other path no longer transmits energy, thus playing a load isolation role, ensuring that there is always a load at any time to form a PT-symmetric MC-WPT system.

Based on the above analysis of the time-sharing control strategy, the multi-frequency multi-load PT-symmetric MC-WPT system can be viewed as a single-frequency single-load PT-symmetric MC-WPT system at any time. Taking the switch S_i conduction at a certain time as an example, the equivalent circuit of the system is shown in Fig. 7.13, then the system has to satisfy the following PT-symmetric conditions:

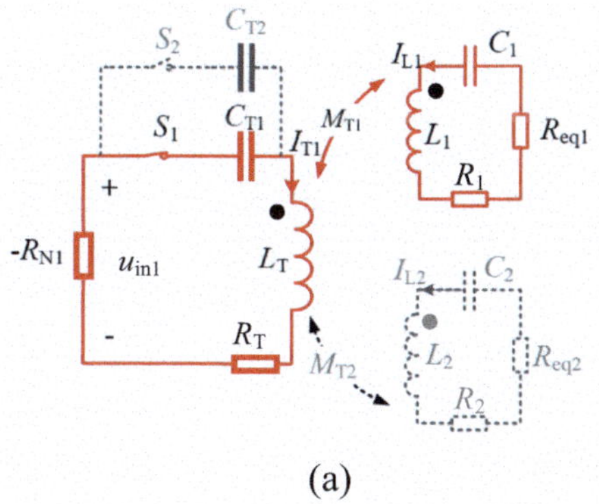

Fig. 7.13 Work modes of dual-frequency dual-load PT-symmetric MC-WPT system. **a** Mode 1. **b** Mode 2

$$\begin{cases} \omega_{Ti} = \omega_{Ri} = \omega_i \\ -\dfrac{-R_{Ni} + R_T}{L_T} = \dfrac{R_{eqi} + R_i}{L_i} \end{cases}, \tag{7.42}$$

where $\omega_{Ti} = 2\pi f_{Ti} = 1/\sqrt{L_T C_{Ti}}$ and $\omega_{Ri} = 2\pi f_{Ri} = 1/\sqrt{L_i C_i}$ are the natural resonant angular frequencies of the transmitter and receiver, respectively. $-R_{Ni}$ is the equivalent negative resistance when the C_{Ti} is selected.

According to the Kirchhoff's voltage law, the circuit model of the system can be written as

$$\begin{bmatrix} \frac{-R_{Ni}+R_T}{L_T} + j\omega - j\frac{\omega_i^2}{\omega} & j\omega k_i\sqrt{\frac{L_i}{L_T}} \\ j\omega k_i\sqrt{\frac{L_T}{L_i}} & \frac{R_{eqi}+R_i}{L_i} + j\omega - j\frac{\omega_i^2}{\omega} \end{bmatrix} \begin{bmatrix} \dot{I}_{Ti} \\ \dot{I}_{Li} \end{bmatrix} = 0, \tag{7.43}$$

where $k_i = M_{Ti}/\sqrt{L_T L_i}$ is the coupling coefficient between the transmitter and i-th receiver.

Then, taking the determinant of (7.43) to be zero and ω to be real, the frequency solution can be derived as

$$f = \begin{cases} f_i, \ 0 < k_i < k_{Ci} \\ \dfrac{f_i\sqrt{b \pm \sqrt{b^2 + 4(k_i^2 - 1)}}}{\sqrt{2(1 - k_i^2)}}, \ k_{Ci} \le k_i < 1 \end{cases}, \tag{7.44}$$

where $b = 2 - \left(\frac{R_{Li}+R_i}{\omega_i L_i}\right)^2$, $k_{Ci} = \sqrt{1 - \frac{b^2}{4}}$ is defined as the critical coupling coefficient of i-th receiver.

From (7.44), i can be observed that the coupling region can be divided into two parts, that is $0 < k_i < k_{Ci}$ and $k_{Ci} \le k_i < 1$, which corresponds to the weak coupling region and strong coupling region, respectively.

Hence, during the time ΔT_i, the switch S_i is turned on and the i-th receiver is working, the output power P_{oi_on} and transfer efficiency η_{i_on} of the corresponding load can be obtained as

$$\begin{cases} P_{oi\_on} = I_{Li\_on}^2 R_L = \dfrac{R_{eqi} U_{in}^2}{\frac{L_T}{L_i}\left(R_{eqi} + R_i\right)^2 + 2R_T\left(R_{eqi} + R_i\right) + \frac{L_i}{L_T} R_T^2} \\ \eta_{i\_on} = \dfrac{P_{oi\_on}}{P_{ini\_on}} = \dfrac{R_{eqi}}{\frac{L_i}{L_T} R_T + R_{eqi} + R_i} \end{cases}, \tag{7.45}$$

where I_{Li_on} is the RMS value of the load current i_{Li} in the on-time ΔT_i.

As can be seen from (7.45), in the on-time ΔT_i where the switch S_i is turned on and the load i is put into operation, as long as the load i works in its own strong coupling region ($k_{Ci} \le k_i < 1$), no matter how the load position changes, the load i can maintain constant output power and constant transfer efficiency.

7.2.4 Power Distribution Control

The duty cycle of the resonant capacitor C_{Ti} is defined as $\alpha_i = \Delta T_i / T$, which is also the duty cycle of the load i. When the on-delay time and dead-zone time of the switches are ignored, $\sum_{i=1}^{n} \alpha_i = 1$, the output power P_{oi} of the i-th load in a control period T satisfies

$$\begin{aligned} P_{oi} &= \frac{1}{T} \int_0^T i^2(t) R_{eqi} dt \\ &= \frac{1}{T}\left(\int_0^{\Delta T_i} i^2(t) R_{eqi} dt + \int_{\Delta T_i}^{T} i^2(t) R_{eqi} dt \right) \\ &= \alpha_i I_{Li\_on}^2 R_{eqi} + (1 - \alpha_i) I_{Li\_off}^2 R_{eqi}, \end{aligned} \tag{7.46}$$

where I_{Li_on} and I_{Li_off} are the RMS values in the turn-on time and turn-off time in this control period, respectively.

When S_i is turned off, the i-th receiver stops working and $I_{Li\_off} \approx 0$, so (7.46) can be further simplified as

$$P_{oi} = \alpha_i P_{oi\_on} = \frac{\alpha_i R_{eqi} U_{in}^2}{\frac{L_T}{L_i}\left(R_{eqi} + R_i\right)^2 + 2R_T\left(R_{eqi} + R_i\right) + \frac{L_i}{L_T} R_T^2}. \tag{7.47}$$

From (7.47), it can be found that the output power of each load only depends on the duty cycle α_i and the inherent circuit parameters, which is independent of the coupling coefficients. In practice, when loads are determined, the inherent circuit parameters of the system are also determined. Therefore, the power distribution among loads can be realized only by controlling the duty cycle α_i of i-th load, and the output power remains constant simultaneously. Especially, if the inherent circuit parameters of all loads are equal, the power distribution ratio among loads will satisfy

$$P_{o1} : P_{o2} : \cdots : P_{on} = \alpha_1 : \alpha_2 : \cdots : \alpha_n. \tag{7.48}$$

Taking the dual-frequency dual-load PT-symmetric MC-WPT system as an example, assuming that the duty cycle of load 1 is set α, then the duty cycle of load 2 is 1-α. According to the above analysis, the output power and transfer efficiency of the whole system and each load in the strong coupling region can be expressed as

$$\begin{cases} P_{o1} = \alpha P_{o1\_on} \\ P_{o2} = (1-\alpha) P_{o2\_on} \\ P_{total} = \sum_{i=1}^{2} P_{oi} = \alpha P_{o1\_on} + (1-\alpha) P_{o2\_on} \end{cases}, \tag{7.49}$$

$$\begin{cases} \eta_1 = \dfrac{R_{eq1}}{\left(\frac{L_1}{L_T}\frac{\alpha}{1-\alpha} + \frac{L_2}{L_T}\right) R_T + \frac{\alpha}{1-\alpha}(R_{eq1} + R_1) + (R_{eq2} + R_2)} \\ \eta_2 = \dfrac{R_{eq2}}{\left(\frac{L_1}{L_T}\frac{\alpha}{1-\alpha} + \frac{L_2}{L_T}\right) R_T + \frac{\alpha}{1-\alpha}(R_{eq1} + R_1) + (R_{eq2} + R_2)} \\ \eta_{total} = \sum_{i=1}^{2} \eta_i \end{cases}. \tag{7.50}$$

From (7.49) and (7.50), it can be seen that the output power and transfer efficiency are only related to the duty cycle α and always keep constant when coupling coefficients change.

Further, the power distribution ratio between two loads can be expressed as

$$\frac{P_{o1}}{P_{o2}} = \frac{\alpha R_{eq1}}{(1-\alpha) R_{eq2}} \cdot \frac{L_1\left[L_T^2 (R_{eq2} + R_2)^2 + 2 L_T L_2 R_T (R_{eq2} + R_2) + L_2^2 R_T^2\right]}{L_2\left[L_T^2 (R_{eq1} + R_1)^2 + 2 L_T L_1 R_T (R_{eq1} + R_1) + L_1^2 R_T^2\right]}. \tag{7.51}$$

When the inherent circuit parameters of load 1 and 2 are identical, (7.51) can be further simplified as

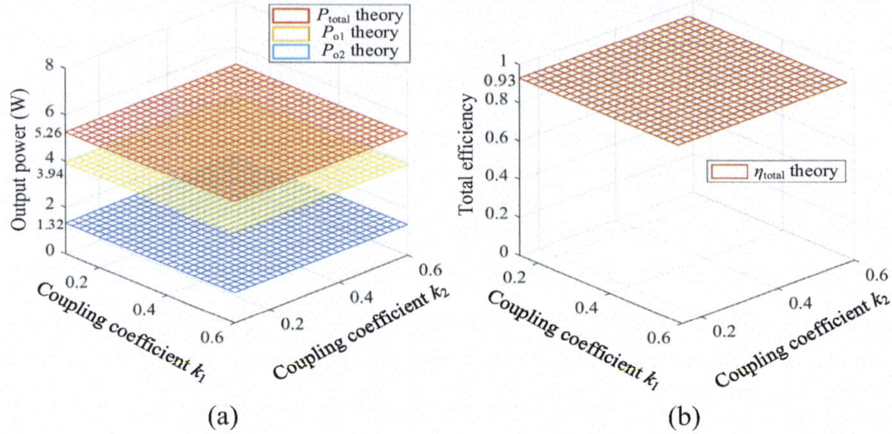

Fig. 7.14 Transfer characteristics of dual-frequency dual-load PT-symmetric MC-WPT system with time-sharing control under $\alpha = 0.75$. **a** Output power. **b** Transfer efficiency

$$\frac{P_{o1}}{P_{o2}} = \frac{\alpha}{1-\alpha}. \tag{7.52}$$

It can be seen that the power distribution relationship between loads is only related to duty cycle α. As shown in Fig. 7.14, when duty cycle $\alpha = 0.75$, the total output power of the system is constant at $P_{total} \approx 5.26$ W, the total transfer efficiency is constant at $\eta_{total} \approx 93\%$, and the output power of load 1 and load 2 are always constant, that is $P_{o1} = 3.94$ W, $P_{o2} = 1.32$ W. Thus, the ratio of output power is P_{o1}:$P_{o2} \approx 3$:1, which is consistent with the theoretical value of. The system simulation parameters are $U_{DC} = 10$ V, $L_T = 191.282$ μH, $L_1 = 240.005$ μH, $L_2 = 241.458$ μH, $R_T = R_1 = R_2 = 0.5$ Ω, $R_{L1} = R_{L2} = 16.67$ Ω, $\alpha = 0.75$.

7.2.5 Comparison with Single-Frequency Multi-load PT-Symmetric MC-WPT System

Table 7.2 describes the comparison of features of the single-frequency multi-load system and the multi-frequency multi-load system. The single-frequency multi-load PT-symmetric MC-WPT system has a simple structure and does not need to add additional control circuits, but needs to satisfy more stringent PT-symmetric conditions. Although the total output power and total transfer efficiency of the system can always be kept constant, it is difficult to maintain a constant output for each load unless $k_1 = k_2$ is always satisfied. In addition, the power distribution between loads is related to k_1 and k_2, which are affected by various factors. Thus, in practical applications, it is difficult to achieve the purpose of power distribution by simply adjusting k_1 and k_2.

Table 7.2 Comparison of multi-load PT-symmetric MC-WPT systems

	Single-frequency multi-load system	Multi-frequency multi-load system
PT-symmetric condition	$\omega_T = \omega_1 = \omega_2, \frac{R_{eq1}}{L_1} = \frac{R_{eq2}}{L_2}, \frac{R_N - R_T}{L_T} = \frac{R_{eq1}+R_1}{L_1} = \frac{R_{eq2}+R_2}{L_2}$	$\omega_{Ti} = \omega_{Ri} = \omega_i, \frac{R_{Ni}-R_T}{L_T} = \frac{R_{eqi}+R_i}{L_i}, (i = 1, 2)$
Critical coupling coefficient	$1 - \frac{1}{4}\left[2 - \left(\frac{R_{eq1}+R_1}{\omega_0 L_1}\right)^2\right]^2$	$1 - \frac{1}{4}\left[2 - \left(\frac{R_{Li}+R_i}{\omega_i L_i}\right)^2\right]^2$
Coupling coefficient condition	$k_C^2 \le k_1^2 + k_2^2 < 1$	$k_{Ci} \le k_i < 1$
Total output power	Constant	Constant
Total transfer efficiency	Constant	Constant
Output power of i-th load	Variation	Constant
Power distribution	$\frac{P_{o1}}{P_{o2}} = \frac{k_1^2}{k_2^2}$	$\frac{P_{o1}}{P_{o2}} = \frac{\alpha}{1-\alpha}$
Number of natural resonant frequencies	Single	Multiple

For the multi-frequency multi-load PT-symmetric MC-WPT system, it can not only achieve constant total output power and constant total transfer efficiency, but also realize constant output of any load. Besides, the power distribution relationship between loads is only related to the duty cycle α, and has nothing to do with the coupling coefficients. Therefore, any proportion of power distribution can be achieved only by controlling the duty cycle α of the switch, which improves the practicability and operability of the system. Meanwhile, the duty cycle α is an open-loop control quantity, which does not require closed-loop control, and the control circuit is simple. However, with the increase of the control circuit and switch array, the cost and switch loss will increase correspondingly.

7.3 System Design and Verification

The schematic diagram of the dual-load PT-symmetric MC-WPT system is shown in Fig. 7.15. For the single-frequency dual-load system, the resonant capacitance C_T is a fixed value. For the dual-frequency dual-load system with a time-sharing control strategy, the resonant capacitance C_T is the variable value.

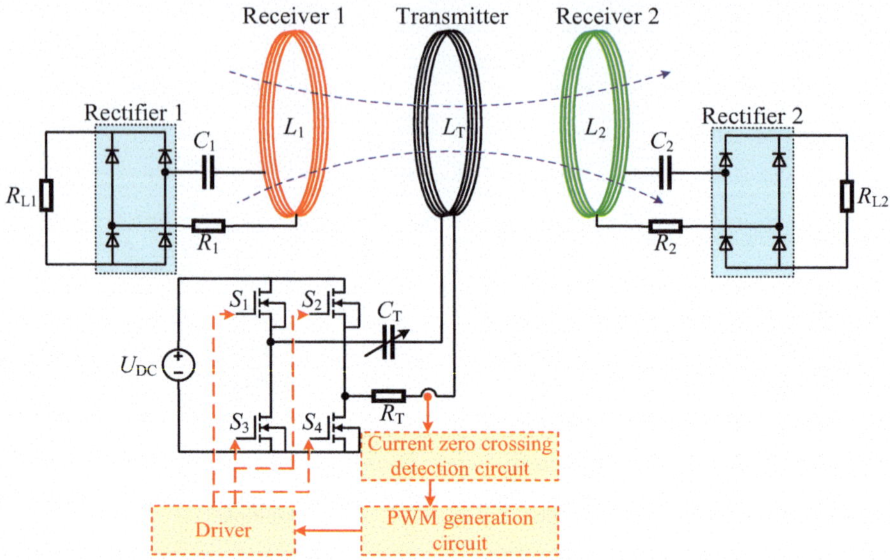

Fig. 7.15 Schematic diagram of the dual-load PT-symmetric MC-WPT system

7.3.1 Negative Resistor

As shown in Fig. 7.15, the negative resistor is constructed by a self-oscillating full-bridge inverter, including a DC power supply, a full-bridge inverter, a current zero-crossing detection circuit, a PWM generation circuit, and a driver. The voltage and current of the negative resistor are required to be in phase, so in order to achieve the output voltage and current of the full-bridge inverter in phase, it is necessary to sample the current signal of the transmitter to generate a self-excited drive signal, without using an additional drive device. The generation of a self-excited drive signal mainly includes the following two key steps:

(1) Current zero crossing detection. First, the current flowing through the transmitting coil is sampled by the current transformer and converted into a voltage signal. Then, the voltage signal is adjusted by a RC phase compensation circuit, and input to the differential amplifier circuit for signal amplification, as shown in Fig. 7.16a. Finally, the voltage amplification signal is input to the zero-crossing comparator to generate a PWM signal, as shown in Fig. 7.16b. Here, the current transformer CU8965 is used to sample the current. The differential amplifier adopts the OPA2690, and the comparator uses the TL3016.
(2) Generate two complementary PWM drive signals. The full-bridge inverter requires two sets of complementary PWM signals, and it is necessary to set a certain dead zone for the drive signals to avoid the phenomenon of "straight through" between the two switch tubes of the same bridge arm. Figure 7.17

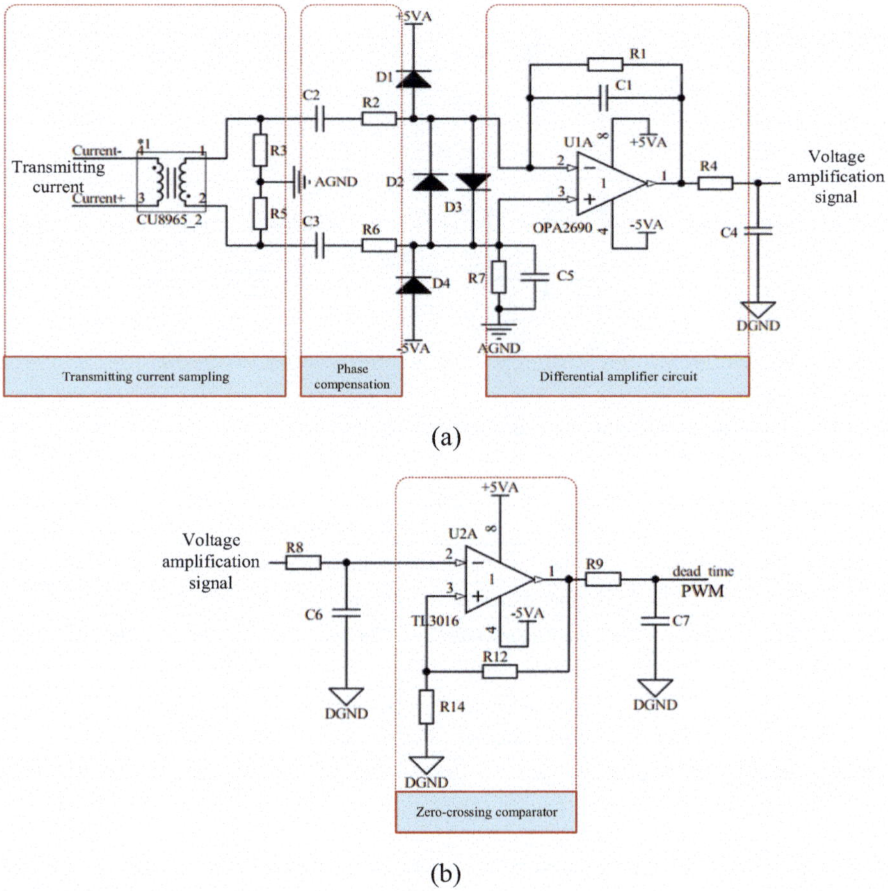

Fig. 7.16 Current sampling and zero-crossing comparator circuits. **a** Current sampling and differential amplification circuit. **b** Zero-crossing comparator

shows the circuit for generating the complementary PWM signal with a dead zone.

7.3.2 Coupling Coil Design

To ensure the PT symmetry of the multi-load MC-WPT system, the structure of the transmitting coil and the receiving coils are constructed as equally as possible. In consideration of skin effect and magnetic field uniformity, the planar spiral transmitting coil and receiving coils made by Litz wire ($\phi = 0.05$ mm $\times$ 400 strands) are chosen. The detailed circuit parameters of the transmitting coil, and receiving coils 1 and 2 coils are listed in Table 7.3. As measured by the precise impedance analyzer

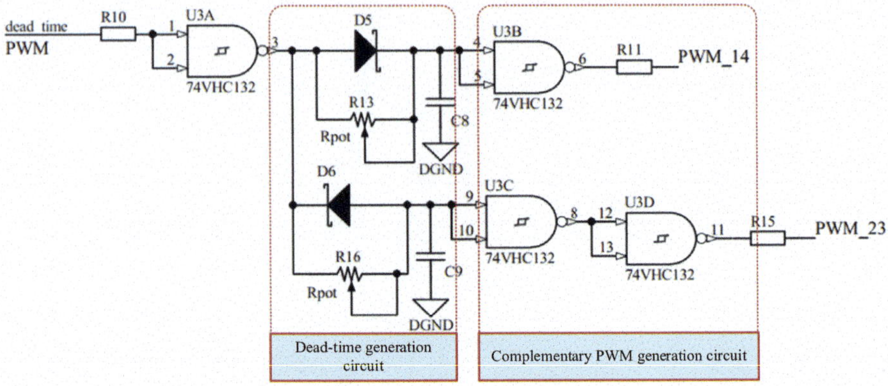

Fig. 7.17 Complementary PWM generation circuit with dead time

Table 7.3 Parameters of coils

Coil	Transmitting coil	Receiving coil
Type	Planar spiral	Planar spiral
Outer diameter	26 cm	26 cm
Inner diameter	19.4 cm	18.2 cm
Turns	22	26
Turn spacing	1.5 mm	1.5 mm
Wire diameter	1.5 mm	1.5 mm
Layers	1	1

(Wayne Kerr 6500B), the inductance of the transmitting coil is about $L_T = 191.282$ μH, and the inductance of the receiving coils 1 and 2 are about $L_1 = 240.005$ μH and $L_2 = 241.458$ μH. Assuming that receiving coils 1 and 2 are placed on both sides of the transmitting coil, the variation of coupling coefficient is shown as Fig. 7.18 when receiving coils 1 and 2 are in axial movement with an offset angle of 0°. It can be seen that the measured values are basically consistent with the calculated results, but it is slightly lower, which may be caused by cross-coupling, measurement environment, etc. As for the series resonant capacitors, multilayer ceramic capacitors are chosen for their low loss and high current-bearing capability at high frequency.

7.3.3 *Controllable Resonant Capacitor Array Design*

For the multi-frequency multi-load PT-WPT system with a time-sharing control strategy, a controllable resonant capacitor array module, including a switch controller (control center) and multiple parallel switching capacitor branches, needs to be added,

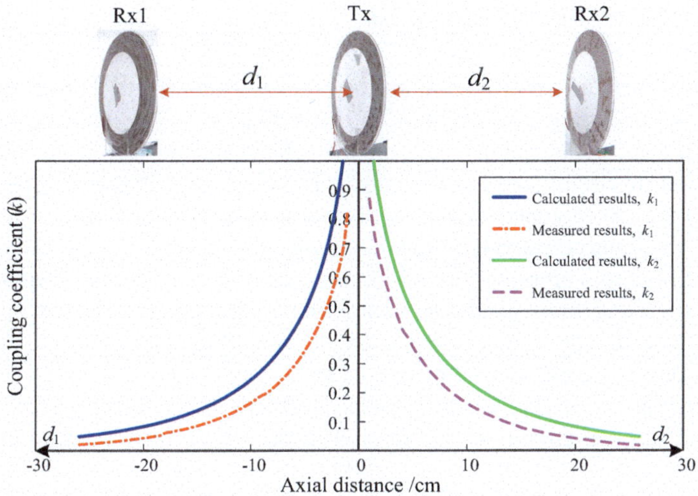

Fig. 7.18 Coupling coefficient between transmitting coil and receiving coils

as shown in Fig. 7.19. The switch controller is not only used to drive the switches, but also to control the power distribution among the loads. Therefore, when selecting a switch controller, the following main functions should be considered: the first is the programmable output of at least two complementary and adjustable duty cycle PWM square wave signals, the second is strong driving ability, and the third is that the sampling frequency is high enough.

Considering various factors, the ARM microcontroller (S9KEAZN8AMFK) is chosen as the main control chip to build the switch controller. To avoid switches S_1 and S_2 from being damaged by excessive instantaneous voltage and current when switches are turned on and off, a current crossing detection circuit is designed to detect the current signal of the transmitter. When the current zero-crossing is detected, the ARM outputs a PWM signal to turn on/off the corresponding switch.

Fig. 7.19 Schematic diagram of controllable switching capacitor

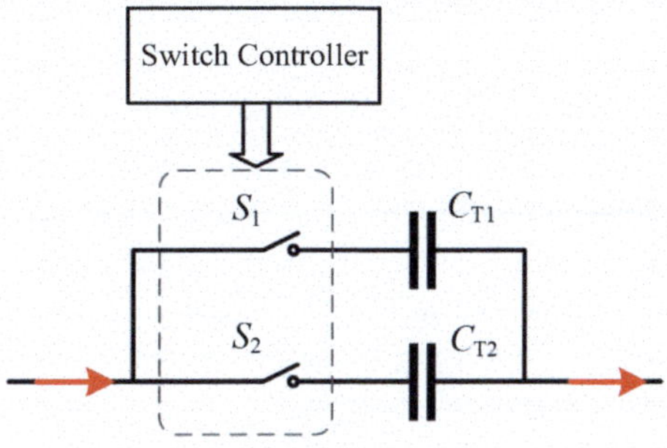

7.3.4 Experimental Verification

Figure 7.20 shows the experimental prototype of the dual-load PT-symmetric MC-WPT system, including a transmitter and two receivers. The main power circuit is shown in Fig. 7.20b, mainly including a self-oscillating full-bridge inverter, a PWM generation circuit, and an over-current protection (OCP) circuit. The working principle of the PWM generation circuit is that the current transformer samples the current i_T (or i_{Ti}) and converts it into a voltage signal, then amplifies the voltage signal through a differential amplifier circuit, finally, generates two complementary PWM signals through a zero-crossing comparator and a logic circuit. The self-oscillating full-bridge inverter (Q_1 ~ Q_4) is driven by the two PWM signals via isolated gate drivers. Through this self-oscillating circuit, the output voltage U_{in} is always in phase with the current i_T (or i_{Ti}), which is consistent with the characteristic of negative resistance. The OCP circuit is used to prevent the circuit from burning out due to the overcurrent caused by the system working in the weak coupling area. The two experiments share a main power circuit: one is the single-frequency dual-load system, and the other is the dual-frequency dual-load system.

Moreover, the controllable resonant capacitor needs to be used in the second experiment, including a time-sharing frequency controller, as shown in Fig. 7.20c, and a controlled switches array, as shown in Fig. 7.20c. The time-sharing frequency controller is composed of an ARM microcontroller (S9KEAZN8AMFK) and a signal isolation chip. Taking into account the current-bearing capability of switch relays, every four relays (Coto 9001–05-01) are connected in parallel to form switches S_1 and S_2. The two switches S_1 and S_2 are also in parallel to form the controlled switches array. Then, the switches array is connected in series with the resonant capacitors array and the transmitting coil to dynamically select the capacitor.

(1) Experimental verification of single-frequency multi-load PT-symmetric MC-WPT system.

The experimental prototype of the single-frequency multi-load PT-symmetric MC-WPT system is shown in Fig. 7.20a. The transmitter and the two receivers have the same natural resonant frequency and are designed to be 300 kHz. Therefore, the transmitter does not need a switching controller and switching capacitor array, and the transmitting coil only needs to be connected in series with a resonant capacitor and tuned to 300 kHz. Other circuit parameters are shown in Table 7.4.

Substituting the parameters of Table 7.4 into $k_C = \sqrt{1 - \frac{1}{4}(2 - Q_{L1}^{-2})^2}$, the critical coupling coefficient $k_C = 0.038$. The experimental waveforms of the output voltage u_{in} of the full-bridge inverter and the input current i_T flowing through the transmitting coil are shown in Fig. 7.21, with a phase difference of 0°, which means generating energy. Therefore, it can be equivalent to negative resistance -R_N, which is satisfied with theoretical analysis.

In the strong coupling region ($k_C^2 \le k_1^2 + k_2^2 < 1$), when the positions of load 1 and load 2 are arbitrarily changed, the operating frequency, total output power, and total transmission efficiency of the system under different coupling coefficients are

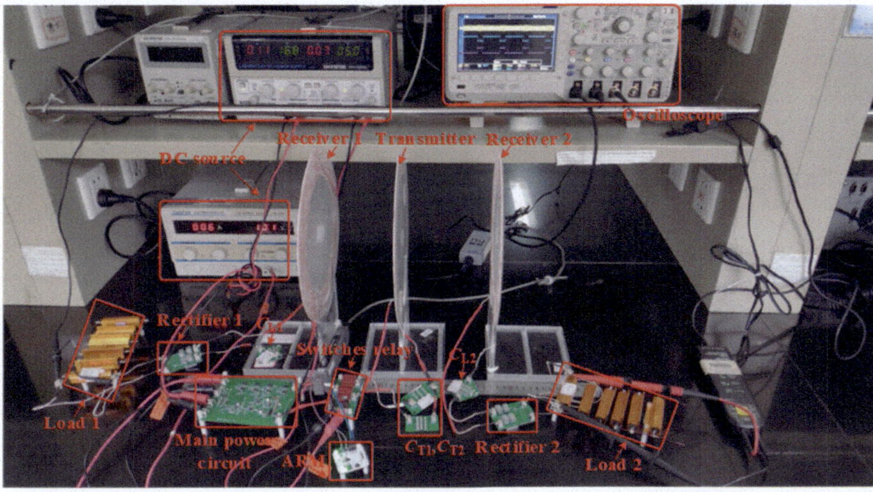

(a)

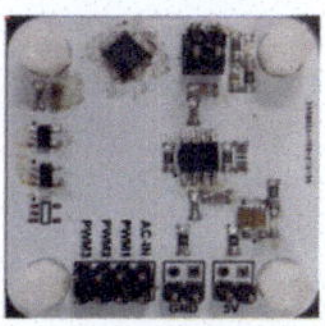

(c)

(b) (d)

Fig. 7.20 Experimental prototype of the dual-load PT-symmetric MC-WPT system. **a** Whole set-up. **b** Main power circuit. **c** Switch controller. **d** Switches relay

measured, as shown in Fig. 7.22. It can be seen that in the strong coupling region, the operating frequency of the system is automatically adjusted as the coupling coefficient k_1 and k_2 change, and the total output power of the system always remains at 36 W with a constant transfer efficiency of 90% when the coupling coefficient k_1 and k_2 changes, which is not affected by the load's position. In addition, within the error range, the theoretical value is consistent with the experimental result.

When $k_1 = k_2$, the power distribution characteristic curves are shown in Fig. 7.23. The blue line and dots represent the total output power and total efficiency of the

Table 7.4 Circuit parameters of single-frequency multi-load PT-symmetric MC-WPT system

Parameters	Values	Parameters	Values
DC input voltage U_{DC}	25 V	Internal resistance of receiver 1 R_1	0.623 Ω
Transmitting coil L_T	191.282 μH	Internal resistance of receiver 2 R_2	0.602 Ω
Transmitting capacitor C_T	1.471 nF	Load resistance 1 R_{L1}	16.67 Ω
Receiving coil 1 L_1	240.005 μH	Load resistance 2 R_{L2}	16.67 Ω
Receiving capacitor 1 C_1	1.173 nF	Natural frequency of transmitter f_T	300 kHz
Receiving coil 2 L_2	241.458 μH	Natural frequency of receiver 1 f_1	300k Hz
Receiving capacitor 2 L_2	1.166 nF	Natural frequency of receiver 2 f_2	300 kHz
Internal resistance of transmitter R_T	0.536 Ω		

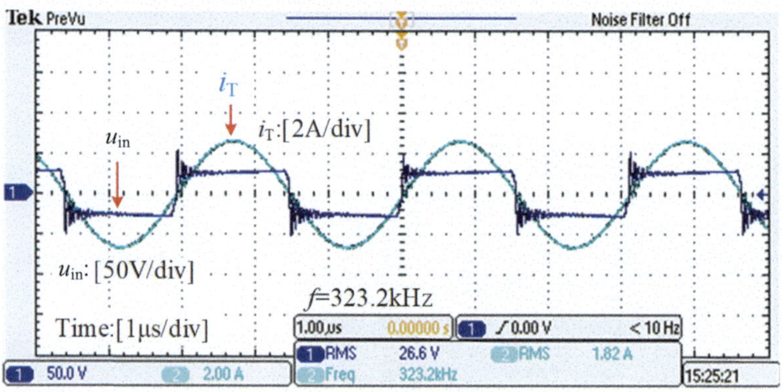

Fig. 7.21 Voltage and current waveforms of full-bridge inverter

system, while the red and green lines and dots represent the output power and efficiency of load 1 and load 2. As shown in Fig. 7.23, it can be seen that in the strong coupling region (the shaded region in Fig. 7.23), the total output power and total transfer efficiency of the system always remain constant at 36 W and 90%, independent of k_1 and k_2. For each load, the output power and transfer efficiency of loads 1 and 2 also remain constant, and $P_{o1} = P_{o2} = 50\% P_{total}$, $\eta_1 = \eta_2 = 50\% \eta_{total}$. Within the error range, the experimental results are consistent with the theoretical analysis.

When $k_1 \neq k_2$ and $k_C^2 \leq k_1^2 + k_2^2 < 1$, the power distribution characteristic curves are shown in Fig. 7.24. The coupling coefficient of load 2 is assumed to be fixed at $k_2 = 0.094$, and the coupling coefficient k_1 of load 1 is changed from 0 to 0.25. From Fig. 7.24, it can be seen that even if $k_1 \neq k_2$, the system always maintains a constant output, which is equal to the total output power $P_{total} = 36$ W when $k_1 = k_2$. However, the output power of each load varies with the coupling coefficient and cannot be kept at a constant value. Furthermore, the greater the coupling coefficient is (that is, the receiver is closer to the transmitter), the greater the output power is.

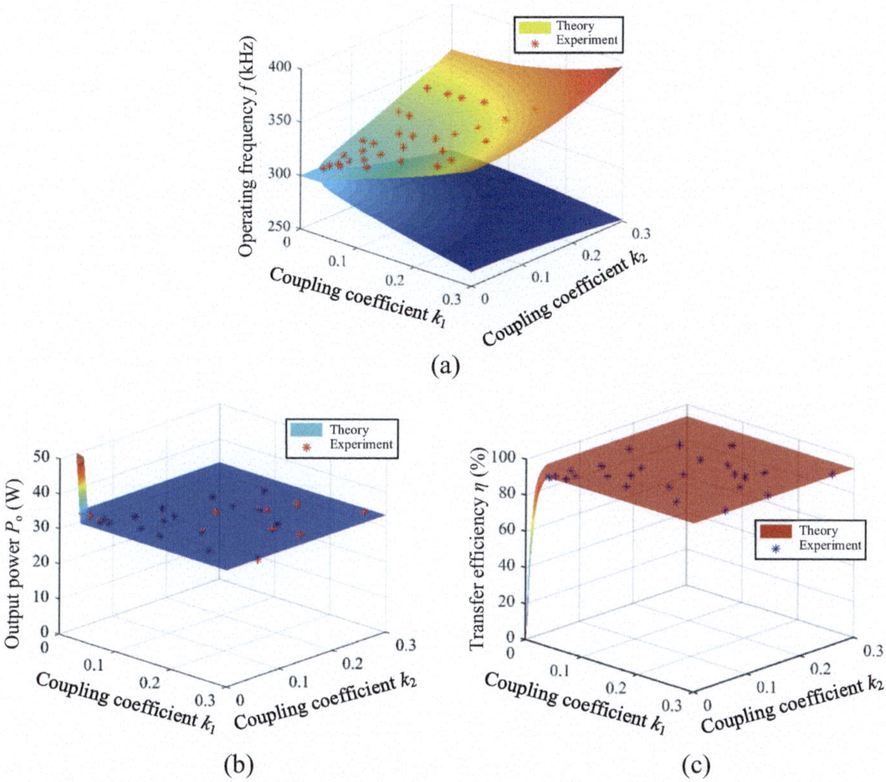

Fig. 7.22 Transfer characteristics of single-frequency dual-load PT-symmetric MC-WPT system. **a** Operating frequency f. **b** Total output power P_{total}. **c** Total transfer efficiency η_{total}

The power distribution relationship between loads is further compared, as shown in Fig. 7.24b, which is consistent with the theoretical analysis that P_{o1}:$P_{o2} = k_1^2$:k_2^2.

In conclusion, for the whole single-frequency multi-load PT-symmetric MC-WPT system, the system always has a constant output characteristic, and the whole system shows the same constant output characteristic as the single-load system. For a specific load, when the coupling coefficients of all loads are always equal, each load also has a constant output characteristic, and the output power is equally distributed to each load without any control. When the coupling coefficients are not equal, the power distribution relationship between the loads meets P_{o1}:P_{o2}:...:$P_{on} = k_1^2$:k_2^2:...:k_n^2. If the control mechanism of the coupling coefficient is added, the proportional relationship between the coupling coefficients is always unchanged during the load movement, then the output power of each load can still be kept constant.

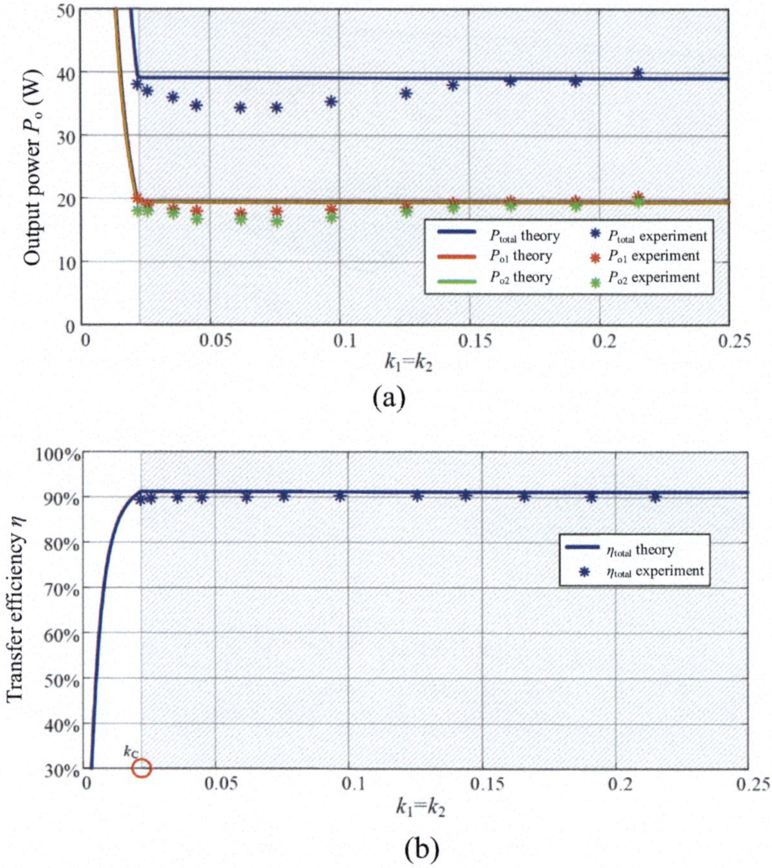

Fig. 7.23 Power distribution and efficiency distribution when $k_1 = k_2$. **a** Power distribution. **b** Efficiency distribution

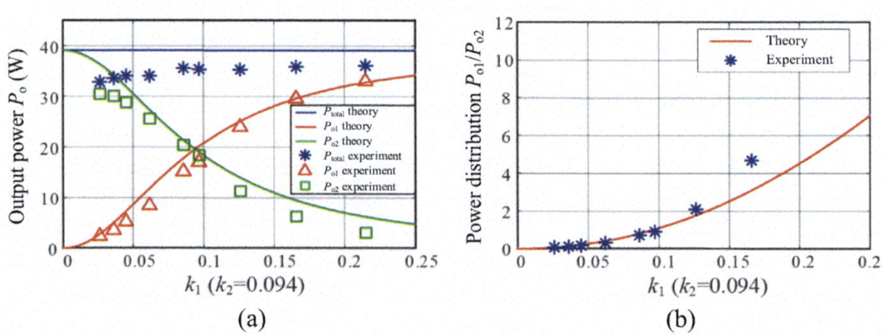

Fig. 7.24 Power distribution characteristic curves when $k_1 \neq k_2$. **a** Power variation. **b** Power distribution relationship

(2) Experimental verification of multi-frequency multi-load PT-symmetric MC-WPT system.

The experimental prototype of this experiment is still shown in Fig. 7.20, but the two load receivers have different natural resonant frequencies. The natural resonant frequencies of load 1 and load 2 are designed to be 130 kHz and 200 kHz, respectively. The corresponding natural resonant frequencies of the two switching capacitor branches in series with the transmitting coil are also tuned to 130 kHz and 200 kHz, respectively. A switch controller and a switch capacitor array are added to the transmitter, as shown in Fig. 7.20c, d. Other circuit parameters of this experiment are shown in Table 7.5.

Substituting the parameters of Table 7.5 into $kc = \sqrt{1 - \frac{1}{4}\left(2 - Q_{\text{L1}}^{-2}\right)^2}$, the critical coupling coefficients of loads 1 and 2 are $k_{\text{C1}} = 0.038$ and $k_{\text{C2}} = 0.057$, respectively. Then, the strong coupling regions corresponding to the two loads are $0.088 \le k_1 < 1$ and $0.057 \le k_2 < 1$, respectively.

Setting the switching frequency $f_{\text{sw}} = 10$ Hz, the duty cycle $\alpha = 0.25$, and the coil distance $d_1 = d_2 = 10$ cm, then the ratio of the on-time of switches S_1 and S_2 is 1:3, that is ΔT_1: $\Delta T_2 = 1$:3. Figure 7.25 shows the waveforms of input current i_{T} and load voltages u_{o1} and u_{o2}. It can be seen that during time ΔT_1, the load 1 is operating with an output voltage of about $U_{\text{o1}} = 8.9$ V and the operating frequency $f_1 = 139.3$ kHz, and the load 2 almost does not work. While during time ΔT_2, the load 2 is working with an output voltage of about $U_{\text{o2}} = 9.0$ V and the operating frequency $f_2 = 213$ kHz, and the load 1 almost has no output voltage. In other words, only one load is working at any time. In addition, in the whole control cycle T, the output power of load 1 is $P_{\text{o1}} \approx 1.19$ W, the output power of load 2 is $P_{\text{o2}} \approx 3.64$ W,

Table 7.5 Circuit parameters of multi-frequency multi-load PT-symmetric MC-WPT system

Parameters	Values	Parameters	Values
DC input voltage U_{DC}	10 V	Internal resistance of receiver 1 R_1	0.623 Ω
Transmitting coil L_{T}	191.282 μH	Internal resistance of receiver 2 R_2	0.602 Ω
Receiving coil 1 L_1	240.005 μH	Load resistance 1 R_{L1}	16.67 Ω
Receiving coil 2 L_2	241.458 μH	Load resistance 2 R_{L2}	16.67 Ω
Transmitting capacitor 1 C_{T1}	7.836 nF	Natural frequency 1 of transmitter f_{T1}	130 kHz
Transmitting capacitor 2 C_{T2}	3.311 nF	Natural frequency 2 of transmitter f_{T2}	200 kHz
Receiving capacitor 1 C_1	6.245 nF	Natural frequency of receiver 1 f_1	130 kHz
Receiving capacitor 2 C_2	2.623 nF	Natural frequency of receiver 2 f_2	200 kHz
Internal resistance of transmitter R_{T}	0.536 Ω	Switching frequency f_{sw}	5 Hz–100 Hz

and the power distribution relationship satisfies P_{o1}:P_{o2} ≈ 1:3, which is consistent with the theoretical ratio α:1-α = 1:3. Further, by moving the positions of the load 1 and 2 to change the coupling coefficients k_1 and k_2, the operating frequencies of load 1 and load 2 are measured, as shown in Fig. 7.26. Both operating frequencies f_1 and f_2 are constantly self-adjusting around the natural resonant frequencies, i.e., 130 kHz and 200 kHz, along with k_1 and k_2, which are consistent with the theoretical value of (7.44).

Then, by changing the duty cycle α (α = 0.25, 0.5, 0.75, 1) and coupling coefficients k_1 and k_2, the output power of load 1 and load 2, the total output power of the system, and the total transfer efficiency of the system are measured, as shown in Fig. 7.27. It can be seen that the experimental results are basically consistent with the theoretical results. As long as each load works in its own strong coupling region, the output power P_{o1} or P_{o2} of loads 1 and 2, as well as the overall output power P_{total} and transfer efficiency η_{total} of the whole system will always remain constant regardless of the coupling coefficient k_1 and k_2. The power distribution relationship is shown in Fig. 7.28, it can be seen that the power distribution of loads 1 and 2 is only related to the switching duty cycle α, which is consistent with the theoretical analysis is only related to the switching duty cycle α, as shown in Fig. 7.15.

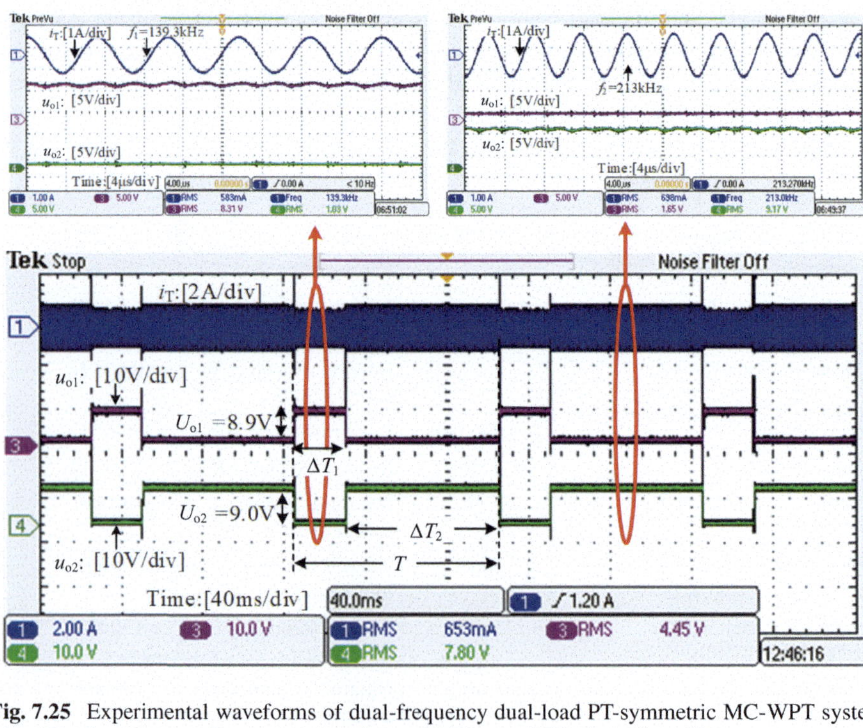

Fig. 7.25 Experimental waveforms of dual-frequency dual-load PT-symmetric MC-WPT system with time-sharing control strategy when f_{sw} = 10 Hz, α = 0.25, d_1 = d_2 = 10 cm

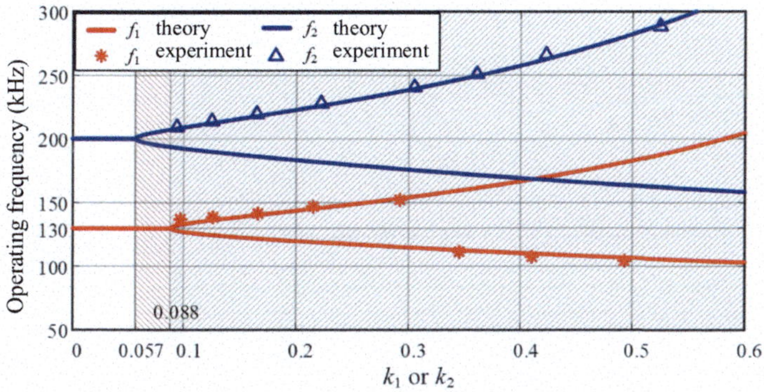

Fig. 7.26 Operation frequencies of load 1 and load 2 ($\alpha = 0.25, f_1 = 130$ kHz, $f_2 = 200$ kHz)

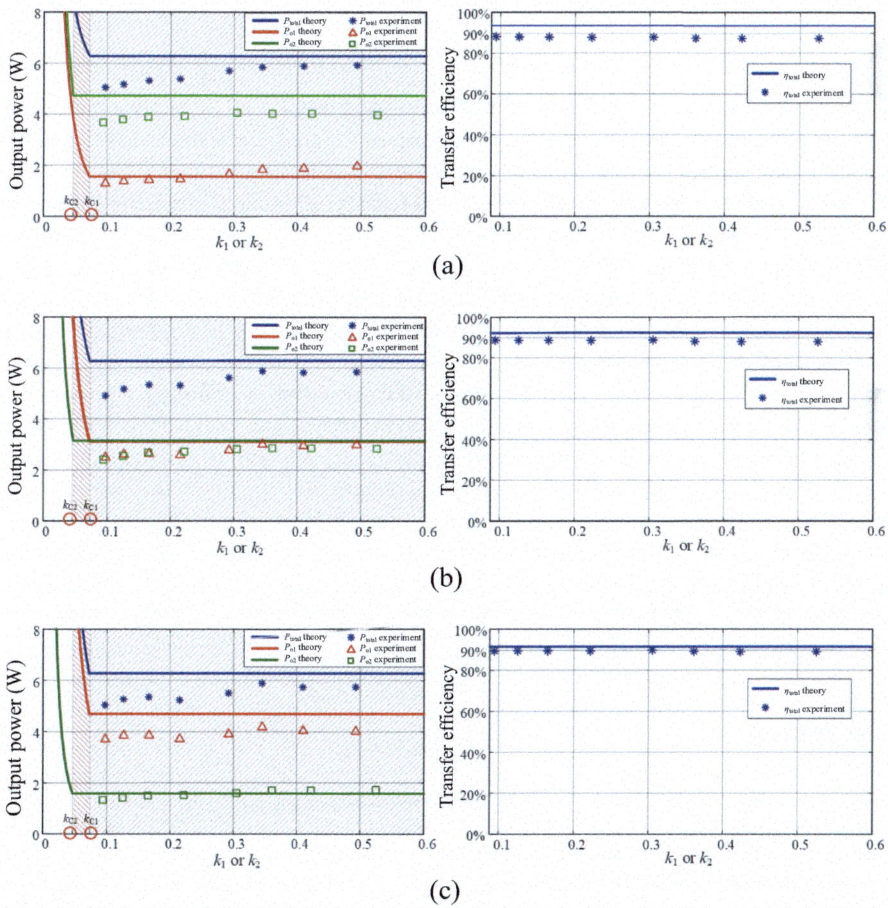

Fig. 7.27 Transfer performance of the dual-frequency dual-load PT-symmetric MC-WPT system in different duty cycles. **a** $\alpha = 0.25$. **b** $\alpha = 0.5$. **c** $\alpha = 0.75$

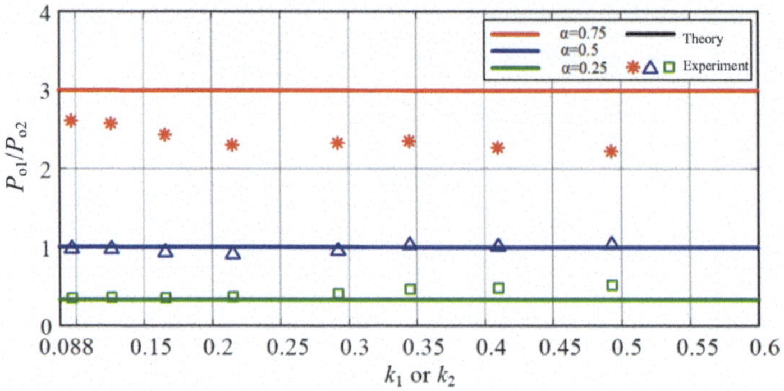

Fig. 7.28 Power distribution under different duty cycles

7.4 Summary

This chapter focuses on PT-symmetric multi-load MC-WPT systems, including the single-frequency multi-load system and the multi-frequency multi-load system. Through the modeling and analysis of the single-frequency multi-load PT-symmetric MC-WPT system, constant output power and transfer efficiency are achieved under the PT-symmetric condition, but the power distribution among loads is related to the coupling coefficients when the coupling coefficients are not identical. In order to solve the problem of uneven power distribution, the multi-frequency multi-load PT-symmetric MC-WPT system is introduced. Based on the time-sharing control strategy, the whole output power and transfer efficiency of the system, and output power of each load remain constant against the variation of coupling coefficients. Finally, the experimental prototype of the PT-symmetric dual-load MC-WPT system is implemented, and the experiments of single-frequency and dual-frequency dua-load systems are carried out. The experimental results show that both the two systems can achieve constant total output power and efficiency. However, for the specific load, the single-frequency multi-load system is still limited by the coupling coefficient, while the multi-frequency multi-load system can completely eliminate the influence of the coupling coefficient.

References

1. Assawaworrarit S, Yu X, Fan S (2017) Robust wireless power transfer using a nonlinear parity-time-symmetric circuit. Nature 546(7658):387–390
2. Schindler J, Lin Z, Lee J, Ramezani H, Ellis F, Kottos T (2012) PT-symmetric electronics. J Phys A: Math Theor 45:1–17
3. Zhou J, Zhang B, Xiao W, Qiu D, Chen Y (2019) Nonlinear parity-time-symmetric model for constant efficiency wireless power transfer: application to a drone-in-flight wireless charging platform. IEEE Trans Industr Electron 66(5):4097–4107

4. Wu L, Zhang B, Zhou J (2020) Efficiency improvement of the parity-time-symmetric wireless power transfer system for electric vehicle charging. IEEE Trans Power Electron 35(11):12497–12508
5. Luo C, Qiu D, Lin M, Zhang B (2020) Circuit model and analysis of multi-load wireless power transfer system based on parity-time symmetry. Energies 13(12):3260
6. Liu F, Yang Y, Jiang D, Ruan X, Chen X (2017) Modeling and optimization of magnetically coupled resonant wireless power transfer system with varying spatial scales. IEEE Trans Power Electron 32(4):3240–3250
7. Kim YJ, Ha D, Chappell WJ, Irazoqui PP (2016) Selective wireless power transfer for smart power distribution in a miniature-sized multiple-receiver system. IEEE Trans Industr Electron 63(3):1853–1862
8. Luo C, Qiu D, Gu W, Zhang B, Chen Y, Xiao W (2022) Multiload wireless power transfer system with constant output power and efficiency. IEEE Trans Ind Appl 58(1):1101–1114
9. Cai W, Ma D, Lai X, Hashmi K, Tang H, Xu J (2020) Time-sharing control strategy for multiple-receiver wireless power transfer systems. Energies 13(3):599

Chapter 8
PT-Symmetric Magnetic-Field Coupled Wireless Power Transfer System with Relay Coils

In the MC-WPT system, the transfer efficiency and the transfer distance are contradictory. Generally speaking, the farther the transfer distance is, the lower the transfer efficiency is. In order to simultaneously realize the requirements of high transfer efficiency and long transfer distance, the MC-WPT system with relay coils is proposed [1–3]. Compared with the common two-coil MC-WPT system, the MC-WPT system with relay coils can effectively improve the power transfer distance, efficiency, and capacity by tuning the relative position of the relay coil. Based on the high-order PT-symmetric principle, the MC-WPT system with relay coils satisfying the PT symmetry can obtain constant output power and transfer efficiency [4–7]. Meanwhile, when the transfer distance varies, the distance of stable power transfer can be extended. Currently, the high-order PT symmetry has been widely applied in ultrasensitive sensing. Compared to the two-coil PT-symmetric MC-WPT system, the PT-symmetric MC-WPT system with relay coils needs additional conditions to meet the high-order PT-symmetric conditions. This chapter first analyzes the PT-symmetric conditions and the transfer characteristics of the PT-symmetric MC-WPT system with a single relay coil, including operating frequency, output power, and transfer efficiency. Then, the operating frequency, output power, and transfer efficiency of the PT-symmetric MC-WPT system with double relay coils are studied, and finally the basic characteristics of the PT-symmetric MC-WPT system with multiple relay coils from the two aspects of odd and even number of relay coils are explored.

8.1 System Structure

The PT-symmetric MC-WPT system with multiple relay coils is shown in Fig. 8.1. M_{12}, M_{23}, …, and $M_{\text{n}-1\ \text{n}}$ are mutual inductances between adjacent coils. Generally, for an MC-WPT system with multiple relay coils, compared with the strong coupling coefficients between two adjacent coils $k_{\text{n-1}}=M_{\text{n-1 n}}/\sqrt{L_{\text{n-1}}L_{\text{n}}}$ (n = 2, 3, …),

B. Zhang et al., *Parity-Time Symmetric Wireless Power Transfer*, CPSS Power Electronics Series, https://doi.org/10.1007/978-981-96-3149-0_8

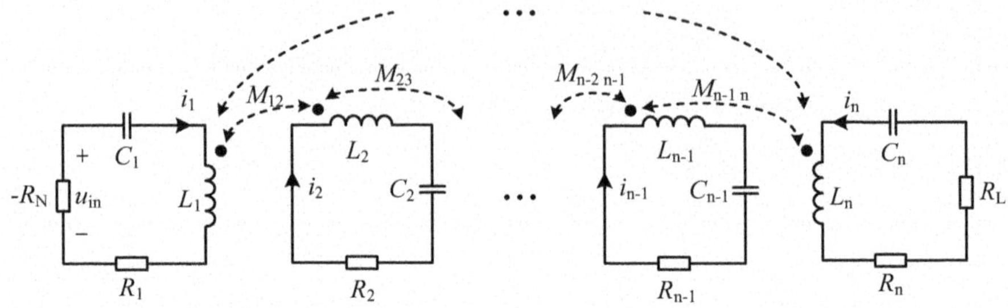

Fig. 8.1 System structure of the PT-symmetric MC-WPT system with multiple relay coils

the coupling coefficients between nonadjacent coils that are far away from each other can be negligible, i.e., $k_{ij} = M_{ij}/\sqrt{L_i L_j} \approx 0$ (i = 1, 2, …, n-2. j = i + 2, i + 3, …, n). $M_{n\text{-}1\ n}$ is the mutual inductance between n-1th and nth adjacent coils, and M_{ij} is the mutual inductance between the ith and jth nonadjacent coils.

Based on the CMT, the dynamic equations of the multiple relay-coil system shown in Fig. 8.1 can be written as

$$\frac{d}{dt}\begin{bmatrix} \boldsymbol{a}_1 \\ \boldsymbol{a}_2 \\ \boldsymbol{a}_3 \\ \vdots \\ \boldsymbol{a}_n \end{bmatrix} = \begin{bmatrix} j\omega_1 + g & -j\kappa_{12} & 0 & \cdots & 0 \\ -j\kappa_{12} & j\omega_2 - \tau_2 & -j\kappa_{23} & \cdots & 0 \\ 0 & -j\kappa_{23} & j\omega_3 - \tau_3 & \cdots & 0 \\ \vdots & \vdots & \vdots & \cdots & \vdots \\ 0 & 0 & 0 & \cdots & j\omega_n - \tau_n \end{bmatrix} \begin{bmatrix} \boldsymbol{a}_1 \\ \boldsymbol{a}_2 \\ \boldsymbol{a}_3 \\ \vdots \\ \boldsymbol{a}_n \end{bmatrix}. \tag{8.1}$$

8.2 Odd Number Relay-Coil System

A single relay-coil system is used as an example. The PT-symmetric MC-WPT system with a single relay coil is shown in Fig. 8.2. Here, $-R_N$ is a negative resistor that is used as a power source to provide input voltage u_{in} to the system, and represents the total gain rate of the system. L_n, C_n, and R_n (n = 1, 2, 3) are inductance, compensation capacitance, and internal resistance of the transmitting coil (n = 1), relay coil (n = 2), and receiving coil (n = 3), respectively. M_{12} is the mutual inductance between the transmitting coil and the relay coil. M_{23} is the mutual inductance between the receiving coil and the relay coil. M_{13} is the mutual inductance between the transmitting coil and the receiving coil. i_n (n = 1, 2, 3) are currents flowing through the transmitting coil (n = 1), relay coil (n = 2) and receiving coil (n = 3), respectively. R_L is the equivalent resistance of the load. Particularly, the coupling coefficients of the mutual inductances M_{12}, M_{23}, and M_{13} are defined as $k_{12}=M_{12}/\sqrt{L_1L_2}$, $k_{23}=M_{23}/\sqrt{L_2L_3}$ and

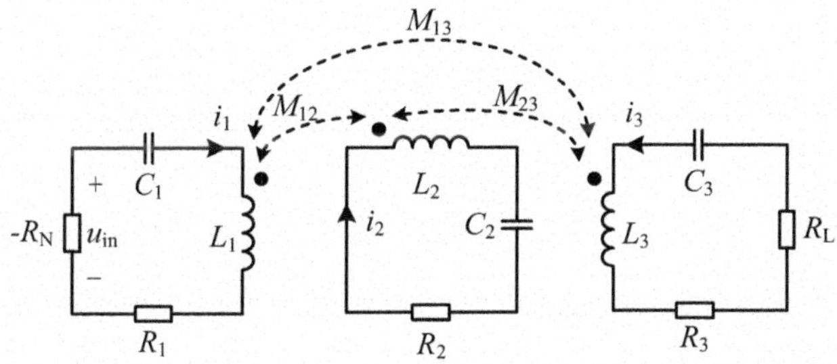

Fig. 8.2 System structure of the PT-symmetric MC-WPT system with single relay coil

$k_{13}=M_{13}/\sqrt{L_1L_3}$, and the natural resonant frequencies of the coils are defined as $\omega_1 = 1/\sqrt{L_1C_1}$, $\omega_2=1/\sqrt{L_2C_2}$ and $\omega_3 = 1/\sqrt{L_3C_3}$, respectively.

Based on the CMT, the dynamic equations of the single relay-coil PT-symmetric MC-WPT system shown in Fig. 8.2 can be obtained by the coupled-mode formalism [6–9] as follows:

$$\frac{d}{dt}\begin{bmatrix} \boldsymbol{a}_1 \\ \boldsymbol{a}_2 \\ \boldsymbol{a}_3 \end{bmatrix} = \begin{bmatrix} j\omega_1 + g & -j\kappa_{12} & -j\kappa_{13} \\ -j\kappa_{12} & j\omega_2 - \tau_2 & -j\kappa_{23} \\ -j\kappa_{13} & -j\kappa_{23} & j\omega_3 - \tau_3 \end{bmatrix} \begin{bmatrix} \boldsymbol{a}_1 \\ \boldsymbol{a}_2 \\ \boldsymbol{a}_3 \end{bmatrix}, \tag{8.2}$$

where $\boldsymbol{a}_n$ (n = 1, 2, 3) is energy mode, and $|\boldsymbol{a}_n|^2$ represents the energy stored in coil n. $g = g_{10}$-τ_{10} is the total gain rate of the transmitting coil with the gain rate $g_{10} = U_{in}/(2\sqrt{L_1}|\boldsymbol{a}_1|)$ and intrinsic loss rate $\tau_{10} = R_1/(2L_1)$, $\tau_2 = R_2/(2L_2)$ is the loss rate of the relay coil, $\tau_3 = \tau_{30} + \tau_L$ is the total loss rate of the receiving coil with the intrinsic loss rate $\tau_{30} = R_3/(2L_3)$ and the load rate $\tau_L = R_L/(2L_3)$, and U_{in} is the RMS value of u_{in}. $\kappa_{12} = (\omega_1 + \omega_2)k_{12}/4$ or $\kappa_{12}=\sqrt{\omega_1\omega_2}\, k_{12}/2$ is the coupling rate between transmitting and relay coils, $\kappa_{23} = (\omega_2 + \omega_3)k_{23}/4$ or $\kappa_{23}=\sqrt{\omega_2\omega_3}\, k_{23}/2$ is the coupling rate between receiving and relay coils, $\kappa_{13} = (\omega_1 + \omega_3)k_{13}/4$ or $\kappa_{13}=\sqrt{\omega_1\omega_3}\, k_{13}/2$ is the coupling rate between transmitting and receiving coils.

In order to satisfy the PT symmetry, the dynamic equation of (8.2) should be invariant under the combined P and T reversal transformation, so as to obtain the PT-symmetric conditions that $\omega_1 \approx \omega_2 \approx \omega_3 = \omega_0$, $\kappa_{12} = \kappa_{23} = \kappa_0$, $\tau_2 = 0$, and $\kappa_{13} = 0$ [6, 7]. Generally, $\omega_1 \approx \omega_2 \approx \omega_3 = \omega_0$ is easily achieved in MC-WPT systems by designing the parameters of the coil and compensated capacitance, and k_{13} is small enough to be neglected too because the mutual inductance M_{13} between nonadjacent coils is much smaller than the mutual inductances M_{12} and M_{23} between adjacent coils [10]. Meanwhile, the relay coil mainly plays the role of widening the transfer distance, its internal resistance R_2 is small enough to be ignored. In addition, $\kappa_{12} = \kappa_{23} = \kappa_0$ can be kept by controlling the variable inductor of the transmitting coil or mechanically rotating the transmitting coil [4, 6]. Hence, the coupled-mode equation of the PT-symmetric MC-WPT system with a single relay coil can be rearranged as

$$\frac{d}{dt}\begin{bmatrix} \boldsymbol{a}_1 \\ \boldsymbol{a}_2 \\ \boldsymbol{a}_3 \end{bmatrix} = \begin{bmatrix} j\omega_0 + g & -j\kappa_0 & 0 \\ -j\kappa_0 & j\omega_0 & -j\kappa_0 \\ 0 & -j\kappa_0 & j\omega_0 - \tau_3 \end{bmatrix}\begin{bmatrix} \boldsymbol{a}_1 \\ \boldsymbol{a}_2 \\ \boldsymbol{a}_3 \end{bmatrix}. \tag{8.3}$$

8.2.1 Operating Frequency

Assuming that the steady solutions of (8.3) are the form $[\boldsymbol{a}_1, \boldsymbol{a}_2, \boldsymbol{a}_3]^T = [A_1, A_2, A_3]^T e^{j\omega t}$, A_n (n = 1, 2, 3) are the amplitudes of $\boldsymbol{a}_n$, ω is the operating frequency of the single relay-coil system, which is determined by the characteristic equation, as follows:

$$\begin{vmatrix} j(\omega - \omega_0) - g & j\kappa_0 & 0 \\ j\kappa_0 & j(\omega - \omega_0) & j\kappa_0 \\ 0 & j\kappa_0 & j(\omega - \omega_0) + \tau_3 \end{vmatrix} = 0. \tag{8.4}$$

By separating the real and imaginary parts of the characteristic equation, we can get

$$\left[(\omega - \omega_0)^2 - \kappa_0^2\right](g - \tau_3) = 0, \tag{8.5}$$

$$(\omega - \omega_0)\left[(\omega - \omega_0)^2 - 2\kappa_0^2 + g\tau_3\right] = 0. \tag{8.6}$$

From (8.5) and, it can be seen that the operating frequency has three forms of solution, as follows:

(1) Exact PT-symmetric state.

In this case, the gain rate g and the operating frequency ω can be deduced as

$$\begin{cases} g = \tau_3 \\ \omega = \omega_0 \pm \sqrt{2\kappa_0^2 - \tau_3^2} \end{cases}. \tag{8.7}$$

From (8.7), it can be seen that the gain and loss rates are automatically balanced, and the operating frequency has two branches. In order to ensure that the operating frequency ω is purely real, the coupling rate κ_0 has to satisfy the following constraint condition

$$\kappa_0 \geq \frac{\tau_3}{\sqrt{2}}. \tag{8.8}$$

Substituting the definition of κ_0 and τ_3 into (8.8), we can get

$$k_0 \geq k_C = \frac{(R_3 + R_L)}{\sqrt{2}\omega_0 L_3}, \tag{8.9}$$

where k_C is defined as the critical coupling coefficient.

From (8.9), it can be seen that the critical coupling coefficient k_C is determined by the load resistance and natural resonant frequency of the coil, and it is $1/\sqrt{2}$ times of the conventional PT-symmetric MC-WPT system expressed by (3.42). Thus, compared with the two-coil PT-symmetric MC-WPT system shown in Fig. 3.2a, the exact PT-symmetric region has been effectively improved.

(2) Broken PT-symmetric state.

When the coupling coefficient k_0 is less than the critical coupling point k_C, the operating frequencies begin to branch into two conjugate pairs, and the PT symmetry is spontaneously broken. In the region ($k_0 < k_C$), the PT symmetry is broken, and the corresponding gain rate g and operating frequency ω are

$$\begin{cases} g = \dfrac{\kappa_0^2}{\tau_3} \\ \omega = \omega_0 \pm \kappa_0 \end{cases}. \tag{8.10}$$

In this case, the balance between the gain rate and loss rate is broken.

(3) Resonant sate.

By solving (8.5) and, we can find a set of solutions that

$$\begin{cases} g = \tau_3 \\ \omega = \omega_0 \end{cases}, \tag{8.11}$$

which is independent of the coupling coefficient, while the gain rate is automatically equal to the loss rate.

Based on Lyapunov theory, the stability of the above steady-state solutions of operating frequency can be judged by referring to the analysis in Sect. 3.3.1. Figure 8.3 shows the operating frequency characteristic curve of the PT-symmetric MC-WPT system with a single relay coil, it can be seen that there are three operating frequencies for each coupling coefficient in the exact PT-symmetric region and broken PT-symmetric region. The circuit parameters are listed in Table 8.1.

8.2.2 *Output Power*

Substituting the solutions of operating frequency into (8.3), and combining the definition of output power that $P_O = 2\tau_L|\boldsymbol{a}_3|^2$ and the negative resistor constructed by a half-bridge inverter $U_{in}=\sqrt{2}\, U_{DC}/\pi$, the output power can be obtained as follows:

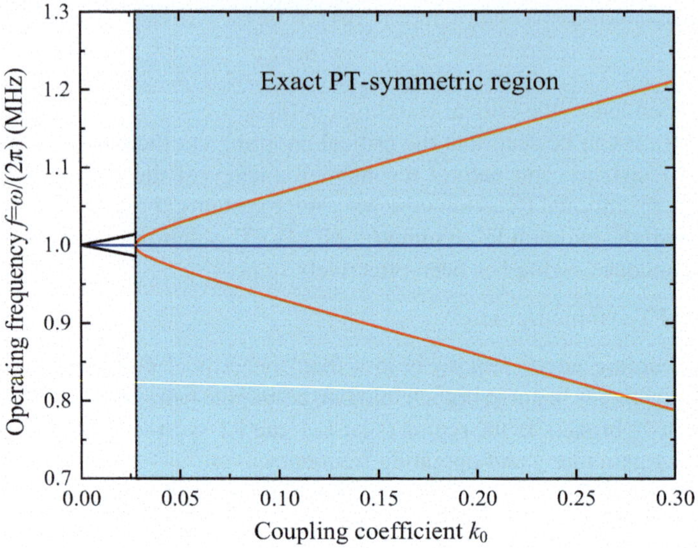

Fig. 8.3 The operating frequency characteristic

Table 8.1 Circuit parameters of the experimental prototype

Description	Symbol	Values
Inductance of transmitting coil	L_1	15.48 μH
Inductance of relay coil 1	L_2	15.47 μH
Inductance of relay coil 2	L_3	15.50 μH
Inductance of receiving coil	L_4	15.52 μH
Compensation capacitance of transmitter	C_1	1.67 nF
Compensation capacitance of relay coil 1	C_2	1.67 nF
Compensation capacitance of relay coil 2	C_3	1.66 nF
Compensation capacitance of receiver	C_4	1.66 nF
Internal resistance of transmitter	R_1	0.246 Ω
Internal resistance of relay coil 1	R_2	0.244 Ω
Internal resistance of relay coil 2	R_3	0.208 Ω
Internal resistance of receiver	R_4	0.222 Ω
DC voltage source	U_{dc}	15 V
Resonant frequency	f_0	995 kHz
Load resistance	R_L	98 Ω

(1) In exact PT-symmetric state and resonant state.

The amplitude of the modes $\boldsymbol{a}_1$ and $\boldsymbol{a}_3$ can be derived as

$$|\boldsymbol{a}_1| = |\boldsymbol{a}_3| = \frac{U_{\text{in}}}{2\sqrt{L_1}(\tau_{10} + \tau_3)}. \tag{8.12}$$

Then, the output power can be obtained as

$$P_{\text{O}} = 2\tau_{\text{L}}|\boldsymbol{a}_3|^2 = \frac{\tau_{\text{L}} U_{\text{in}}^2}{2L_1(\tau_{10} + \tau_3)^2} = \frac{\tau_{\text{L}} U_{\text{DC}}^2}{\pi^2 L_1(\tau_{10} + \tau_3)^2}. \tag{8.13}$$

(2) In broken PT-symmetric state.

The amplitude of the modes $\boldsymbol{a}_1$ and $\boldsymbol{a}_3$ can be deduced as

$$|\boldsymbol{a}_1| = \frac{\tau_3 U_{\text{in}}}{2\sqrt{L_1}\left(\tau_{10}\tau_3 + \kappa_0^2\right)}, \tag{8.14}$$

$$|\boldsymbol{a}_3| = \frac{\kappa_0 U_{\text{in}}}{2\sqrt{L_1}\left(\tau_{10}\tau_3 + \kappa_0^2\right)}. \tag{8.15}$$

Then, the output power can be obtained as

$$P_{\text{O}} = 2\tau_{\text{L}}|\boldsymbol{a}_3|^2 = \frac{\tau_{\text{L}}\kappa_0^2 U_{\text{in}}^2}{2L_1\left(\tau_{10}\tau_3 + \kappa_0^2\right)^2} = \frac{\tau_{\text{L}}\kappa_0^2 U_{\text{DC}}^2}{\pi^2 L_1\left(\tau_{10}\tau_3 + \kappa_0^2\right)^2}. \tag{8.16}$$

From the above analysis, it can be known that the expression of the output power in the exact PT-symmetric state and resonant state is the same, and it is irrelevant to the coupling coefficient, meaning that the output power remains constant against the variations of coupling coefficient, as shown in Fig. 8.4.

8.2.3 *Transfer Efficiency*

According to the definition of transfer efficiency (the ratio of the active power absorbed by the load to the total active power consumed by all the resistors of the system), that is

$$\eta = \frac{2\tau_{\text{L}}|\boldsymbol{a}_3|^2}{2\tau_{10}|\boldsymbol{a}_1|^2 + 2\tau_3|\boldsymbol{a}_3|^2}. \tag{8.17}$$

The transfer efficiency can be obtained as follows:

(1) In exact PT-symmetric state and resonant state.

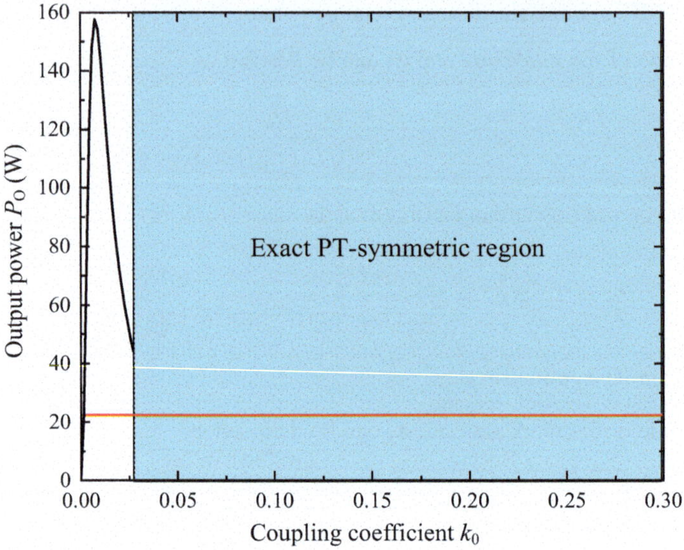

Fig. 8.4 The output power characteristic

$$\eta = \frac{\tau_L}{\tau_{10} + \tau_3}. \tag{8.18}$$

(2) In broken PT-symmetric state.

$$\eta = \frac{\kappa_0^2 \tau_L}{\tau_3 \left(\tau_{10}\tau_3 + \kappa_0^2\right)}. \tag{8.19}$$

It can be seen from (8.18) and that the transfer efficiency in the exact PT-symmetric state and resonant state are constant with the variation of the coupling coefficient, as shown in Fig. 8.5.

From the above analysis, it can be seen that the PT-symmetric MC-WPT system with a single relay coil has three working states, and the system can automatically select a certain working state for stable operation. In the exact PT-symmetric state and resonant state, the output power and transfer efficiency are constant, which are independent of the coupling coefficient.

The same analysis can be extended to more than one relay coil. For the systems with repeater numbers of 3, 5, 7, …, there are still three operating states, and output power and transfer efficiency of the system can remain constant against the variation of transfer distance in the exact PT-symmetric and resonant states. Generally, in the systems with an odd number of repeaters, it is not desirable to operate in the broken PT-symmetric state owing to the variable output power and transfer efficiency related to the coupling coefficient.

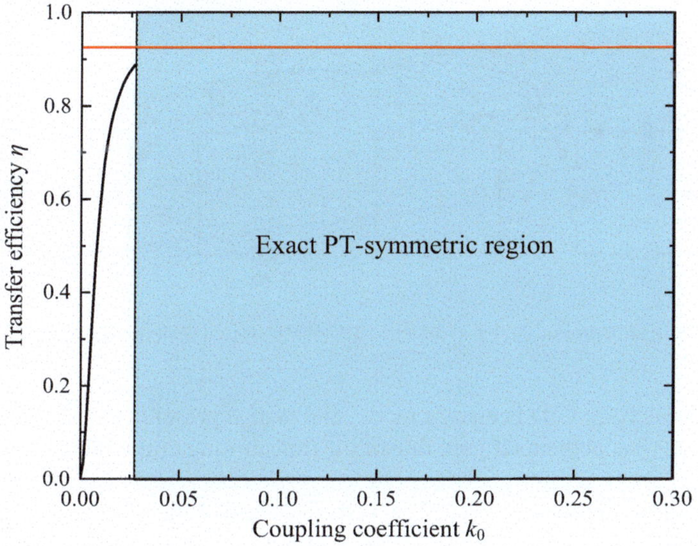

Fig. 8.5 The transfer efficiency characteristic

8.3 Even Number Relay-Coil System

A double relay-coil system is used as an example. Figure 8.6 gives the PT-symmetric MC-WPT system with double relay coils. Here, -R_N is the negative resistance that is used as a power source to provide input voltage u_{in} to the system. n = 1, 2, 3, 4 represent the transmitting coil, relay coil one, relay coil two, and receiving coil. L_n, C_n, and R_n are inductance, compensation capacitance and internal resistance of the coils. M_{12}, M_{23}, and M_{34} are mutual inductances between adjacent coils. M_{13}, M_{14}, and M_{24} are mutual inductances between non-adjacent coils. i_n are currents flowing through the coils. R_L is the equivalent resistance of the load. In particular, k_{12}=M_{12}/$\sqrt{L_1L_2}$, k_{23}=M_{23}/$\sqrt{L_2L_3}$ and k_{34}=M_{34}/$\sqrt{L_3L_4}$ are coupling coefficients between adjacent coils. k_{13}=M_{13}/$\sqrt{L_1L_3}$, k_{14}=M_{14}/$\sqrt{L_1L_4}$ and k_{24}=M_{24}/$\sqrt{L_2L_4}$ are coupling coefficients between non-adjacent coils. $\omega_1 = 1/\sqrt{L_1C_1}$, $\omega_2 = 1/\sqrt{L_2C_2}$, ω_3=$1/\sqrt{L_3C_3}$ and ω_4=$1/\sqrt{L_4C_4}$ are the natural resonant frequencies of the coils.

Based on the CMT, the dynamic equations of the double relay-coil system shown in Fig. 8.6 can be written as

$$\frac{d}{dt}\begin{bmatrix} \boldsymbol{a}_1 \\ \boldsymbol{a}_2 \\ \boldsymbol{a}_3 \\ \boldsymbol{a}_4 \end{bmatrix} = \begin{bmatrix} j\omega_1 + g & -j\kappa_{12} & -j\kappa_{13} & -j\kappa_{14} \\ -j\kappa_{12} & j\omega_2 - \tau_2 & -j\kappa_{23} & -j\kappa_{24} \\ -j\kappa_{13} & -j\kappa_{23} & j\omega_3 - \tau_3 & -j\kappa_{34} \\ -j\kappa_{14} & -j\kappa_{24} & -j\kappa_{34} & j\omega_4 - \tau_4 \end{bmatrix} \begin{bmatrix} \boldsymbol{a}_1 \\ \boldsymbol{a}_2 \\ \boldsymbol{a}_3 \\ \boldsymbol{a}_4 \end{bmatrix}, \tag{8.20}$$

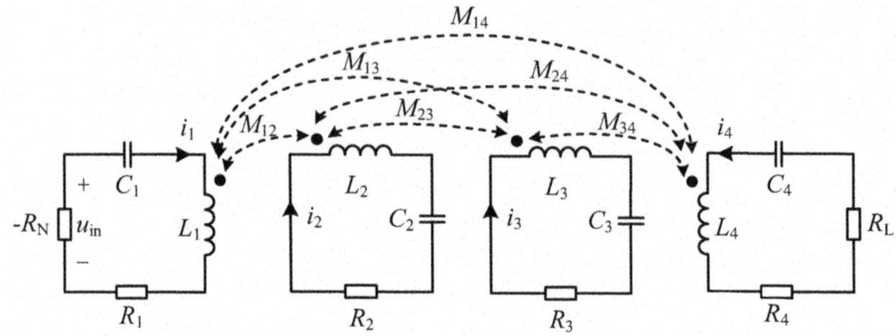

Fig. 8.6 System structure of the PT-symmetric MC-WPT system with double relay coils

where $\boldsymbol{a}_n$ (n = 1, 2, 3, 4) is energy mode, and $|\boldsymbol{a}_n|^2$ represents the energy stored in coil n. $g = g_{10}$-τ_{10} is the total gain rate of the transmitting coil with the gain rate $g_{10} = U_{in}/(2\sqrt{L_1}|\boldsymbol{a}_1|)$ and intrinsic loss rate $\tau_{10} = R_1/(2L_1)$, $\tau_2 = R_2/(2L_2)$ is the loss rate of the relay coil one, $\tau_3 = R_3/(2L_3)$ is the loss rate of the relay coil two, $\tau_4 = \tau_{40} + \tau_L$ is the total loss rate of the receiving coil with the intrinsic loss rate $\tau_{40} = R_4/(2L_4)$ and the load rate $\tau_L = R_L/(2L_4)$, and U_{in} is the RMS value of u_{in}. $\kappa_{12} = (\omega_1 + \omega_2)k_{12}/4$ or $\kappa_{12}=\sqrt{\omega_1\omega_2}\,k_{12}/2$, $\kappa_{23} = (\omega_2 + \omega_3)k_{23}/4$ or $\kappa_{23}=\sqrt{\omega_2\omega_3}\,k_{23}/2$, $\kappa_{34} = (\omega_3 + \omega_4)k_{34}/4$ or $\kappa_{34}=\sqrt{\omega_3\omega_4}\,k_{34}/2$ are coupling rate between adjacent coils. $\kappa_{13} = (\omega_1 + \omega_3)k_{13}/4$ or $\kappa_{13}=\sqrt{\omega_1\omega_3}\,k_{13}/2$, $\kappa_{14} = (\omega_1 + \omega_4)k_{14}/4$ or $\kappa_{14}=\sqrt{\omega_1\omega_4}\,k_{14}/2$, $\kappa_{24} = (\omega_2 + \omega_4)k_{24}/4$ or $\kappa_{24}=\sqrt{\omega_2\omega_4}\,k_{24}/2$ are coupling rate between nonadjacent coils.

According to the definition of PT symmetry, we can get the PT-symmetric conditions of the double relay-coil PT-symmetric MC-WPT system, that is $\omega_1 \approx \omega_2 \approx \omega_3 \approx \omega_4 = \omega_0$, $\kappa_{12} = \kappa_{34} = \kappa_0$, $\tau_2 = \tau_3 = 0$ and $\kappa_{13} = \kappa_{24} = \kappa_{14} = 0$ [6, 7]. General speaking, $\omega_1 \approx \omega_2 \approx \omega_3 \approx \omega_4 = \omega_0$ is easily achieved in MC-WPT systems by designing the parameters of the coils and compensated capacitances, k_{13}, k_{24}, and k_{14} are small enough to be neglected too because the mutual inductances between nonadjacent coils are much smaller than the mutual inductances between adjacent coils [10]. Meanwhile, the relay coils mainly play the role of widening the transfer distance, the internal resistances R_2 and R_3 are small enough to be ignored. In addition, $\kappa_{12} = \kappa_{34} = \kappa_0$ can be kept by controlling the variable inductor of the transmitting coil or mechanically rotating the transmitting coil [4, 6]. Hence, the coupled-mode equation of the PT-symmetric MC-WPT system with double relay coils can be rearranged as

$$\frac{d}{dt}\begin{bmatrix} \boldsymbol{a}_1 \\ \boldsymbol{a}_2 \\ \boldsymbol{a}_3 \\ \boldsymbol{a}_4 \end{bmatrix} = \begin{bmatrix} j\omega_0 + g & -j\kappa_0 & 0 & 0 \\ -j\kappa_0 & j\omega_0 & -j\kappa_{23} & 0 \\ 0 & -j\kappa_{23} & j\omega_0 & -j\kappa_0 \\ 0 & 0 & -j\kappa_0 & j\omega_0 - \tau_4 \end{bmatrix} \begin{bmatrix} \boldsymbol{a}_1 \\ \boldsymbol{a}_2 \\ \boldsymbol{a}_3 \\ \boldsymbol{a}_4 \end{bmatrix}. \tag{8.21}$$

8.3.1 Operating Frequency

Assuming that the steady solutions of (8.21) are the form $[\boldsymbol{a}_1, \boldsymbol{a}_2, \boldsymbol{a}_3, \boldsymbol{a}_4]^T = [A_1, A_2, A_3, A_4]^T e^{j\omega t}$, A_n (n = 1, 2, 3, 4) are the amplitudes of $\boldsymbol{a}_n$, ω is the operating frequency of the double relay-coils PT-symmetric MC-WPT system, which is determined by the characteristic equation. Thus, the characteristic equation can be obtained as

$$\begin{vmatrix} j(\omega-\omega_0)-g & j\kappa_0 & 0 & 0 \\ j\kappa_0 & j(\omega-\omega_0) & j\kappa_{23} & 0 \\ 0 & j\kappa_{23} & j(\omega-\omega_0) & j\kappa_0 \\ 0 & 0 & j\kappa_0 & j(\omega-\omega_0)+\tau_4 \end{vmatrix} = 0. \tag{8.22}$$

By arranging (8.22) and separating the real and imaginary parts, we can get

$$\begin{cases} (\omega-\omega_0)^4 - (\omega-\omega_0)^2\left[\left(2\kappa_0^2+\kappa_{23}^2\right)+\tau_4(\tau_1-g)\right]+\kappa_{23}^2\tau_4(\tau_1-g)+\kappa_0^4=0 \\ (\omega-\omega_0)\left[(\omega-\omega_0)^2-\left(\kappa_0^2+\kappa_{23}^2\right)\right](\tau_1+\tau_4-g)=0. \end{cases} \tag{8.23}$$

Then, the operating frequency can be derived as follows:

(1) Exact PT-symmetric state.

In this case, the gain rate g and the operating frequency ω can be deduced as

$$\begin{cases} g=\tau_1+\tau_4 \\ \omega=\omega_0\pm\sqrt{\kappa_0^2+\frac{1}{2}\left(\kappa_{23}^2-\tau_4^2\right)\pm\frac{1}{2}\sqrt{4\kappa_0^2\left(\kappa_{23}^2-\tau_4^2\right)+\left(\kappa_{23}^2+\tau_4^2\right)^2}} \end{cases}. \tag{8.24}$$

From, it can be seen that the operating frequency has four branches, and the gain and loss rates are automatically balanced. In order to ensure that all four operating frequencies are purely real, the coupling rate κ_{23} has to satisfy the constraint condition $\kappa_{CL} \le \kappa_{23} \le \kappa_{CH}$. κ_{CH} and κ_{CL} are the critical coupling rates, which determine the exact PT-symmetric region that the system can transfer power stably, can be described as

$$\begin{cases} \kappa_{CH}=\dfrac{\kappa_0^2}{\tau_4} \\ \kappa_{CL}=\sqrt{-2\kappa_0^2-\tau_4^2+2\kappa_0\sqrt{\kappa_0^2+2\tau_4^2}} \end{cases}. \tag{8.25}$$

where κ_0 is can be set to a fixed value in advance, it has to satisfy $\kappa_0 \ge \tau_4/2$ to ensure the real κ_{CL}. If the coupling coefficients between adjacent coils are equal $\kappa_{12} = \kappa_{23} = \kappa_{34}$, the constraint condition of the coupling coefficient will degenerate into $\kappa_0 \ge \tau_4$.

Substituting the definition of κ_0 and τ_4 into, we can get the critical coupling coefficients k_{CH} and k_{CL} as follows

$$\begin{cases} k_{CH} = \dfrac{\omega_0 L_4}{R_4 + R_L} k_0^2 \\ k_{CL} = k_0 \sqrt{2\sqrt{1 + \dfrac{2(R_4 + R_L)^2}{\omega_0^2 L_4^2 k_0^2}} - 2 - \dfrac{(R_4 + R_L)^2}{\omega_0^2 L_4^2 k_0^2}} \end{cases}. \tag{8.26}$$

As can be seen from, it can be seen that the critical coupling coefficients k_{CH} and k_{CL} are determined by the predetermined coupling coefficient k_0, the load resistance R_L, and the natural resonant frequency of the coil ω_0. Remarkably, k_{CL} determines the farthest boundary of the exact PT-symmetric region, so the exact PT-symmetric region can be widened by adjusting the coupling coefficient k_0, which is beneficial for applications with long-distance requirements.

(2) Broken PT-symmetric state.

In this case, the constraint conditions of k_{23} are $k_{23} < k_{CL}$ and $k_{23} > k_{CH}$ when $k_0 \neq k_{23}$, and if $k_0 = k_{23}$, the restricted condition of k_{23} will degenerate to $k_{23} < (R_4 + R_L)/(\omega_0 L_4)$. The following analysis mainly takes into account $k_0 = k_{23}$.

The gain rate g and operating frequency ω can be obtained by (8.23) as

$$\begin{cases} g = \tau_1 + \dfrac{\kappa_{23}^2}{\tau_4} \\ \omega = \omega_0 \pm \sqrt{\kappa_0^2 + \kappa_{23}^2} \end{cases}, \tag{8.27}$$

where the balance between gain rate and loss rate is broken.

(3) Resonant sate.

According to (8.23), it can be seen that $\omega = \omega_0$ is a solution for the operating frequency, and the gain rate can be derived as

$$g = \tau_1 + \frac{\kappa_0^4}{\kappa_{23}^2 \tau_4}, \tag{8.28}$$

where the balance between the gain rate and loss rate is broken, although the operating frequency is located at $\omega = \omega_0$, which is different from the delay-coil system.

Let $k_0 = 0.1$, Fig. 8.7 shows the operating frequency of the PT-symmetric MC-WPT system with double relay coils, it can be seen that there are three operating frequencies for each coupling coefficient in the exact PT-symmetric region and broken PT-symmetric region.

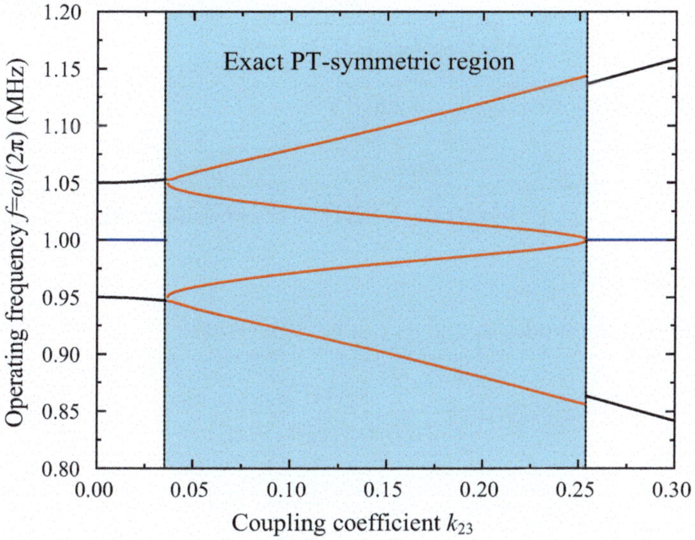

Fig. 8.7 The operating frequency characteristic

8.3.2 Output Power

Substituting the solutions of operating frequency into (8.21), and combining the definition of output power that $P_O = 2\tau_L|\boldsymbol{a}_4|^2$, the output power can be obtained as follows:

(1) In exact PT-symmetric state.

The amplitude of the modes $\boldsymbol{a}_1$ and $\boldsymbol{a}_4$ can be derived as

$$|\boldsymbol{a}_1| = |\boldsymbol{a}_4| = \frac{U_{in}}{2\sqrt{L_1}(\tau_{10} + \tau_4)}. \tag{8.29}$$

Then, the output power can be obtained as

$$P_O = 2\tau_L|\boldsymbol{a}_4|^2 = \frac{\tau_L U_{in}^2}{2L_1(\tau_{10} + \tau_4)^2} = \frac{\tau_L U_{DC}^2}{\pi^2 L_1(\tau_{10} + \tau_4)^2}. \tag{8.30}$$

(2) In broken PT-symmetric state.

The amplitude of the modes $\boldsymbol{a}_1$ and $\boldsymbol{a}_4$ can be deduced as

$$|\boldsymbol{a}_1| = \frac{\tau_4 U_{in}}{2\sqrt{L_1}\left(\tau_{10}\tau_4 + \kappa_{23}^2\right)}, \tag{8.31}$$

$$|\boldsymbol{a}_4| = \frac{\kappa_{23} U_{\text{in}}}{2\sqrt{L_1}\left(\tau_{10}\tau_4 + \kappa_{23}^2\right)}. \tag{8.32}$$

Then, the output power can be obtained as

$$P_{\text{O}} = 2\tau_{\text{L}}|\boldsymbol{a}_4|^2 = \frac{\tau_{\text{L}}\kappa_{23}^2 U_{\text{in}}^2}{2L_1\left(\tau_{10}\tau_4 + \kappa_{23}^2\right)^2} = \frac{\tau_{\text{L}}\kappa_{23}^2 U_{\text{DC}}^2}{\pi^2 L_1\left(\tau_{10}\tau_4 + \kappa_{23}^2\right)^2} \tag{8.33}$$

(3) In resonant state.

The amplitude of the modes $\boldsymbol{a}_1$ and $\boldsymbol{a}_4$ can be deduced as

$$|\boldsymbol{a}_1| = \frac{\kappa_{23}^2 \tau_4 U_{\text{in}}}{2\sqrt{L_1}\left(\kappa_{23}^2\tau_{10}\tau_4 + \kappa_0^4\right)}, \tag{8.34}$$

$$|\boldsymbol{a}_4| = \frac{\kappa_{23}\kappa_0^2 U_{\text{in}}}{2\sqrt{L_1}\left(\kappa_{23}^2\tau_{10}\tau_4 + \kappa_0^4\right)}. \tag{8.35}$$

Then, the output power can be obtained as

$$P_{\text{O}} = 2\tau_{\text{L}}|\boldsymbol{a}_4|^2 = \frac{\tau_{\text{L}}\kappa_{23}^2\kappa_0^4 U_{\text{in}}^2}{2L_1\left(\kappa_{23}^2\tau_{10}\tau_4 + \kappa_0^4\right)^2} = \frac{\tau_{\text{L}}\kappa_{23}^2\kappa_0^4 U_{\text{DC}}^2}{\pi^2 L_1\left(\kappa_{23}^2\tau_{10}\tau_4 + \kappa_0^4\right)^2}. \tag{8.36}$$

From the above analysis, it can be known that the output power in the exact PT-symmetric state is independent of the coupling coefficient, meaning that the output power remains constant against the variations of the coupling coefficient, as shown in Fig. 8.8.

8.3.3 *Transfer Efficiency*

According to the above analysis, the transfer efficiency can also be divided into three states, as follows:

(1) In exact PT-symmetric state.

The transfer efficiency is

$$\eta = \frac{2\tau_{\text{L}}|\boldsymbol{a}_4|^2}{2\tau_{10}|\boldsymbol{a}_1|^2 + 2\tau_4|\boldsymbol{a}_4|^2} = \frac{\tau_{\text{L}}}{\tau_{10} + \tau_4}. \tag{8.37}$$

(2) In broken PT-symmetric state.

The transfer efficiency is

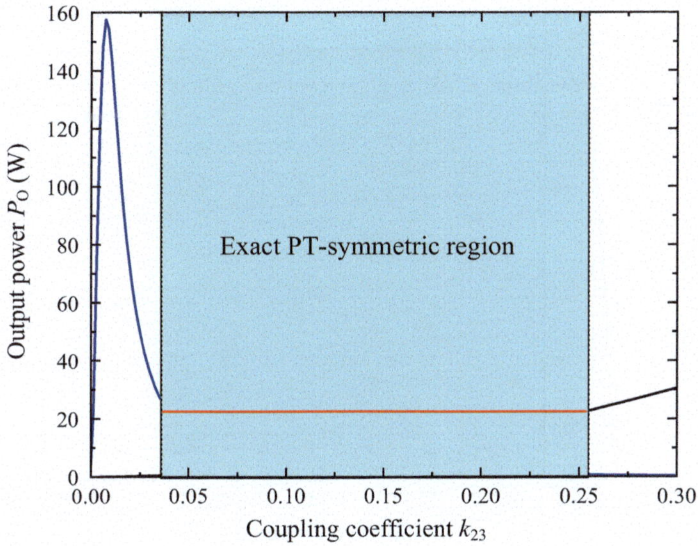

Fig. 8.8 The output power characteristic

$$\eta = \frac{2\tau_L|\boldsymbol{a}_4|^2}{2\tau_{10}|\boldsymbol{a}_1|^2 + 2\tau_4|\boldsymbol{a}_4|^2} = \frac{\kappa_{23}^2\tau_L}{\left(\tau_{10}\tau_4 + \kappa_{23}^2\right)\tau_4}. \tag{8.38}$$

(3) In resonant state.

The transfer efficiency is

$$\eta = \frac{2\tau_L|\boldsymbol{a}_4|^2}{2\tau_{10}|\boldsymbol{a}_1|^2 + 2\tau_4|\boldsymbol{a}_4|^2} = \frac{\kappa_0^4\tau_L}{\left(\tau_{10}\kappa_{23}^2\tau_4 + \kappa_0^4\right)\tau_4}. \tag{8.39}$$

As can be seen from (8.37), and (8.39), only the transfer efficiency in exact PT-symmetric state is constant with the variation of the coupling coefficient. Figure 8.9 gives the transfer efficiency versus the coupling coefficient.

From the above analysis, it can be seen that the PT-symmetric MC-WPT system with double relay coils has three working states, and the system can automatically select a certain working state for stable operation. Only in the exact PT-symmetric state, the output power and transfer efficiency are constant, which are independent of the coupling coefficient.

The same analysis can be extended to the system with repeater numbers of 6, 8, 10, ...,. In general, in systems with an even number of repeaters, it is not desirable to operate in the broken PT-symmetric state and resonant state due to the variable output power and transfer efficiency related to the coupling coefficient.

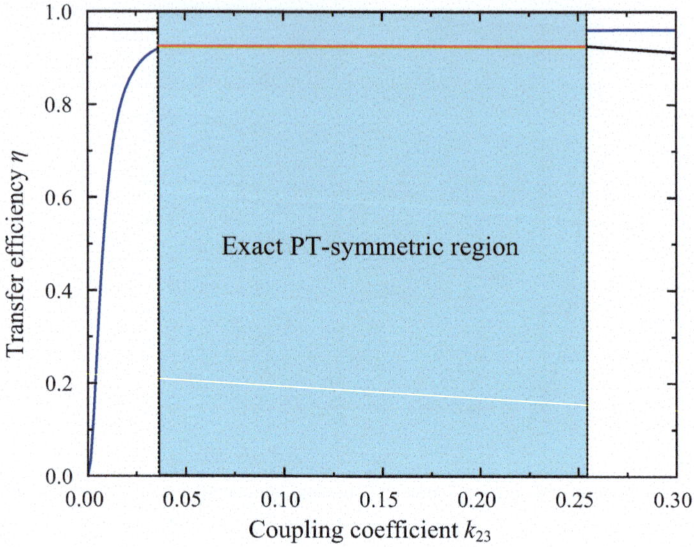

Fig. 8.9 The transfer efficiency characteristic

8.4 System Design and Verification

Figure 8.10 shows the experimental prototype of the PT-symmetric MC-WPT system with relay coils. The size of each coil is 320 mm × 320 mm, and the coils are fabricated by using Φ0.05 mm*800 silk-covered wire with a turn spacing of 1.4 mm. All circuit parameters of the prototype are listed in Table 8.1, which are obtained by the precise impedance analyzer (Wayne Kerr 6500B). In addition, the implementation of the negative resistance is described in the Sect. 3.4.1, the difference is the full-bridge inverter used here, so the fundamental component of the output voltage can be derived as $U_{in} = 2\sqrt{2}\,U_{dc}/\pi$, which is twice as much as the half-bridge inverter.

Based on the theoretical analysis in Sects. 8.2 and 8.3, for the one-repeater system, the critical coupling coefficient k_C can be calculated as 0.0748. Similarly, for the two-repeater system, the coupling coefficient k_1 has to satisfy $k_1 \geq 0.0529$. Here, considering that the exact PT-symmetric region ($k_{CL} \leq k_2 \leq k_{CH}$) is determined by k_1, we assume that k_1 is 0.12, then the critical coupling coefficients k_{CL} and k_{CH} can be calculated as 0.0777 and 0.1362, respectively. Therefore, the effective range that the system maintains constant output power and constant transfer efficiency is limited to $0.0777 \leq k_2 \leq 0.1362$. Figures 8.11 and 8.12 show the transfer performance of the one-relay-coil and double-relay-coil prototypes, respectively.

As can be seen from Fig. 8.11, the output power and transfer efficiency of the one-repeater system remain constant within a range of about 420 mm. In this region, the output power of the one-repeater prototype is nearly stable at 15 W, and the transfer efficiency is constant near 91%. When the transfer distance exceeds 420 mm, the output power and transfer efficiency of the prototype are dependent on the variation

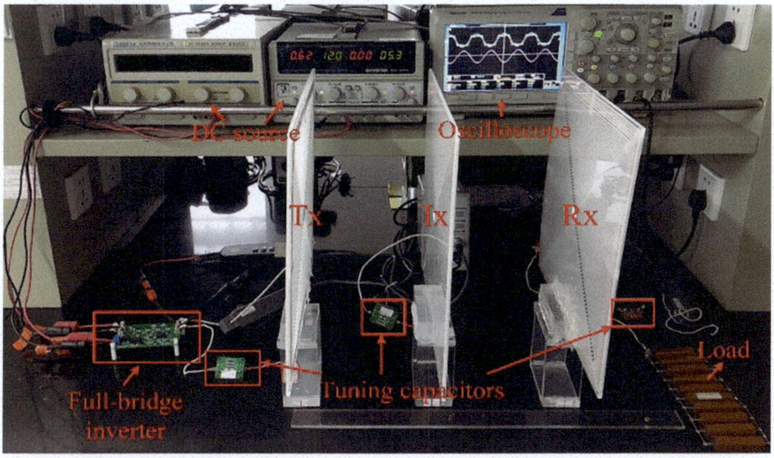

Fig. 8.10 Experimental prototype of the PT-symmetric MC-WPT system with a repeater

of the transfer distance and do not remain constant, which is not desirable. Besides, the operating frequency of the one-repeater system has three real solutions in the exact PT-symmetric region and broken PT-symmetric region. Here, in the range of $d_{13} >$ 420 mm, the one-repeater prototype automatically works in the exact PT-symmetric state, but for the range of $d_{13} <$ 420 mm, the prototype operates in the broken PT-symmetric state, the measured operating frequencies are in good agreement with the calculated results. To sum up, for the one-repeater prototype, the output power and transfer efficiency always remain constant against the variation of the coupling coefficient in the exact PT-symmetric region, no matter whether the system works in the exact PT-symmetric state or the resonant state.

Similarly, as illustrated in Fig. 8.12, the two-repeater prototype achieves a stable power transfer of about 15 W with a constant transfer efficiency of approximately 89% within a range of about 420–500 mm air gap. In this region, the operating frequency of the system has five real solutions, and the system works on a certain operating frequency branch. While in the broken PT-symmetric region ($d_{14} >$ 500 mm and $d_{14} <$ 420 mm), the prototype works in the broken PT-symmetric state or the resonant state, in which both output power and transfer efficiency vary with different transfer distances, which is not desirable. Therefore, for the two-repeater system, it is desired that the system operates in the PT-symmetric region, in which the output power and transfer efficiency can maintain constant against the variation of the transfer distance.

Fig. 8.11 Transfer performance of the one-repeater prototype for the coaxial aligned case. **a** The operating frequency $\omega/(2\pi)$ versus the transfer distance d_{13}. **b** The output power P_o versus the transfer distance d_{13}. **c** The transfer efficiency η versus the transfer distance d_{13}

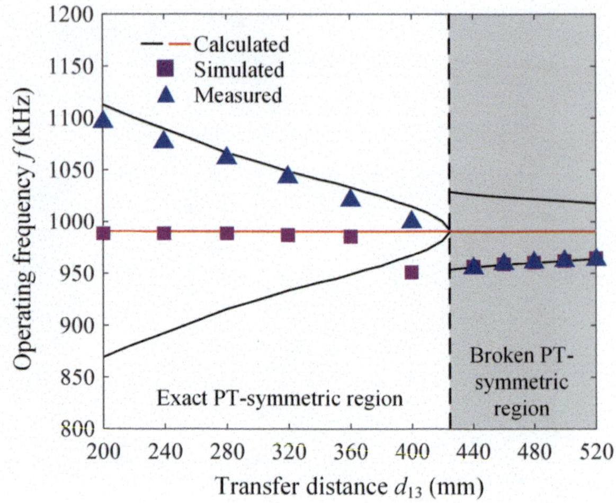

(a)

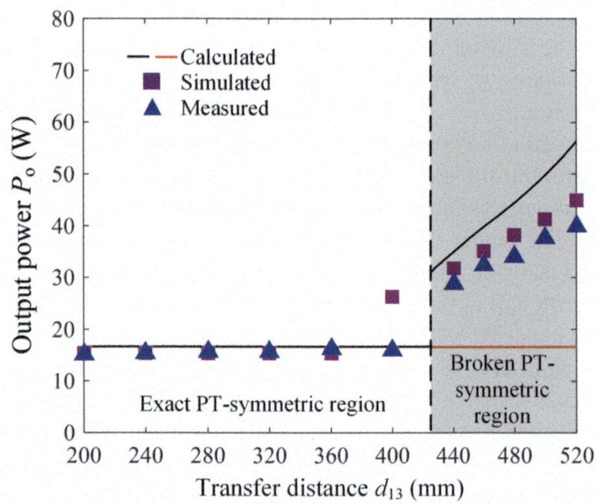

(b)

8.5 Summary

This chapter focuses on PT-symmetric MC-WPT systems with relay coils, analyzes the operating frequency, output power, and transfer efficiency characteristics of PT-symmetric MC-WPT systems with single relay coil and double relay coils successively, and summarizes the operating frequency and fundamental characteristics of PT-symmetric multi-relay-coil MC-WPT system from two aspects of even and odd number of relay coils. It can be found that the critical coupling coefficient of the PT-symmetric MC-WPT system with any number of relay coils is reduced, which

Fig. 8.11 (continued)

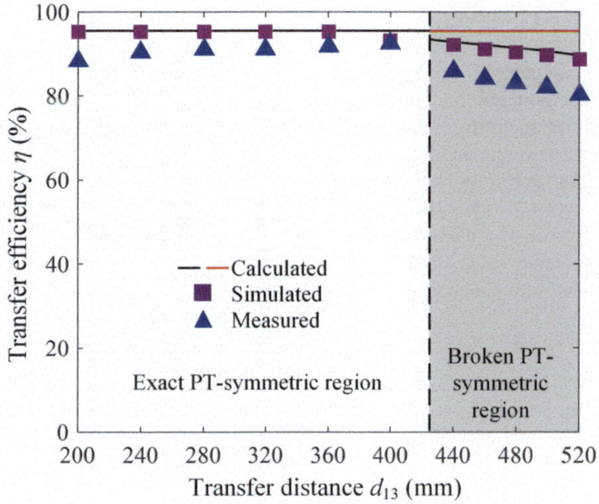

(c)

means that the maximum transfer distance for stable and efficient transfer power is improved. For the even number of relay coils, the output power and transfer efficiency can remain constant against the variation of the coupling coefficient in the exact PT-symmetric state, which indicates that the system is robust against the misalignments and variations in the transfer distance. For the odd number of relay coils, there is also a resonant state, at which the operating frequency of the system is fixed, and the output power and transfer efficiency can be kept constant in the near-field range. Therefore, the multi-relay-coil MC-WPT system with PT symmetry can achieve stable output power and constant transfer efficiency that are independent of transfer distance variation in the exact PT-symmetric region.

Fig. 8.12 Transfer performance of the two-repeater prototype for the coaxial aligned case. **a** The operating frequency $\omega/(2\pi)$ versus the transfer distance d_{13}. **b** The output power P_o versus the transfer distance d_{13}. **c** The transfer efficiency η versus the transfer distance d_{13}

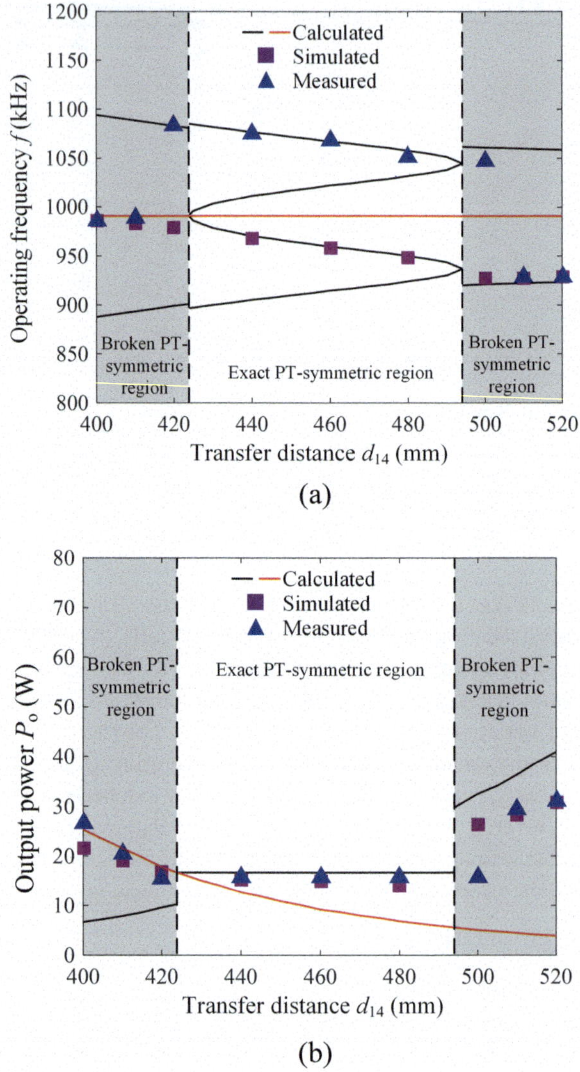

References

1. Tran DH, Vu VB, Choi W (2018) Design of a high-efficiency wireless power transfer system with intermediate coils for the on-board chargers of electric vehicles. IEEE Trans Power Electron 33(1):175–187
2. Wen F, Chu X, Li Q, Li R, Liu L, Jing F (2020) Optimization on three-coil long-range and dimension-asymmetric wireless power transfer system. IEEE Trans Electromagn Compat 62(5):1859–1868
3. Liu X, Wang G (2016) A novel wireless power transfer system with double intermediate resonant coils. IEEE Trans Industr Electron 63(4):2174–2180

Fig. 8.12 (continued)

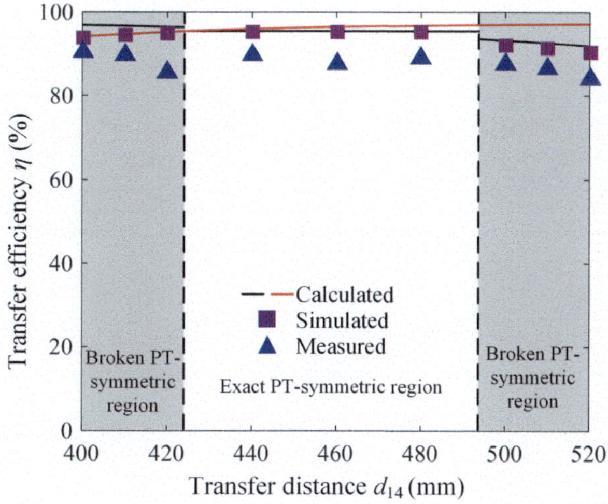

(c)

4. Wang Z, Qiu D, Zhang B, Xiao W, Chen Y, Xie F (2021) Extended-range wireless power transfer system based on high-order pt symmetric principle. In: 2021 IEEE 1st international power electronics and application symposium (PEAS), pp 1–6
5. Guo Z et al (2023) Rotation manipulation of high-order PT-symmetry for robust wireless power transfer. Fundamental Res
6. Sakhdari M, Hajizadegan M, Chen P-Y (2020) Robust extended-range wireless power transfer using a higher-order PT-symmetric platform. Phys Rev Res 2(1):013152
7. Shu X, Zhang B, Wei Z, Rong C, Sun S (2021) Extended-distance wireless power transfer system with constant output power and transfer efficiency based on parity-time-symmetric principle. IEEE Trans Power Electron 36(8):8861–8871
8. Haus HAJP-H (1984) Waves and fields in optoelectronics. Prentice Hall
9. L. W. H, Coupled mode and parametric electronics. Wiley, New York (1960)
10. Kim J, Son HC, Kim KH, Park YJ (2011) Efficiency analysis of magnetic resonance wireless power transfer with intermediate resonant coil. IEEE Antennas Wirel Propag Lett 10:389–392

Chapter 9
Techniques for Improving Performances of PT-Symmetric Magnetic-Field Wireless Power Transfer System

Although the MC-WPT system based on PT symmetry has been proved to be a promising WPT system due to its characteristics of constant output power, constant transfer efficiency, constant voltage, and constant current, the system performances are limited by PT-symmetric range, switching frequency, switching loss, single-phase power supply, and so on. Therefore, it is significantly important to explore techniques to improve the performance of the PT-symmetric MC-WPT system. Firstly, aiming at the limited PT-symmetric region, which is the operating range of constant and efficient transfer, this chapter analyzes the main factors affecting the PT-symmetric region, then introduces several methods to improve the transfer distance of the stable and efficient operation of the system, and compares them with the traditional PT-symmetric MC-WPT system introduced in Chap. 3. Then, focusing on the whole efficiency of the PT-symmetric MC-WPT system with a negative resistor constructed by a power electronic converter, the methods of reducing the switching frequency and switching losses of the negative resistor are introduced. Meanwhile, based on the universality of the three-phase power supply, a three-phase PT-symmetric MC-WPT system is built, and the transfer characteristics of the three-phase system are analyzed. Finally, a quasi-PT-symmetric MC-WPT system based on gain–loss modulation is introduced, which can enhance either the maximum transfer efficiency or distance alternatively.

9.1 Improvement of Transfer Distance

9.1.1 Analysis of Influencing Factors

According to the analysis in Chap. 3, when the coupling coefficient is greater than the critical value, the output power and transfer efficiency of the PT-symmetric MC-WPT system are independent of the coupling coefficient, which means that the system

B. Zhang et al., *Parity-Time Symmetric Wireless Power Transfer*, CPSS Power Electronics Series, https://doi.org/10.1007/978-981-96-3149-0_9

has strong robustness. Therefore, it is necessary to analyze the methods to reduce the critical coupling coefficient, so as to increase the transfer distance of the PT-symmetric MC-WPT system.

Assuming that the inductance, resonant capacitance, and internal resistance of the transmitting coil are L_t, C_t, and R_t, the inductance, resonant capacitance, and internal resistance of the receiving coil are L_r, C_r, and R_r, negative resistance and load resistance are $-R_N$ and R_L, and the natural resonant angular frequencies of transmitting and receiving coils are defined as ω_t and ω_r, respectively, then the transfer performance of PT-symmetric MC-WPT system with SS, PP, SP, and PS structures can be listed in Table 9.1 according to the coupled-mode model.

As can be observed from Table 9.1, when the circuit structure of the receiver is the same, the expression of the critical coupling coefficient of the PT-symmetric MC-WPT system is the same, that is, the PT-symmetric regions of SS and PS structures are identical, and PT-symmetric regions of PS and PP structures are identical. Besides, the critical coupling coefficient of the PT-symmetric MC-WPT system is related to the inductance, internal resistance, compensation capacitance of the receiving coil, and load resistance, that is, the critical coupling coefficient of the PT-symmetric MC-WPT system is only related to the circuit parameters of the receiver.

Figure 9.1 shows the relationship curves between the critical coupling coefficient of different structures and the inductance of the receiving coil, it can be found that the critical coupling coefficient of SS and PS structures decreases with the increase of receiving-coil inductance, which indicates that the critical transfer distance of SS and PS structures can be increased by increasing the receiving-coil inductance. However, for SP and PP structures, the critical coupling coefficient first decreases sharply and then increases slowly with the increase of receiving-coil inductance, and there is a minimum point, which can be obtained by

$$\frac{dk_C}{dL_r} = -\frac{\sqrt{C_r}R_r}{\sqrt{L_r^3}} + \frac{1}{\sqrt{L_r C_r} R_L}. \tag{9.1}$$

Let $dk_C/dL_r = 0$, the extreme point of the critical coupling coefficient can be obtained. When $L_r = C_r R_L R_r$, the minimum critical coupling coefficient of SP and

Table 9.1 Comparison of different PT-symmetric structures

Structure	PT-symmetric condition	PT-symmetric region	Transfer efficiency
SS	$\omega_t = \omega_r$, $\frac{R_N - R_t}{L_t} = \frac{R_r + R_L}{L_r}$	$k \geq k_C = \frac{\sqrt{C_r}}{\sqrt{L_r}}(R_r + R_L)$	$\eta = \frac{R_L L_t}{L_r R_t + (R_r + R_L) L_t}$
SP	$\omega_t = \omega_r$, $\frac{R_N - R_t}{L_t} = \frac{R_r}{L_r} + \frac{1}{R_L C_r}$	$k \geq k_C = \frac{\sqrt{C_r} R_r}{\sqrt{L_r}} + \frac{\sqrt{L_r}}{\sqrt{C_r} R_L}$	$\eta = \frac{1}{\frac{R_t R_L}{\omega_t^2 L_t L_r} + \frac{R_r R_L}{\omega_t^2 L_r^2} + 1}$
PS	$\omega_t = \omega_r$, $\frac{1}{R_N C_t} - \frac{R_t}{L_t} = \frac{R_r + R_L}{L_r}$	$k \geq k_C = \frac{\sqrt{C_r}}{\sqrt{L_r}}(R_r + R_L)$	$\eta = \frac{R_L L_t}{L_r R_t + (R_r + R_L) L_t}$
PP	$\omega_t = \omega_r$, $\frac{1}{R_N C_t} - \frac{R_t}{L_t} = \frac{R_r}{L_r} + \frac{1}{R_L C_r}$	$k \geq k_C = \frac{\sqrt{C_r} R_r}{\sqrt{L_r}} + \frac{\sqrt{L_r}}{\sqrt{C_r} R_L}$	$\eta = \frac{1}{\frac{R_t R_L}{\omega_t^2 L_t L_r} + \frac{R_r R_L}{\omega_t^2 L_r^2} + 1}$

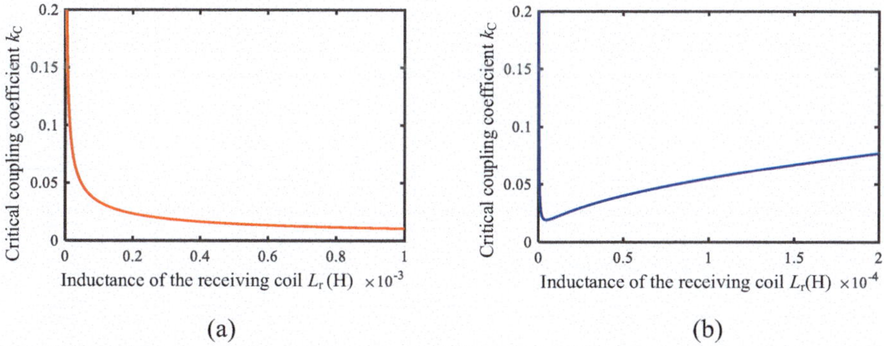

Fig. 9.1 The relationship between the critical coupling coefficient and the inductance of receiving coil. **a** SS and PS structures. **b** SP and PP structures

PP structures can be obtained as $k_{\mathrm{Cmin}} = 2\sqrt{R_r/R_L}$. Therefore, in practical application, the farthest PT-symmetric wireless power transfer distance can be obtained by making the receiving-coil inductance equal to $C_rR_LR_r$. Compared with SS and PS PT-symmetric MC-WPT systems, the smaller critical coupling coefficient can be obtained because of the smaller receiving-coil inductance of SP and PP structures, which means that the SP and PP structures are more suitable for applications requiring compact and lightweight receivers, such as wireless charging for mobile phones and implantable medical devices.

Similarly, the relationship curves between the critical coupling coefficient of different structures and the compensated capacitance of the receiver can be described in Fig. 9.2, it can be found that the critical coupling coefficients of SS and PS structures decrease with the decrease of the compensation capacitance on the receiver, that is, the PT-symmetric wireless power transfer distance can be increased by reducing the compensation capacitance on the receiver. However, when the inductance of the receiving coil is determined, the natural resonance frequency increases with the decrease of the compensation capacitance on the receiver, resulting in the increase of the operating frequency $\omega = \omega_t \pm \sqrt{(\omega_0\kappa)^2/2\text{-}[(R_r + R_L)/(2L_r)]^2}$. Consequently, the current skin effect and switching loss caused by high frequency limit the method of increasing the transfer distance of SS and PS structures by reducing the compensation capacitance on the receiver. In addition, for SP and PP structures, with the increase of the compensated capacitance of the receiver, the critical coupling coefficient decreases first and then increases, and there is a minimum point $k_{\mathrm{Cmin}} = 2\sqrt{R_r/R_L}$ when $C_r = L_r/(R_LR_r)$, which demonstrates that the designed system has a long critical transfer distance when the compensation capacitance of the receiver is equal to or close to $L_r/(R_LR_r)$ in practical application.

Figure 9.3 gives the relationship between the critical coupling coefficient of different structures and load resistance, it can be seen that the critical coupling coefficient of SS and PS structures increases with the increase of the load resistance, but the transfer efficiency of the system will be low when the load resistance is small.

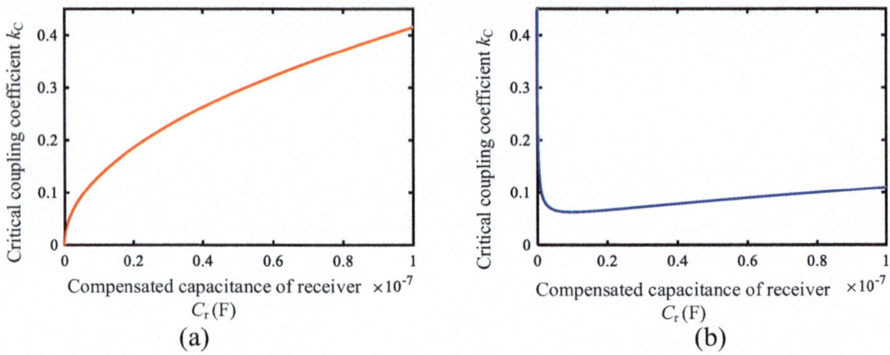

Fig. 9.2 The relationship between the critical coupling coefficient and the compensated capacitance of receiver. **a** SS and PS structures. **b** SP and PP structures

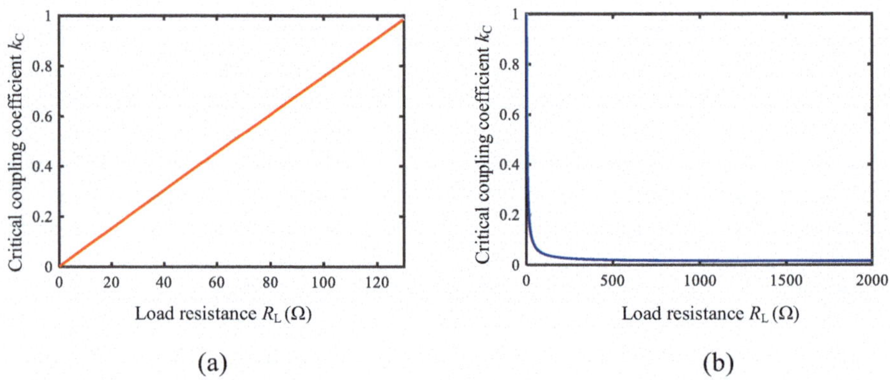

Fig. 9.3 The relationship between the critical coupling coefficient and load resistance. **a** SS and PS structures. **b** SP and PP structures

Therefore, the method of increasing the transfer distance of SS and PS structures by reducing the load resistance has certain limitations. Meanwhile, for SP and PP structures, the critical coupling coefficient decreases with the increase of load resistance, but when the load resistance is large, the efficiency of SP and PP systems will be low, so the method of increasing the load resistance to improve the critical transfer distance of SP and PP structures will be limited in practical application. In other words, SS and PS systems are suitable for occasions where the load resistance is small, and SP and PP systems are suitable for occasions where the load resistance is large.

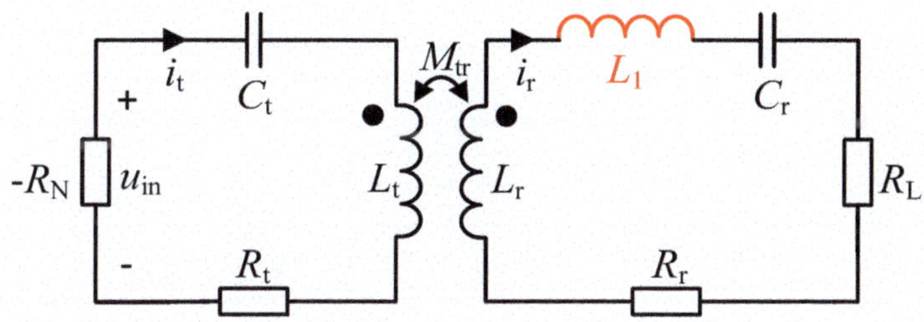

Fig. 9.4 The schematic diagram of the PT-symmetric MC-WPT system with an inductor added on the receiver

9.1.2 Adding Inductive Element L on the Receiver

The schematic diagram of the PT-symmetric MC-WPT system with an inductor added on the receiver is shown in Fig. 9.4, in which $-R_N$ is a negative resistance and is used as a power source to provide energy to the whole system, L_t, C_t, and R_t are the coil inductance, compensation capacitance, and internal resistance of the transmitter, L_r, C_r, and R_r are the coil inductance, compensation capacitance, and internal resistance of the receiver, respectively. It is noted that R_r represents the sum of the internal resistance of the receiving coil and the internal resistance of the additional inductor. R_L is the load resistance, M_{tr} is the mutual inductance between coils, and L_1 is the inductance of the inductor added on the receiver [1].

The state equation of the PT-symmetric MC-WPT system shown in Fig. 9.4 can be derived as

$$\begin{cases} u_{in} = L_t \dfrac{di_t}{dt} + M_{tr} \dfrac{di_r}{dt} + u_{ct} + i_t R_t \\ 0 = M_{tr} \dfrac{di_t}{dt} + (L_r + L_1) \dfrac{di_r}{dt} + u_{cr} + i_r (R_r + R_L) \\ i_t = C_t \dfrac{du_{ct}}{dt} \\ i_r = C_r \dfrac{du_{cr}}{dt} \end{cases}, \quad (9.2)$$

where $u_{in} = \text{sgn}(i_t) U_{DC}$ is the AC input voltage, U_{DC} is the DC input voltage of the full-bridge inverter described in Chap. 4. i_t and i_r are currents through transmitting and receiving coils, respectively. u_{ct} and u_{cr} are voltages across the compensation capacitors of the transmitter and receiver, respectively.

The coupled modes a_t and a_r can be represented as [2]

$$\begin{cases} \boldsymbol{a}_t = A_t e^{j(\omega t+\theta_t)} = \sqrt{\frac{L_t}{2}} i_t + j\sqrt{\frac{C_t}{2}} u_{ct} \\ \boldsymbol{a}_r = A_r e^{j(\omega t+\theta_r)} = \sqrt{\frac{L_r+L_1}{2}} i_r + j\sqrt{\frac{C_r}{2}} u_{cr} \end{cases}, \tag{9.3}$$

where ω is the operating angular frequency, A_t and A_r are the amplitudes of the coupled modes, θ_t and θ_r are the phases of the coupled modes.

By representing i_t, i_r, u_{ct}, and u_{cr} with A_t, A_r, θ_t, and θ_r, there is

$$\begin{cases} u_{ct} = A_t\sqrt{\frac{2}{C_t}} \sin(\omega t + \theta_t) \\ i_t = A_t\sqrt{\frac{2}{L_t}} \cos(\omega t + \theta_t) \\ u_{cr} = A_r\sqrt{\frac{2}{C_r}} \sin(\omega t + \theta_r) \\ i_r = A_r\sqrt{\frac{2}{L_r+L_1}} \cos(\omega t + \theta_r) \end{cases}. \tag{9.4}$$

Referring to the derivation of the coupled-mode equation in Chap. 3, the dynamic coupled-mode equation of the system can be obtained as follows:

$$\frac{d}{dt}\begin{bmatrix} \boldsymbol{a}_t \\ \boldsymbol{a}_r \end{bmatrix} = \begin{bmatrix} j\omega_t + \frac{1}{A_t}\frac{2U_{DC}}{\pi\sqrt{2L_t}} - \tau_{tt} & -j\frac{\omega_r}{2}k' \\ -j\frac{\omega_t}{2}k' & j\omega_r - (\tau_{rr}+\tau_L) \end{bmatrix}\begin{bmatrix} \boldsymbol{a}_t \\ \boldsymbol{a}_r \end{bmatrix}, \tag{9.5}$$

where $\omega_t = 1/\sqrt{L_t C_t}$ and $\omega_r = 1/\sqrt{L_r C_r}$ represent the natural resonant angular frequencies of the transmitter and receiver respectively. $k' = M_{tr}/\sqrt{L_t(L_r + L_1)}$ is a variable related to mutual inductance. In addition, the coupling coefficient of the system is $k = M_{tr}/\sqrt{L_t L_r}$. $\tau_{tt} = R_t/(2L_t)$ and $\tau_{rr} = R_r/(2L_r)$ are the inherent loss rates of the transmitter and receiver, respectively. $\tau_L = R_L/[2(L_t + L_r)]$ is the load loss rates, and $\tau_r = \tau_{rr} + \tau_L = (R_r + R_L)/[2(L_t + L_r)]$ is the total loss rate of the receiver. Compared with the SS structure, the total loss rate of the receiving circuit with additional inductive elements is smaller.

Then, considering that $\omega_t = \omega_r$, the characteristic equation of the system can be deduced as

$$\begin{vmatrix} j(\omega - \omega_t) - g_t & j\kappa \\ j\kappa & j(\omega - \omega_r)+\tau_r \end{vmatrix} = 0, \tag{9.6}$$

where $g_t = 2U_{DC}/(\pi A_t \sqrt{2L_t})$-$\tau_{tt}$ is defined as the total gain rate of the transmitter, $\kappa = \omega_t k'/2$ is defined as the coupling rate of the system.

Thus, the operating angular frequency can be derived as

$$\omega = \omega_t + \chi = \omega_t + j\frac{g_t - \tau_r}{2} \pm \sqrt{\kappa^2 - \left(\frac{g_t + \tau_r}{2}\right)^2}, \tag{9.7}$$

where χ is the real number.

As can be seen from (9.7), when $\kappa^2 = g_t\tau_r$, the operating angular frequency is $\omega = \omega_t$, and the frequency solution is unstable in the exact PT-symmetric region [3]. When $g_t = \tau_r$ and $\kappa > \tau_r$, the operating angular frequency is $\omega = \omega_t \pm \sqrt{\kappa^2 + \tau_r^2}$, and the system works in the exact PT-symmetric region. Then, the critical coupling rate can be expressed as $\kappa_C = \tau_r$, and the corresponding critical coefficient is

$$k_C = \frac{R_r + R_L}{\omega_t\sqrt{L_r(L_r + L_1)}}. \tag{9.8}$$

From (9.8), it can be seen that the critical coupling coefficient can be reduced by adding an inductor to the receiving circuit, and thus the corresponding critical transfer distance can be increased.

According to (9.5), the amplitudes of the coupled modes can be derived as

$$|\boldsymbol{a}_t| = |\boldsymbol{a}_r| = \frac{2U_{DC}}{\pi\sqrt{2L_t}}\frac{1}{\tau_{tt} + \tau_r}. \tag{9.9}$$

Then the output power and transfer efficiency can be obtained as

$$P_o = 2\tau_L|\boldsymbol{a}_r|^2 = \frac{4\tau_L U_{DC}^2}{\pi^2 L_t(\tau_{tt} + \tau_r)^2}, \tag{9.10}$$

$$\eta = \frac{2\tau_L|\boldsymbol{a}_r|^2}{2\tau_{tt}|\boldsymbol{a}_t|^2 + 2\tau_r|\boldsymbol{a}_r|^2} = \frac{R_L L_t}{(L_r{+}L_1)R_t + (R_r{+}R_L)L_t}, \tag{9.11}$$

From (9.10) and (9.11), it can be seen that the output power and transfer efficiency of the system are independent of the coupling coefficient, i.e., in the PT-symmetric region, the output power and transfer efficiency of the system remain unchanged when the transfer distance between the transmitting and receiving coils changes. Figure 9.5 depicts the transfer characteristics of the PT-symmetric MC-WPT system with an additional inductor, including operating frequency, output power, and transfer efficiency, it can be seen that when the coupling coefficient is greater than the critical coupling coefficient, the system works in the PT-symmetric region (i.e., strong coupling region), the operating frequency has two solutions, the output power and transfer efficiency remain constant with the change of the coupling coefficient. When the coupling coefficient is less than the critical coupling coefficient, the system works in the weak coupling region, (i.e., broken PT-symmetric region), the operating frequency has only one solution, the output power and transfer efficiency change with the change of the coupling coefficient.

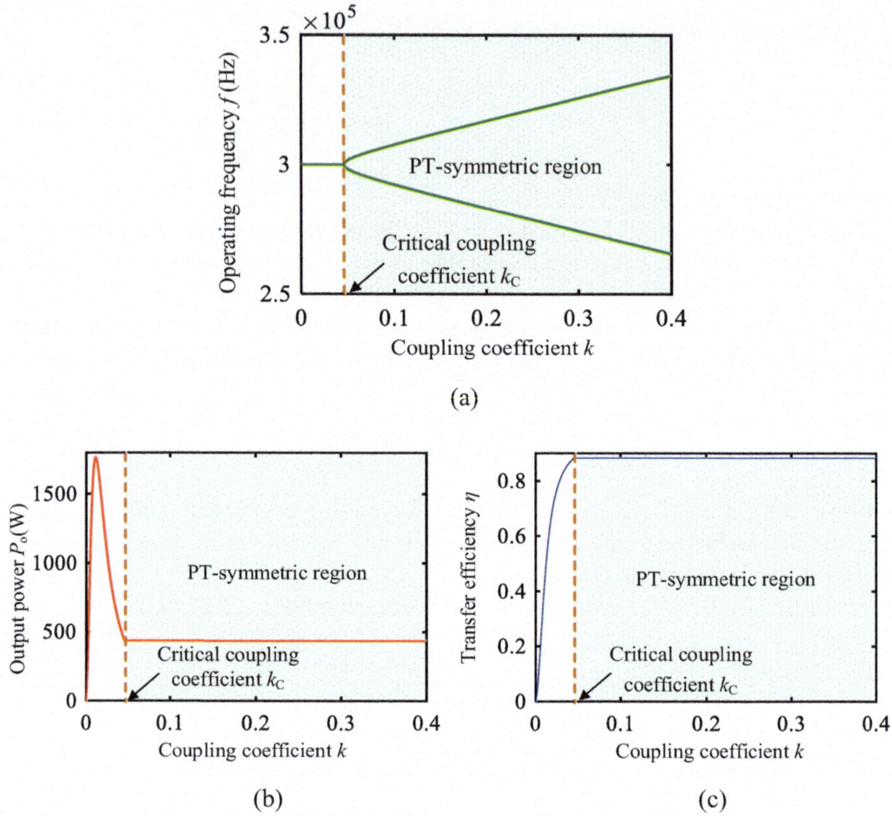

Fig. 9.5 Transfer characteristics of additional inductive elements on the receiving circuit. **a** Operating frequency. **b** Output power. **c** Transfer efficiency

9.1.3 Adding Capacitive Element C on the Receiver

The schematic diagram of the PT-symmetric MC-WPT system with a capacitor added on the receiver is shown in Fig. 9.6, in which $-R_N$ is a negative resistance and is used as a power source to provide energy to the whole system, L_t, C_t and R_t are the coil inductance, compensation capacitance and internal resistance of the transmitter, L_r, C_r and R_r are the coil inductance, compensation capacitance and internal resistance of the receiver, respectively. R_L is the load resistance, M_{tr} is the mutual inductance between coils, and C_1 is the capacitance of the capacitor added on the receiver [4].

According to the Kirchhoff's voltage and current laws, the circuit equation of the system shown in Fig. 9.6 can be written as

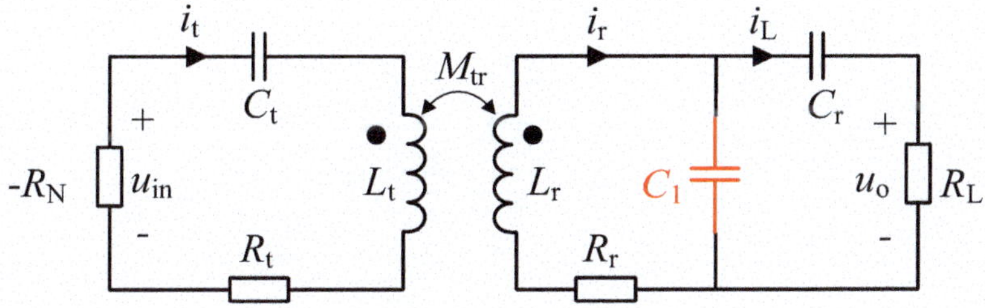

Fig. 9.6 The schematic diagram of the PT-symmetric MC-WPT system with a capacitor added on the receiver

$$\begin{bmatrix} -R_N + R_t + \frac{1}{j\omega C_t} + j\omega L_t & j\omega M_{tr} & 0 \\ j\omega M_{tr} & j\omega L_r + R_r & \frac{1}{j\omega C_r} + R_L \\ 0 & -\frac{1}{j\omega C_1} & \frac{1}{j\omega C_1} + \frac{1}{j\omega C_r} + R_L \end{bmatrix} \begin{bmatrix} \dot{I}_t \\ \dot{I}_r \\ \dot{I}_L \end{bmatrix} = 0, \tag{9.12}$$

where ω is the operating angular frequency, $\dot{I}_t$ is the RMS value of the current flowing through the transmitting coil i_t, $\dot{I}_r$ is the RMS value of the current flowing through the receiving coil i_r, $\dot{I}_L$ is the RMS value of the current flowing through the load i_L.

Then the ratio of the output current to the input current can be derived as

$$\gamma_I = \frac{I_L}{I_t} = \frac{M_{tr}}{C_1\sqrt{\left(\frac{L_r}{C_1} + \frac{L_t}{C_r} + R_r R_L - \frac{1}{\omega^2 C_r C_1}\right)^2 + \left(\omega L_r R_L - \frac{R_r + R_L}{\omega C_1} - \frac{R_r}{\omega C_r}\right)^2}}, \tag{9.13}$$

And the input impedance can be expressed as

$$Z_{in} = R_t + j\omega L_t + \frac{1}{j\omega C_t} + \frac{\omega^2 M_{tr}^2}{R_r + j\omega L_r + \left(\frac{1}{j\omega C_r} + R_L // \frac{1}{j\omega C_1}\right)}. \tag{9.14}$$

Since the system is powered by the negative resistor, the imaginary part of the input impedance is zero, resulting in the following equation:

$$\omega L_t - \frac{1}{\omega C_t} - \frac{\omega^2 M_{tr}^2(\omega L_r - X_2)}{(R_r + X_1)^2 + (\omega L_r - X_2)^2} = 0, \tag{9.15}$$

where $X_1 = \frac{R_L C_r^2}{\omega^2 C_r^2 C_1^2 R_L^2 + (C_r + C_1)^2}$, $X_2 = \frac{C_r + C_1 + \omega^2 C_1 C_r^2 R_L^2}{\omega^3 C_r^2 C_1^2 R_L^2 + \omega (C_r + C_1)^2}$.

Thus, the expression of the negative resistance can be written as

$$-R_N = -R_t - \frac{(R_r + X_1)(\omega^2 L_t C_t - 1)}{\omega^2 L_r C_t - \omega C_t X_2}. \tag{9.16}$$

And the RMS values of the currents flowing through the transmitting and receiving coils are

$$I_t = \frac{U_{in}}{R_t + \frac{(R_r + X_1)(\omega^2 L_t C_t - 1)}{\omega^2 L_r C_t - \omega C_t X_2}}, \tag{9.17}$$

$$I_L = \gamma_1 I_t = \frac{\gamma_1 U_{in}}{R_t + \frac{(R_r + X_1)(\omega^2 L_t C_t - 1)}{\omega^2 L_r C_t - \omega C_t X_2}}, \tag{9.18}$$

where U_{in} is the RMS value of the voltage across the negative resistor.

The output voltage is

$$U_o = I_L R_L = \frac{\gamma_1 U_{in} R_L}{R_t + \frac{(R_r + X_1)(\omega^2 L_t C_t - 1)}{\omega^2 L_r C_t - \omega C_t X_2}}. \tag{9.19}$$

where U_o is the RMS value of the voltage across the load resistor.

The output power and transfer efficiency are

$$P_o = \frac{U_o^2}{R_L}, \tag{9.20}$$

$$\eta = \frac{P_o}{P_{in}} = \frac{U_o^2}{I_t^2 R_L \left[R_t + \frac{(R_r + X_1)(\omega^2 L_t C_t - 1)}{\omega^2 L_r C_t - \omega C_t X_2} \right]}. \tag{9.21}$$

By arranging (9.15), the frequency characteristic equation can be obtained, which is a one-variable equation of degree with respect to ω, and an analytical expression cannot be solved. Therefore, the equivalent impedance theory is used to transform the PS compensated capacitor network of the receiver into S compensated network [4], as shown in Fig. 9.7. The equivalent system parameters are as follows:

$$\begin{cases} C_R = \dfrac{(C_r + C_1)^2 + \omega^2 C_r^2 C_1^2 R_L^2}{(C_r + C_1) + \omega^2 C_r^2 C_1 R_L^2} \\ R_{Leq} = \dfrac{C_r^2 R_L}{(C_r + C_1)^2 + \omega^2 C_r^2 C_1^2 R_L^2} \end{cases}. \tag{9.22}$$

According to the parameters of the conventional MC-WPT system, the following simplification conditions can be obtained:

$$\begin{cases} \omega^2 C_r^2 C_1 R_L^2 << C_r + C_1 \\ \omega^2 C_r^2 C_1^2 R_L^2 << (C_r + C_1)^2 \end{cases}. \tag{9.23}$$

Then, $\omega^2 C_r C_1^2 R_L^2$ and $\omega^2 C_r^2 C_1^2 R_L^2$ can be ignored, and (9.22) can be simplified as

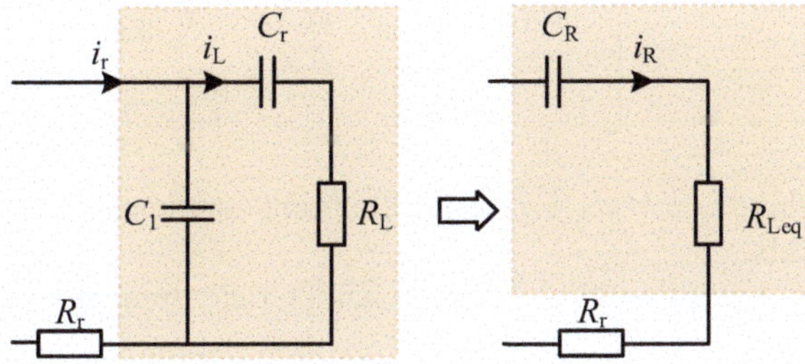

Fig. 9.7 Equivalent transformation of the receiver

$$\begin{cases} C_R = C_r + C_1 \\ R_{Leq} = \dfrac{C_r^2 R_L}{(C_r + C_1)^2} \end{cases}. \tag{9.24}$$

Defining $\alpha = C_r/C_R$ as the capacitance distribution ratio, there is $C_1 = (1–\alpha)C_R$, and the equivalent load resistance can be expressed as

$$R_{Leq} = \alpha^2 R_L. \tag{9.25}$$

Thus, the PT-symmetric MC-WPT system with S-PS compensation network shown in Fig. 9.6 can be equivalent to the PT-symmetric MC-WPT system with SS compensation network, as shown in Fig. 9.8.

The circuit equation of the equivalent circuit shown in Fig. 9.8 can be written as

$$\begin{bmatrix} -R_N + R_t + j\omega L_t - j\frac{1}{\omega C_t} & j\omega M_{tr} \\ j\omega M_{tr} & R_{Leq} + R_r + j\omega L_r - j\frac{1}{\omega C_R} \end{bmatrix} \begin{bmatrix} \dot{I}_t \\ \dot{I}_r \end{bmatrix} = 0. \tag{9.26}$$

Defining ω_t=$1/\sqrt{L_t C_t}$ and ω_R=$1/\sqrt{L_r C_R}$ as the natural resonant angular frequencies of the transmitter and receiver respectively, and considering the PT-symmetric condition of $\omega_t = \omega_R = \omega_0$, (9.26) can be reorganized into

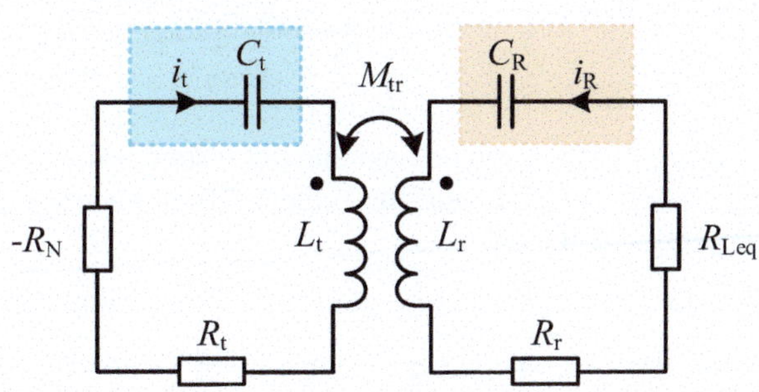

Fig. 9.8 Equivalent circuit of the PT-symmetric MC-WPT system with S-PS compensation network

$$\begin{bmatrix} \frac{-R_N+R_t}{L_t} + j\left(\omega - \frac{\omega_0^2}{\omega}\right) & j\omega k\sqrt{\frac{L_t}{L_r}} \\ j\omega k\sqrt{\frac{L_r}{L_t}} & \frac{R_{Leq}+R_r}{L_r} + j\left(\omega - \frac{\omega_0^2}{\omega}\right) \end{bmatrix} \begin{bmatrix} \dot{I}_t \\ \dot{I}_R \end{bmatrix} = 0, \tag{9.27}$$

where $k = M_{tr}/\sqrt{L_t L_r}$ is the coupling coefficient between transmitting and receiving coils.

According to (9.27), the characteristic frequency equation can be deduced as

$$\left[Q_t^{-1} + j\left(\frac{\omega}{\omega_0} - \frac{\omega_0}{\omega}\right)\right]\left[Q_{Req}^{-1} + j\left(\frac{\omega}{\omega_0} - \frac{\omega_0}{\omega}\right)\right] + \left(\frac{\omega}{\omega_0}k\right)^2 = 0, \tag{9.28}$$

where $Q_t = \omega_0 L_t/(-R_N + R_t)$ and $Q_{Req} = \omega_0 L_r/(R_{Leq} + R_r)$ are defined as the quality factor of the transmitter and the receiver, respectively.

Separating the real and imaginary parts of the characteristic frequency equation, there are

$$Q_t^{-1} Q_{Req}^{-1} - \left(\frac{\omega}{\omega_0} - \frac{\omega_0}{\omega}\right)^2 + \left(\frac{\omega}{\omega_0}k\right)^2 = 0, \tag{9.29}$$

$$\left(\frac{\omega}{\omega_0} - \frac{\omega_0}{\omega}\right)^2 \left(Q_t^{-1} + Q_{Req}^{-1}\right) = 0. \tag{9.30}$$

By solving (9.29) and (9.30), the operating frequency can be derived as

$$\omega = \begin{cases} \omega_0, \ 0 < k < k_C \\ \omega_0 \sqrt{\dfrac{2 - Q_{Req}^{-2} \pm \sqrt{\left(2 - Q_{Req}^{-2}\right)^2 + 4(k^2 - 1)}}{2(1 - k^2)}}, \ k_C \le k < 1 \end{cases}, \tag{9.31}$$

where k_C is defined as the critical coupling coefficient, that is

$$k_C = \frac{1}{Q_{Req}} \sqrt{1 - \frac{1}{4Q_{Req}^2}} \tag{9.32}$$

From (9.31), it can be seen that the system is divided into the exact PT-symmetric region ($k_C \le k < 1$) and the broken PT-symmetric region ($k < k_C$) by the critical coupling coefficient.

(1) Exact PT-symmetric region (i.e., strong coupling region $k_C \le k < 1$).

In this case, f_{o1} and f_{o2} are the operating frequencies of the system, and the system will operate at f_{o1} or f_{o2} automatically.

$$f_{o1,o2} = f_0\sqrt{\frac{2 - Q_{Req}^{-2} \pm \sqrt{\left(2 - Q_{Req}^{-2}\right)^2 + 4\left(k^2 - 1\right)}}{2\left(1 - k^2\right)}}, \tag{9.33}$$

Due to $\omega \neq \omega_0$ and (9.30), the negative resistance can be described as

$$Q_t^{-1} + Q_{Req}^{-1} = 0, \tag{9.34}$$

that is

$$-R_N = -\frac{L_t}{L_r}\left(\alpha^2 R_L + R_r\right) - R_t. \tag{9.35}$$

By substituting (9.29) and (9.34) into (9.27), the current ratio of the receiver and the transmitter can be derived as

$$\frac{I_R}{I_t} = \left|\frac{\dot{I}_R}{\dot{I}_t}\right| = \sqrt{\frac{L_t}{L_r}}. \tag{9.36}$$

Then, the RMS value of the current flowing through the load resistance can be obtained as

$$I_L = \alpha I_R = \alpha I_t\sqrt{\frac{L_t}{L_r}}. \tag{9.37}$$

Due to $U_o = R_L I_L$ and $U_{in} = R_N I_t$, the voltage ratio of the output voltage and input voltage can be written as

$$\frac{U_o}{U_{in}} = \frac{I_L R_L}{I_t R_N} = \sqrt{\frac{L_t}{L_r}}\frac{\alpha R_L}{\frac{L_t}{L_r}\left(\alpha^2 R_L + R_r\right) + R_t}. \tag{9.38}$$

Thus, the output power and transfer efficiency can be obtained as

$$P_o = \frac{U_o^2}{R_L} = \frac{\alpha^2 R_L U_{in}^2}{\frac{L_t}{L_r}\left(\alpha^2 R_L + R_r\right)^2 + 2R_t\left(\alpha^2 R_L + R_r\right) + \frac{L_r}{L_t}R_t^2}, \tag{9.39}$$

$$\eta = \frac{I_L^2 R_L}{I_t^2 R_t + I_R^2 R_r + I_L^2 R_L} = \frac{\alpha^2 R_L}{\alpha^2 R_L + R_r + \frac{L_r}{L_t}R_t}. \tag{9.40}$$

From (9.39) and (9.40), it can be seen that the output power and transfer efficiency in the strong coupling region are independent of the coupling coefficient. Consequently, the output power and transfer efficiency can remain constant even when moving the receiving coil in the strong coupling region.

(2) Broken PT-symmetric region (i.e., weak coupling region $k < k_C$).

In this case, the operating frequency of the system is not bifurcated and automatically tracks the natural resonant frequency of the system f_0. By substituting $\omega = \omega_0$ into (9.29), the negative resistance can be deduced as

$$-R_N = -\frac{k^2\omega_0^2 L_t L_r}{\alpha^2 R_L + R_r} - R_t. \tag{9.41}$$

Then, the current ratio of the receiver and the transmitter can be derived as

$$\frac{I_R}{I_t} = \left|\frac{\dot{I}_R}{\dot{I}_t}\right| = \frac{\omega_0 k\sqrt{L_t L_r}}{\alpha^2 R_L + R_r}. \tag{9.42}$$

The current flowing through the load resistance can be written as

$$I_L = \alpha I_R = \alpha I_t \frac{\omega_0 k\sqrt{L_t L_r}}{\alpha^2 R_L + R_r}. \tag{9.43}$$

The voltage ratio can be derived as

$$\frac{U_o}{U_{in}} = \frac{I_L R_L}{I_t R_N} = \frac{\alpha R_L \omega_0 k\sqrt{L_t L_r}}{k^2\omega_0^2 L_t L_r + r_T(\alpha^2 R_L + R_r)}. \tag{9.44}$$

Therefore, the output power and transfer efficiency can be expressed as

$$P_o = \frac{U_o^2}{R_L} = \frac{\alpha^2 R_L U_{in}^2 \omega_0^2 k^2 L_t L_r}{\left[k^2\omega_0^2 L_t L_r + R_t(\alpha^2 R_L + R_r)\right]^2}, \tag{9.45}$$

$$\eta = \frac{I_L^2 R_L}{I_t^2 r_t + I_R^2 R_r + I_L^2 R_L} = \frac{\alpha^2 R_L \omega_0^2 k^2 L_t L_r}{r_t(\alpha^2 R_L + R_r)^2 + (\alpha^2 R_L + R_r)\omega_0^2 k^2 L_t L_r}. \tag{9.46}$$

From (9.45) and (9.46), it can be seen that the output power and transfer efficiency of the system are related to the coupling coefficient in the weak coupling region. Hence, it is necessary to avoid the system operating in the weak coupling region, and the critical coupling coefficient of the PT-symmetric MC-WPT system should be designed as small as possible.

Figure 9.9 shows the frequency bifurcation curves of the system with different capacitance ratios. With the decrease of the capacitance distribution ratio, the critical coupling coefficient is constantly decreased, which indicates that the strong coupling region is significantly enlarged. Therefore, the PT-symmetric MC-WPT system with S-PS compensation topology can significantly improve the transfer distance and degree of freedom. Moreover, the S-PS topology is completely consistent with the SS topology when $\alpha = 1$.

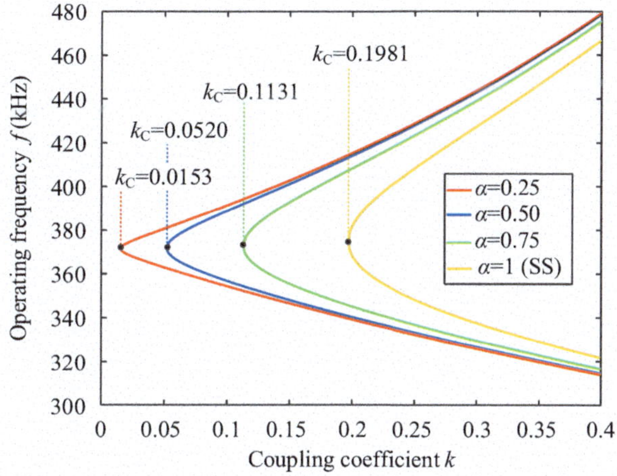

Fig. 9.9 Frequency bifurcation curves of the system in the exact PT-symmetric region

Figure 9.10 shows the output power and transfer efficiency versus coupling coefficient under different capacitance distribution ratios, it can be seen that the output power increases significantly with the decrease of the capacitance distribution ratio, and the system can remain constant output power within its corresponding exact PT-symmetric region, but the cost of extending the transfer distance and increasing the output power by reducing the capacitance distribution ratio is the reduction of transfer efficiency, so the PT-symmetric MC-WPT system with S-PS compensation topology needs to choose the appropriate α value according to different application scenarios to balance the contradiction between transfer distance and transfer efficiency.

9.1.4 Adding Inductive and Capacitive Elements LC on the Receiver

The schematic diagram of the PT-symmetric MC-WPT system with inductive and capacitive elements LC added on the receiver is shown in Fig. 9.11, in which $-R_N$ is a negative resistance and is used as a power source to provide energy to the whole system, L_t, C_t and R_t are the coil inductance, compensation capacitance and internal resistance of the transmitter, L_r, C_r and R_r are the coil inductance, compensation capacitance and internal resistance of the receiver, respectively. It is noted that R_r represents the sum of the internal resistances of the receiving coil and its series additional inductor. R_L is the load resistance, M_{tr} is the mutual inductance between coils, L_1 and C_1 are the inductance and capacitance of the inductive and capacitive elements LC added on the receiver.

According to the Kirchhoff's voltage and current laws, the circuit equation of the system shown in Fig. 9.11 can be written as

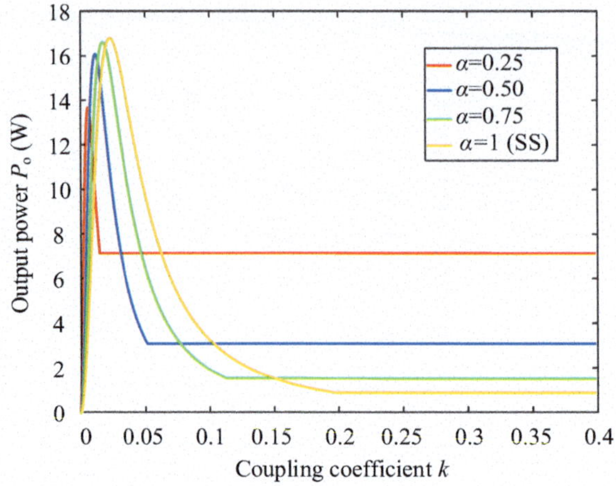

(a)

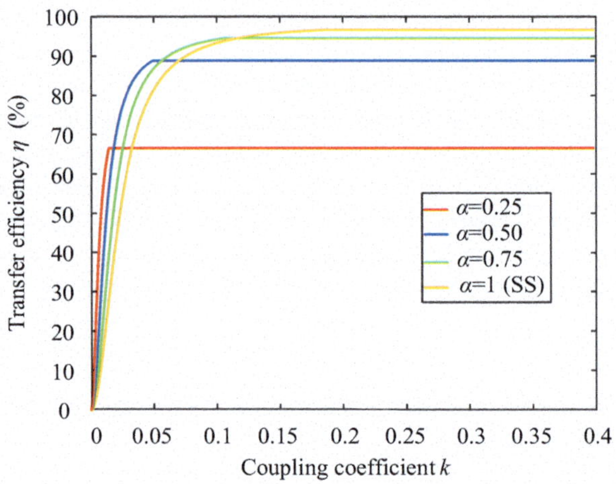

(b)

Fig. 9.10 Transfer characteristics under different capacitance distribution ratios. **a** Output power. **b** Transfer efficiency

$$\begin{bmatrix} -R_N + R_t + \frac{1}{j\omega C_t} + j\omega L_t & j\omega M_{tr} & 0 \\ j\omega M_{tr} & j\omega(L_r + L_1) + R_r' & \frac{1}{j\omega C_r} + R_L \\ 0 & -\frac{1}{j\omega C_1} & \frac{1}{j\omega C_1} + \frac{1}{j\omega C_r} + R_L \end{bmatrix} \begin{bmatrix} \dot{I}_t \\ \dot{I}_r \\ \dot{I}_L \end{bmatrix} = \begin{bmatrix} 0 \\ 0 \\ 0 \end{bmatrix}. \tag{9.47}$$

Then, the current ratio of output current and input current can be derived as

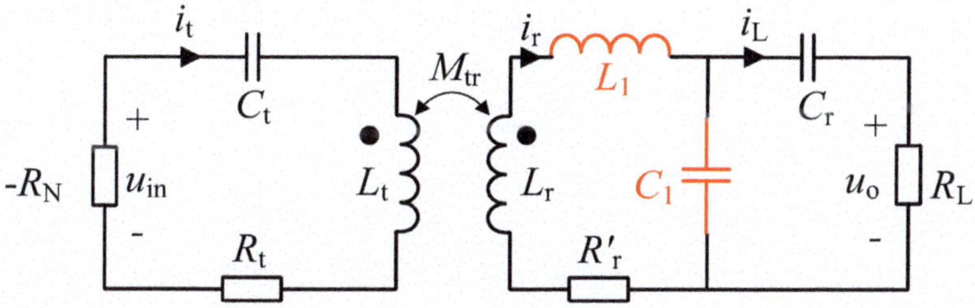

Fig. 9.11 The schematic diagram of the PT-symmetric MC-WPT system with inductive and capacitive elements *LC* added on the receiver

$$\frac{I_L}{I_t} = \frac{M_{tr}}{C_1\sqrt{X_3}}, \tag{9.48}$$

where $X_3 = \left[\frac{\omega^2(L_r+L_1)(C_r+C_1)-1}{\omega^2 C_r C_1} + R'_r R_L\right]^2 + \left[\omega(L_r+L_1)R_L - \frac{R'_r+R_L}{\omega C_1} - \frac{R'_r}{\omega C_r}\right]^2$.

The input impedance of the system is

$$Z_{in} = R_t + j\omega L_t + \frac{1}{j\omega C_t} + \frac{\omega^2 M_{tr}^2}{R'_r + j\omega(L_r+L_1) + \left(\frac{1}{j\omega C_r} + R_L\right)//\frac{1}{j\omega C_1}}. \tag{9.49}$$

When the system is PT-symmetric, the imaginary part of the input impedance is zero, and the following equation can be derived from (9.49):

$$\omega L_t - \frac{1}{\omega C_t} - \frac{\omega^2 M_{tr}^2[\omega(L_r+L_1) - X_2]}{(R'_r+X_1)^2 + [\omega(L_r+L_1) - X_2]^2} = 0. \tag{9.50}$$

The negative resistance can be expressed as

$$-R_N = -R_t - \frac{(R'_r+X_1)(\omega^2 L_t C_t - 1)}{\omega^2(L_r+L_1)C_t - \omega C_t X_2}. \tag{9.51}$$

The current flowing through the transmitting coil can be represented as

$$I_t = \frac{U_{in}}{R_t + \frac{(R'_r+X_1)(\omega^2 L_t C_t - 1)}{\omega^2(L_r+L_1)C_t - \omega C_t X_2}}. \tag{9.52}$$

The voltage across the load resistance can be derived as

$$U_o = \frac{U_{in} R_L \sqrt{\left[(R'_r+X_1)^2 + (\omega(L_r+L_1) - X_2)^2\right](\omega^2 L_t C_t - 1)}}{\omega C_1 R_N \sqrt{X_3\left[\omega^2(L_r+L_1)C_t - \omega C_t X_2\right]}}. \tag{9.53}$$

Therefore, the output power and transfer efficiency can be obtained as

$$P_{\mathrm{o}} = \frac{U_{\mathrm{o}}^2}{R_{\mathrm{L}}}, \tag{9.54}$$

$$\eta = \frac{P_{\mathrm{o}}}{P_{\mathrm{in}}} = \frac{U_{\mathrm{o}}^2}{I_{\mathrm{t}}^2 R_{\mathrm{L}} \left[R_{\mathrm{t}} + \frac{(R'_{\mathrm{r}} + X_1)(\omega^2 L_{\mathrm{t}} C_{\mathrm{t}} - 1)}{\omega^2 (L_{\mathrm{r}} + L_1) C_{\mathrm{t}} - \omega C_{\mathrm{t}} X_2} \right]}. \tag{9.55}$$

In addition, the operating frequency of the system always satisfies (9.50), which can be derived to obtain a one-variable tenth equation about ω, so the operating frequency has no analytical solution and can only be analyzed by numerical methods. Figure 9.12 shows the relationship between the operating frequency and coupling coefficient, it can be seen that when the coupling coefficient is less than the critical coupling coefficient k_{C}, the system has a real frequency solution, while when the coupling coefficient is greater than the critical coupling coefficient k_{C}, the system has three real frequency solutions, defined as high-frequency f_{H}, medium-frequency f_{M}, and low-frequency f_{L}. The system is stable at high-frequency f_{H} and low-frequency f_{L}, and is unstable at medium-frequency f_{M}, which can be seen from the Hamel locus of the transmitting current shown in Fig. 9.13 [5]. Here, the circuit parameters are $U_{\mathrm{in}} = 43$ V, $L_{\mathrm{t}} = 70$ μH, $L_{\mathrm{r}} = 70$ μH, $L_1 = 140$ μH, $C_{\mathrm{t}} = 4.02$ nF, $C_{\mathrm{r}} = 1$ nF, $C_1 = 0.34$ nF, $R_{\mathrm{t}} = 0.25$ Ω, $R'_{\mathrm{r}} = 0.56$Ω, $R_{\mathrm{L}} = 10$ Ω.

The output power and transfer efficiency of the PT-symmetric MC-WPT system with inductive and capacitive elements *LC* added on the receiver is described in Fig. 9.14, it can be seen that when the coupling coefficient is less than the critical

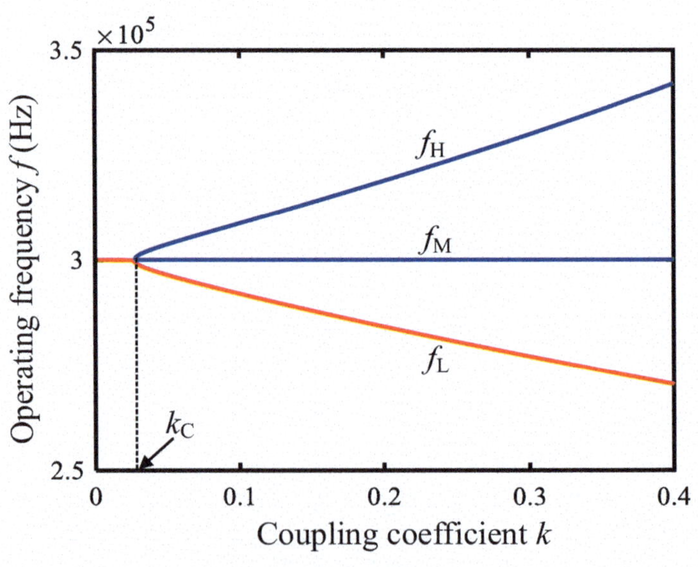

Fig. 9.12 Operating frequency versus coupling coefficient

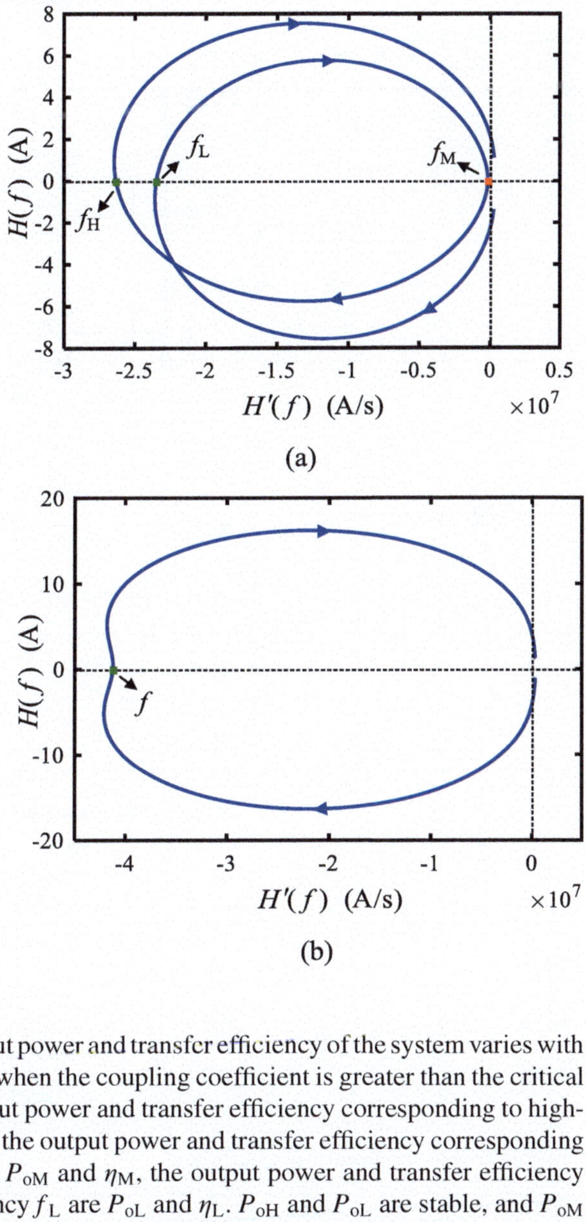

Fig. 9.13 Hamel locus of the transmitting current. **a** In the strong coupling region ($k_C \leq k < 1$). **b** In the weak coupling region ($k < k_C$)

coupling coefficient, the output power and transfer efficiency of the system varies with the coupling coefficient, and when the coupling coefficient is greater than the critical coupling coefficient, the output power and transfer efficiency corresponding to high-frequency f_H are P_{oH} and η_H, the output power and transfer efficiency corresponding to medium-frequency f_M are P_{oM} and η_M, the output power and transfer efficiency corresponding to low-frequency f_L are P_{oL} and η_L. P_{oH} and P_{oL} are stable, and P_{oM} is unstable. The output power P_{oH} and P_{oL} and transfer efficiency η_H and η_L of the system in the strong coupling region are basically unchanged with the change of coupling coefficient. Hence, when the distance between the coils is changed or the coils are misaligned, the output voltage and transfer efficiency of the system in the strong coupling region remain basically unchanged, and the PT-symmetric MC-WPT system with inductive and capacitive elements LC added on the receiver has a

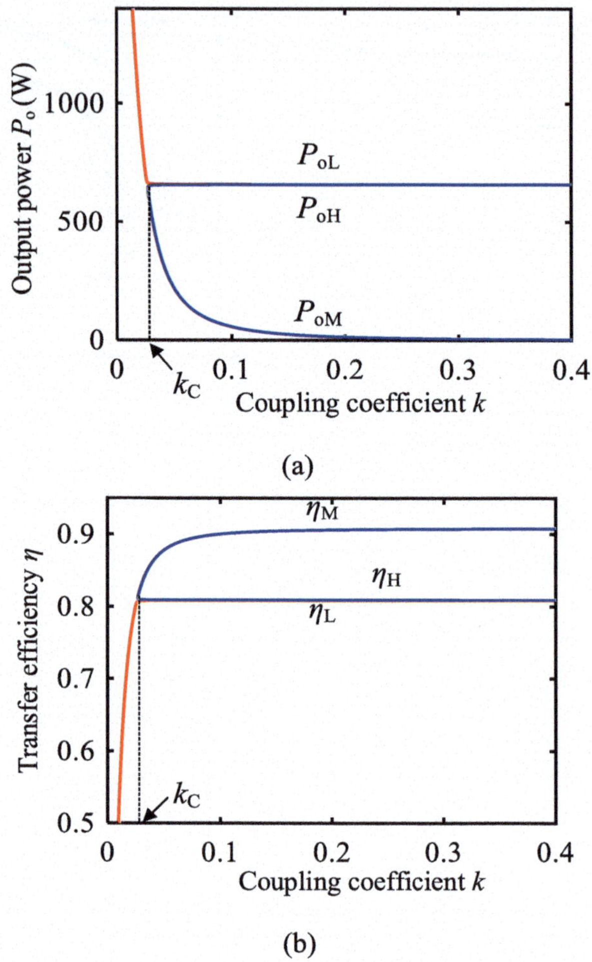

Fig. 9.14 Transfer characteristics of the PT-symmetric MC-WPT system with inductive and capacitive elements *LC* added on the receiver. **a** Output power. **b** Transfer efficiency

high misalignment tolerance. It should be noted that when the coupling coefficient changes, the system will automatically track f_H or f_L to ensure that the phase of the inverter output voltage is the same as that of the inverter output current. At this time, the output power and transfer efficiency of the system in the strong coupling region is insensitive to the coupling coefficient.

Figures 9.15 and 9.16 show the effect of the additional inductance L_1 and capacitance C_1 on the output power and transfer efficiency of the system. From Fig. 9.15, it can be seen that with the increase of the additional inductance L_1 on the receiver, the critical coupling coefficient k_C of the system decreases, making the constant and efficient transfer distance increase. In the strong coupling region, the output power P_{oH} and P_{oL} of the system increase with the increase of the additional inductance L_1, while the transfer efficiency η_H and η_L of the system decreases with the increase of the additional inductance L_1. Similarly, as can be seen from Fig. 9.16, with the increase

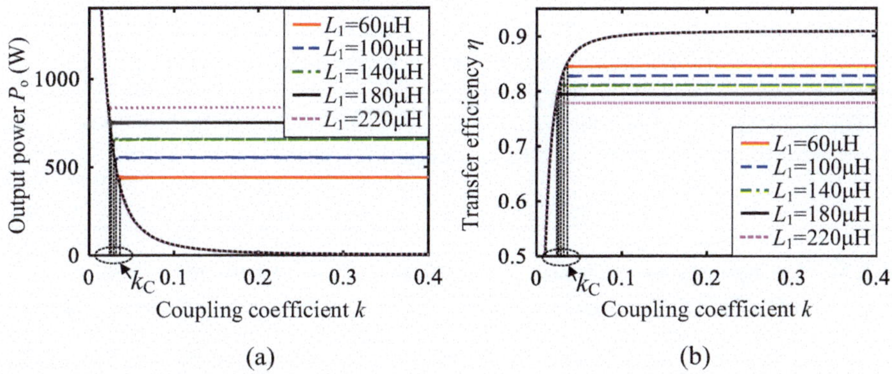

Fig. 9.15 Effect of additional inductance L_1 on transfer performance. **a** Output power. **b** Transfer efficiency

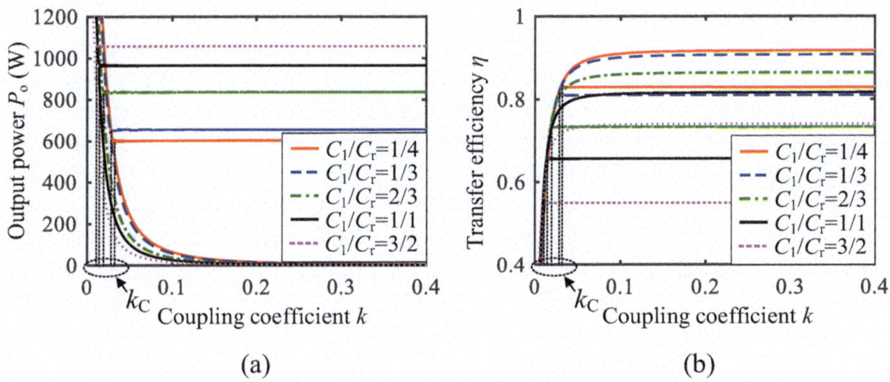

Fig. 9.16 Effect of additional capacitance ratio C_1/C_r on transfer performance

of the ratio of C_1 to C_r, the critical coupling coefficient of the system decreases, thus increasing the stable transfer distance. In the strong coupling region, with the increase of the ratio of C_1 to C_r, the output power P_{oH} and P_{oL} of the system increase, while the transfer efficiency η_H and η_L of the system decrease. Therefore, the PT-symmetric MC-WPT system with LCC compensation, in which the inductive and capacitive elements LC are added to the receiving circuit, is suitable for occasions where the transfer distance and anti-misalignment ability are required. Meanwhile, in order to meet the design requirements of transfer distance and transfer efficiency in practical applications, the parameter design process of the PT-symmetric MC-WPT system with LCC compensation mainly consists of the following steps:

The first step is to determine the maximum inductance value of the receiving coil according to the reserved space of the receiving coil.

The second step is to adjust the critical coupling coefficient of the system by changing the additional inductance L_1 on the receiver and the ratio of C_1 to C_r to meet the requirements of the transfer distance.

The last step is to adjust the transfer efficiency by changing the inductance L_t of the transmitting coil to meet the efficiency requirements.

9.1.5 Comparison with the Traditional PT-Symmetric MC-WPT System

Figure 9.17 shows the output power and transfer efficiency of PT-symmetric MC-WPT systems with different compensation topologies, including traditional SS compensation topology shown in Fig. 3.3a, adding an inductive element *L* in the receiving circuit shown in Fig. 9.4, adding a capacitive element *C* in the receiving circuit (i.e., PS compensation topology on the receiver) shown in Fig. 9.6, adding inductive and capacitive elements *LC* in the receiving circuit (i.e., LCC compensation network on the receiver) shown in Fig. 9.11. Here, the input voltage, inductance of the transmitting coil, internal resistance of the transmitting coil, compensation capacitance of the transmitting coil, inductance of the receiving coil, and load resistance of each system are identical, respectively. That is, $U_{in} = 43$ V, $L_t = 70$ μH, $R_t = 0.25$ Ω, $C_t = 4.02$ nF, $L_r = 70$ μH, $R_L = 10$ Ω. Moreover, in the traditional PT-symmetric MC-WPT system with SS compensation topology, the other parameters are $C_r = 4.02$ nF, $R_r = 0.25$ Ω. in the PT-symmetric MC-WPT system with an additional inductance, there are $L_1 = 140$ μH, $C_r = 1.34$ nF, $R'_r = 0.56$ Ω. In the PS compensation network on receiver, there are $C_r = 3.01$ nF, $C_1 = 1.01$ nF, $R_r = 0.25$ Ω. In the LCC compensation network on receiver, there are $L_1 = 140$ μH, $C_r = 1$ nF, $C_1 = 0.34$ nF, $R'_r = 0.56$ Ω. Table 9.2 lists the transmission performance of the four systems.

From Fig. 9.17, it can be seen that the critical coupling coefficients of SS compensation, additional inductive element on the receiving circuit, PS compensation on the receiving circuit, and LCC compensation on the receiving circuit are 0.0777, 0.0462, 0.0466, and 0.0275, respectively. Among these four systems, the critical coupling coefficient of the LCC compensation on the receiving circuit is the smallest. That is, adding inductive and capacitive elements *LC* on the receiving circuit can improve the transfer distance of the PT-symmetric MC-WPT system, and the critical transfer distance of LCC compensation on the receiving circuit is longer than that of PS compensation or additional inductance on the receiving circuit. Besides, in the strong coupling region, the output power of LCC compensation on the receiving circuit is greater than that of the three systems, but the transfer efficiency is lower than that of the three systems, so the PT-symmetric MC-WPT system with LCC compensation is more suitable for the application of long-distance wireless power transmission.

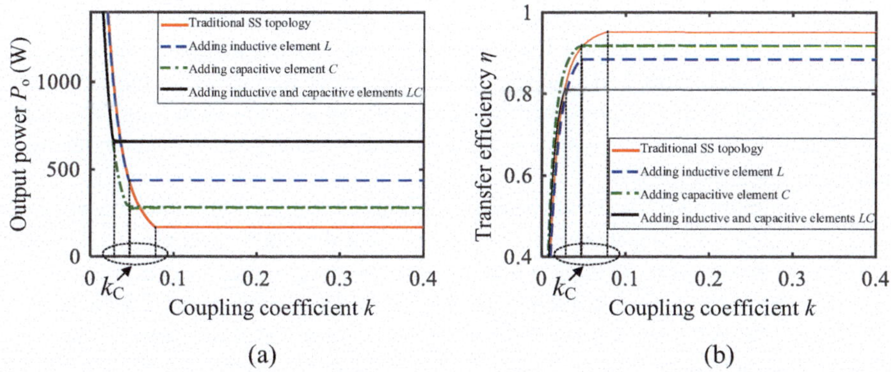

Fig. 9.17 Comparison of transfer characteristics of different PT-symmetric MC-WPT systems. **a** Output power. **b** Transfer efficiency

Table 9.2 Transfer characteristics of four systems

Structure	Critical coupling coefficient	Output power	Transfer efficiency (%)
SS	0.0777	169.4	95.2
Additional inductance on the receiver	0.0462	438.0	88.4
PS compensation on the receiver	0.0466	280.9	91.8
LCC compensation on the the receiver	0.0275	659.3	81.0

9.2 Reduction of Switching Frequency

For the existing PT-symmetric MC-WPT systems using fundamental harmonic operation (FHO) [3, 6, 7], a high resonance frequency means that the inverter needs to operate at a high switching frequency, which will lead to an increase in switching losses. In addition, power semiconductor switches with a fast switching speed and high-speed gate drivers are demanded to match the high switching frequency. As a result, the cost of the system will inevitably increase. Besides, fast switching speed usually brings in bigger electromagnetic interference noise at a high-frequency range. It is valuable to reduce the switching frequency without affecting the stable power transfer range of the PT-symmetric MC-WPT system.

The series-series (SS) compensation has been widely used in MC-WPT systems due to its simplicity. For SS-based MC-WPT systems, only the current around resonance frequency is allowed to flow to the receiver because of the band-pass feature of the resonant network. Therefore, the harmonic component of an ac square voltage near resonance frequency can be designed as a power carrier [8]. When the nth harmonic of the ac square-wave voltage is used as the power carrier (that is, the frequency of the nth harmonic is set equal to the frequency of the resonant current), the corresponding fundamental frequency is $1/n$ of the frequency of the resonant current. Compared with FHO (where the fundamental frequency of the ac square voltage is equal to the frequency of the resonant current), the switching frequency of a system with the same resonant frequency can be reduced to $1/n$ under high harmonic operation (HHO).

In PT-symmetric MC-WPT systems, the output of the inverter should be controlled to be equivalent to a negative resistor (the current direction through the negative resistor is opposite to its voltage when the associated reference direction is chosen [6]). Therefore, the key to reducing the switching frequency of the SS-based PT-symmetric MC-WPT system based on HHO is to ensure that the phase difference between the nth harmonic of the inverter's output voltage and the resonant current is maintained at 180°.

9.2.1 System Modeling and Analysis

Considering the band-pass filter feature of the series resonant compensation network, only the harmonic component with frequency around the resonant frequency is allowed to pass through, while all other frequency components are filtered out by large impedance. Therefore, when the frequency of the selected nth harmonic is around the resonant, it can be assumed that the actual voltage and current signals in the resonant compensation network contain only the selected nth harmonic. Besides, in the proposed system, by using the phase synchronization method (PSM) (the detailed control process is presented in Sect. 9.2.2), the selected harmonic of u_P is always in phase with i_P. Therefore, the system can be seen as a PT-symmetric system powered by the negative resistor, and the equivalent circuit of the system when selecting the nth-harmonic is shown in Fig. 9.18.

Here, R_P (R_S) is the internal resistance of the transmitter (receiver), which includes the equivalent series resistances of coil and tuning capacitors. L_P (L_S) and C_P (C_S) are the self-inductance and tuning capacitance of the transmitter (receiver), respectively. $k_{PS} = M_{PS}/\sqrt{L_P L_S}$ is the coupling coefficient. The phasor of the nth harmonic component of the inverter's output voltage is denoted as $\dot{U}_{Pn}$ (due to the band-pass

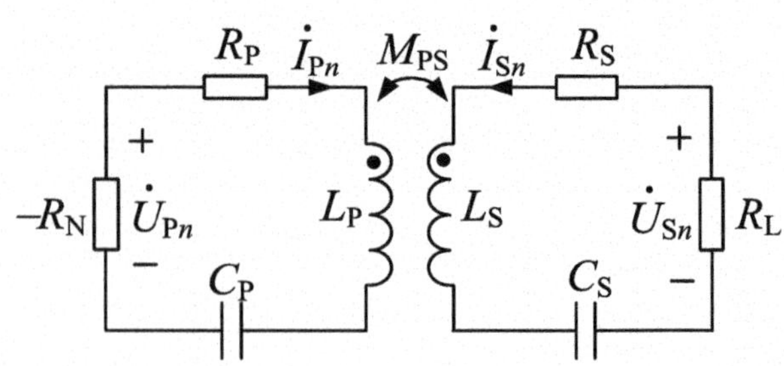

Fig. 9.18 The equivalent circuit of the proposed system when selecting the nth-harmonic

filter feature of the resonant compensation network, only the harmonic around the resonance frequency contributes to power transmission). $\dot{I}_{\mathrm{P}n}$ and $\dot{I}_{\mathrm{S}n}$ represent the phasor of the nth harmonic of transmitting and receiving currents, respectively. $\dot{U}_{\mathrm{S}n}$ is the nth harmonic output voltage across the load resistor. R_{L} is the equivalent load resistance. According to Kirchhoff's Voltage Law (KVL), the system can be expressed as

$$\begin{bmatrix} \left[\frac{-R_{\mathrm{N}}+R_{\mathrm{P}}}{L_{\mathrm{P}}}+j\left(\omega_n-\frac{\omega_0^2}{\omega_n}\right)\right] & j\omega_n k_{\mathrm{PS}}\sqrt{\frac{L_{\mathrm{S}}}{L_{\mathrm{P}}}} \\ j\omega_n k_{\mathrm{PS}}\sqrt{\frac{L_{\mathrm{P}}}{L_{\mathrm{S}}}} & \left[\frac{R_{\mathrm{S}}+R_{\mathrm{L}}}{L_{\mathrm{S}}}+j\left(\omega_n-\frac{\omega_0^2}{\omega_n}\right)\right] \end{bmatrix}\begin{bmatrix}\dot{I}_{\mathrm{P}n}\\ \dot{I}_{\mathrm{S}n}\end{bmatrix}=0, \tag{9.56}$$

where $\omega_n = n\omega$ is the angular frequency of voltage and current signals in the resonant compensation network when selecting the nth harmonic, ω is the switching angular frequency of the inverter and n is an integer representing the order of harmonics. The resonant angular frequency is denoted by $\omega_0 = 1/\sqrt{L_{\mathrm{P}}C_{\mathrm{P}}}=1/\sqrt{L_{\mathrm{S}}C_{\mathrm{S}}}$. It can be found that when the electrical parameters of the system are fixed, ω_n will be determined regardless of the order of the selected harmonic (n), that is, the angular frequencies of voltage and current in the resonant compensation network are equal regardless of whether the fundamental or the nth harmonic is selected.

To obtain a steady-state solution with a purely real ω, the determinant of (9.56) should be equal to zero, that is

$$\begin{cases} \left(n\omega-\frac{\omega_0^2}{n\omega}\right)\left(\frac{R_{\mathrm{L}}+R_{\mathrm{S}}}{L_{\mathrm{S}}}+\frac{-R_{\mathrm{N}}+R_{\mathrm{P}}}{L_{\mathrm{P}}}\right)=0 \\ \frac{-R_{\mathrm{N}}+R_{\mathrm{P}}}{L_{\mathrm{P}}}\frac{R_{\mathrm{L}}+R_{\mathrm{S}}}{L_{\mathrm{S}}}-\left(n\omega-\frac{\omega_0^2}{n\omega}\right)^2+(n\omega)^2k_{\mathrm{PS}}^2=0 \end{cases}. \tag{9.57}$$

Then, the solutions of ω can be deduced from (9.57) as follows:

(1) In the exact PT-symmetric region.

$$\omega=\omega_{\mathrm{H,L}}=\frac{\omega_0}{n}\sqrt{\frac{2-\gamma^2\pm\sqrt{\left[2-\gamma^2\right]^2-4\left(1-k_{\mathrm{PS}}^2\right)}}{2\left(1-k_{\mathrm{PS}}^2\right)}},\ k_{\mathrm{PS}}\geq k_{\mathrm{C}}, \tag{9.58}$$

where $\gamma = (R_{\mathrm{L}} + R_{\mathrm{S}})/(\omega_{\mathrm{L}}L_{\mathrm{S}})$ and $k_{\mathrm{C}}=\sqrt{\gamma^2-\gamma^4/4}$ is the critical coupling coefficient. ω_{H} that corresponds to the + sign in (9.58) denotes the high-frequency branch. The $-$sign gives ω_{L}, which represents the low-frequency branch.

(2) In the broken PT-symmetric region.

$$\omega=\frac{\omega_0}{n},\ k_{\mathrm{PS}}<k_{\mathrm{C}}. \tag{9.59}$$

According to (9.58) and (9.59), the use of higher harmonics (i.e., $n > 1$) can reduce the switching frequency to $1/n$ compared to using the fundamental (i.e., $n = 1$). It should be emphasized that as long as the negative resistor $-R_{\mathrm{N}}$ is implemented, ω can be tracked automatically without knowing k_{PS} and R_{L}. Besides, the expression of k_{C} is consistent with the traditional PT-symmetric MC-WPT system [3], which means the PT-symmetric region is not affected.

Considering the exact PT-symmetric region, the current gain can be deduced by inserting (9.57) into (9.56):

$$\left.\begin{array}{l} jn\omega k_{\mathrm{PS}}\sqrt{\dfrac{L_{\mathrm{P}}}{L_{\mathrm{S}}}}\dot{I}_{\mathrm{P}n} + \left[\dfrac{R_{\mathrm{S}}+R_{\mathrm{L}}}{L_{\mathrm{S}}} + j\left(n\omega - \dfrac{\omega_0^2}{n\omega}\right)\right]\dot{I}_{\mathrm{S}n} = 0 \\ (n\omega)^2 k_{\mathrm{PS}}^2 = \left(\dfrac{R_{\mathrm{L}}+R_{\mathrm{S}}}{L_{\mathrm{S}}}\right)^2 + \left(n\omega - \dfrac{\omega_0^2}{n\omega}\right)^2 \end{array}\right\} \Rightarrow$$
$$\frac{I_{\mathrm{S}n}}{I_{\mathrm{P}n}} = \left|\frac{\dot{I}_{\mathrm{S}n}}{\dot{I}_{\mathrm{P}n}}\right| = \frac{n\omega k_{\mathrm{PS}}\sqrt{L_{\mathrm{P}}/L_{\mathrm{S}}}}{\sqrt{[(R_{\mathrm{S}}+R_{\mathrm{L}})/L_{\mathrm{S}}]^2 + \left(n\omega - \omega_0^2/n\omega\right)^2}} = \sqrt{\frac{L_{\mathrm{P}}}{L_{\mathrm{S}}}}. \tag{9.60}$$

According to (9.57) and (9.60), the voltage gain of the resonant compensation network can be expressed as

$$\left.\begin{array}{l} \dfrac{R_{\mathrm{L}}+R_{\mathrm{S}}}{L_{\mathrm{S}}} + \dfrac{-R_{\mathrm{N}}+R_{\mathrm{P}}}{L_{\mathrm{P}}} = 0 \\ U_{\mathrm{S}n} = R_{\mathrm{L}}I_{\mathrm{S}n} \text{ and } U_{\mathrm{P}n} = R_{\mathrm{N}}I_{\mathrm{P}n} \end{array}\right\} \Rightarrow$$
$$\frac{U_{\mathrm{S}n}}{U_{\mathrm{P}n}} = \left|\frac{\dot{U}_{\mathrm{S}n}}{\dot{U}_{\mathrm{P}n}}\right| = \frac{R_{\mathrm{L}}\sqrt{L_{\mathrm{P}}L_{\mathrm{S}}}}{(R_{\mathrm{L}}+R_{\mathrm{S}})L_{\mathrm{P}} + R_{\mathrm{P}}L_{\mathrm{S}}}, \tag{9.61}$$

where $I_{\mathrm{P}n}$, $I_{\mathrm{S}n}$, $U_{\mathrm{P}n}$, and $U_{\mathrm{S}n}$ are the RMS values of the $\dot{I}_{\mathrm{P}n}$, $\dot{I}_{\mathrm{S}n}$, $\dot{U}_{\mathrm{P}n}$ and $\dot{U}_{\mathrm{S}n}$, respectively.

Using (9.60), the transfer efficiency (coil-coil) is given as:

$$\eta_{\mathrm{T}} = \frac{I_{\mathrm{S}n}^2 R_{\mathrm{L}}}{I_{\mathrm{P}n}^2 R_{\mathrm{P}} + I_{\mathrm{S}n}^2 R_{\mathrm{S}} + I_{\mathrm{S}n}^2 R_{\mathrm{L}}} = \frac{R_{\mathrm{L}}}{\frac{L_{\mathrm{S}}}{L_{\mathrm{P}}}R_{\mathrm{P}} + R_{\mathrm{S}} + R_{\mathrm{L}}}. \tag{9.62}$$

It can be noted that the transfer efficiency is independent of the coupling coefficient, that is, the transfer efficiency can be maintained constant under different transfer distances and misalignments when the higher harmonics are used in the PT-symmetric MC-WPT system.

9.2.2 Content and Influence of the Fundamental and Unused Harmonics

According to the Fourier series, the periodic function can be expressed as

$$u_{\mathrm{P}}(\omega t) = c_0 + \sum_{n=1}^{\infty} [a_n \cos(n\omega t) + b_n \sin(n\omega t)], \tag{9.63}$$

where

$$\begin{cases} c_0 = \dfrac{1}{T}\displaystyle\int_0^T u_{\mathrm{P}}(t)dt = \dfrac{1}{2\pi}\displaystyle\int_0^{2\pi} u_{\mathrm{P}}(\omega t)d(\omega t) \\ a_n = \dfrac{2}{T}\displaystyle\int_0^T u_{\mathrm{P}}(t)\cos(n\omega t)dt = \dfrac{1}{\pi}\displaystyle\int_0^{2\pi} u_{\mathrm{P}}(\omega t)\cos(n\omega t)d(\omega t) \\ b_n = \dfrac{2}{T}\displaystyle\int_0^T u_{\mathrm{P}}(t)\sin(n\omega t)dt = \dfrac{1}{\pi}\displaystyle\int_0^{2\pi} u_{\mathrm{P}}(\omega t)\sin(n\omega t)d(\omega t) \end{cases}. \tag{9.64}$$

Based on (9.63) and (9.64), the inverter square-wave output with duty cycle D_{S} can be expressed in the Fourier series

$$u_{\mathrm{P}} = \sum_{n=1}^{\infty} u_{\mathrm{P}n}, \tag{9.65}$$

where $u_{\mathrm{P}n}$ is the nth harmonic component of inverter output voltage

$$u_{\mathrm{P}n} = \frac{4U_{\mathrm{dc}}}{n\pi}\sin\left(\frac{nD_{\mathrm{S}}\pi}{2}\right)\sin\left(\frac{n\pi}{2}\right)\sin(n\omega t) \tag{9.66}$$

According to (9.65) and (9.66), it can be noted that the square wave contains only the fundamental and the odd-order ($2n$–1)th harmonics, whereas all the even-order $2n$th harmonics are zero. The amplitude of the fundamental and the main harmonics of u_{P} varies with the duty cycle D_{S} as presented in Fig. 9.19.

In this work, the power transferred to the receiver side is expected to be carried by the harmonics to reduce the switching frequency. Therefore, it is necessary to eliminate the effect of the fundamental component and the unused harmonics as much as possible. From Fig. 9.19, the amplitude of the fundamental decreases monotonically as the duty cycle D_{S} decreases, while the maximum amplitude of the harmonics occurs at different duty cycles. Thus, the duty cycle can be limited to a range of zero to a small value to reduce the amplitude of the fundamental component while

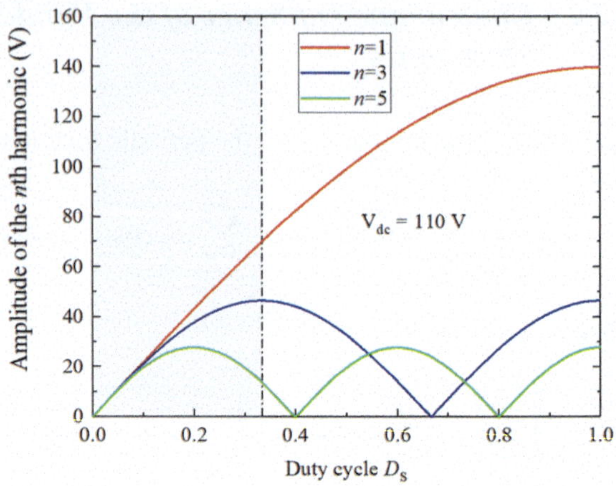

Fig. 9.19 Amplitude of the fundamental and the main harmonics of u_P under different duty cycles

ensuring that the amplitude of the selected harmonic is monotonic. For example, if the third harmonic is selected to deliver power, the duty cycle can be chosen at $0 \leq D_S \leq 1/3$; or if the fifth harmonic is used, the duty cycle can be limited to $0 \leq D_S \leq 1/5$.

Next, the influence of the fundamental component and unused harmonics of u_P on the transmitter current i_P will be examined. The amplitude of the nth harmonic components of i_P depends on the amplitude of the nth harmonic of u_P and the magnitude of the transmitter-side input impedance at the frequency of nth harmonic. Based on (9.66), the nth harmonic component of the transmitting side current (i_{Pn}) can be deduced as

$$i_{Pn} = \frac{4U_{dc}}{Z_{in}(jn\omega)n\pi}\sin\left(\frac{nD_S\pi}{2}\right)\sin\left(\frac{n\pi}{2}\right)\sin(n\omega t), \tag{9.67}$$

where $Z_{in}(jn\omega)$ denotes the transmitter-side input impedance at the frequency of nth harmonic

$$Z_{in}(jn\omega) = \left(\frac{1}{jn\omega C_P} + jn\omega L_P\right) + \frac{(n\omega M_{PS})^2}{R_L + jn\omega L_S + 1/jn\omega C_S}, \tag{9.68}$$

and the transmitting current i_P is expressed as

$$i_P = \sum_{n=1}^{\infty} i_{Pn}. \tag{9.69}$$

Assuming that the third harmonics of u_P is used for power transmission (that is, the switching angular frequency is set as $\omega = \omega_{H,L}|_{n=3}$). Due to the band-pass feature of the resonant compensation network and the small content of the higher harmonics ($n > 5$) in the inverter output voltage u_P, only the fundamental, third, and fifth harmonics of i_P are discussed in this paper, that is, $i_P(t) \approx i_{P1}(t) + i_{P3}(t) + i_{P5}(t)$.

By submitting $\omega = \omega_{H,L}|_{n=3}$ into (9.67), the percentage content of the fundamental and fifth harmonic in the current i_P at ω_L and ω_H under different coupling coefficients are shown in Fig. 9.20a, b, respectively, and Fig. 9.21a, b show the corresponding magnitude of the transmitter-side input impedance at the frequency of the first, third, and fifth harmonic components when $\omega = \omega_L$ and $\omega = \omega_H$, respectively. It can be seen from Fig. 9.20a that when $\omega = \omega_L$, the content of the fifth harmonic increases to 23% at $k_{PS} = 0.47$ due to the smaller value of $|Z_{in}(j5\omega_L)|$. When $\omega = \omega_H$, the content of the fundamental and fifth harmonic is always below 8% due to the large input impedance at the frequency of the fundamental and fifth harmonic components shown in Fig. 9.21b. Hence, to guarantee the performance and controllability of the proposed system, $\omega = \omega_H$ is chosen in this work.

According to (9.65), (9.67), and (9.69), the theoretical waveforms of i_P, i_{P3}, and the corresponding output voltage of the inverter (u_P) at $\omega = \omega_H$ when $k_{PS} = 0.35$ and $k_{PS} = 0.65$ are plotted in Fig. 9.22, it can be seen that $i_P \approx i_{P3}$, and the ripple of transmitting current i_P is acceptable. Besides, it can be found that when $D_S = 1/3$, the soft-switching or zero-phase-angle (ZPA) operation can be realized, which can further improve the system efficiency.

9.2.3 Implementation of Negative Resistor Using the nth Harmonic

In this work, a full-bridge inverter is used to construct the negative resistor, as shown in Fig. 9.23. The switches of each leg of the inverter operate in a complementary state, and the phase shift between the two legs (φ_P) controls the output voltage u_P (which is a quasi-square waveform with rich harmonics).

Here, the third harmonic operation (THO) (that is, the third harmonic of u_P makes a major contribution to the output power) is presented as an example. According to Fig. 9.21b, the content of i_{P1} and i_{P5} in transmitting current will be small due to the large input impedance at the frequency of the third and fifth harmonic components, and the amplitude of the fundamental component of u_P is smaller when $0 \leq D_S \leq 1/3$, which is beneficial to further reduce the amplitude of i_{P1}. For simplicity, suppose i_P only contains the third harmonic, i.e., $i_P \approx i_{P3}$, which is reasonable and acceptable. The desired drive signals, inverter's output voltage together with its harmonics, and the current of transmitting coil i_P in THO mode are shown in Fig. 9.24a.

If the relationship between the driver signals and the current i_P shown in Fig. 9.24a (including the phase and period) is maintained in real-time, the third harmonic of the inverter's output voltage (u_{P3}) will have the same frequency as the current i_P and be always in phase with i_P. In this case, the output characteristic of the inverter can be regarded as a negative resistance (here, the non-associated reference direction is chosen, so u_{P3} should be in phase with i_P). Besides, the switching frequency of the

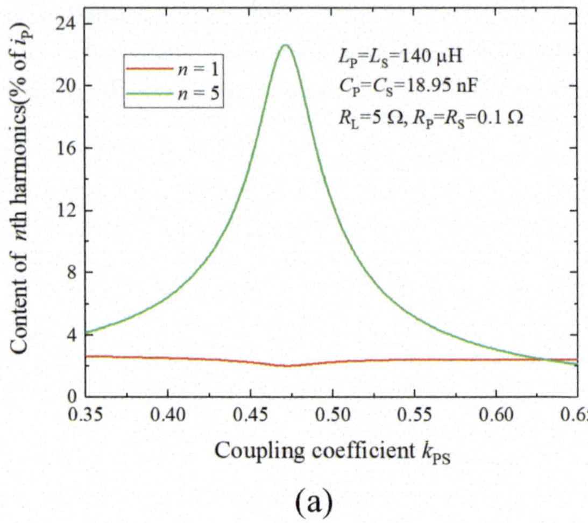

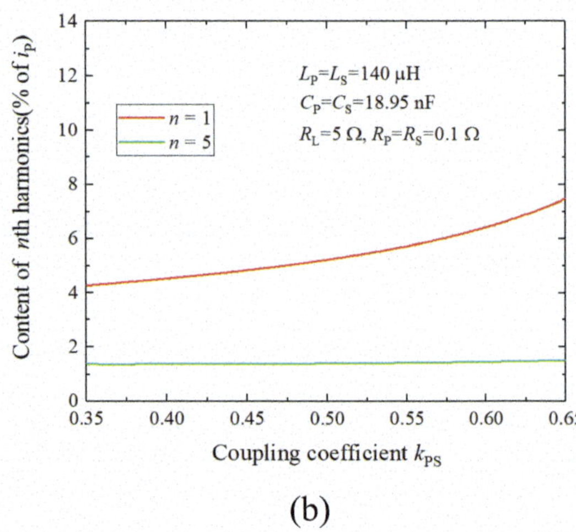

Fig. 9.20 Content of the fundamental and fifth harmonic in the current i_P **a** at $\omega = \omega_L$ and **b** at $\omega = \omega_H$

inverter is updated by measuring the frequency of the current i_P whether in FHO or THO (the control process will be shown in Fig. 9.25). In FHO, the switching frequency in FHO is set to equal to the frequency of i_P [9], as shown in Fig. 9.24b, while the switching frequency of THO is set to 1/3 of the frequency of i_P. According to (9.56), the frequencies of i_P under FHO and THO are equal. Therefore, it can be concluded that the switching frequency of the inverter can be reduced to 1/3 compared to the negative resistor based on FHO.

Fig. 9.21 Magnitude of the transmitter-side input impedance at the frequency of the first, third, and fifth harmonic components when **a** $\omega = \omega_L$ and **b** $\omega = \omega_H$

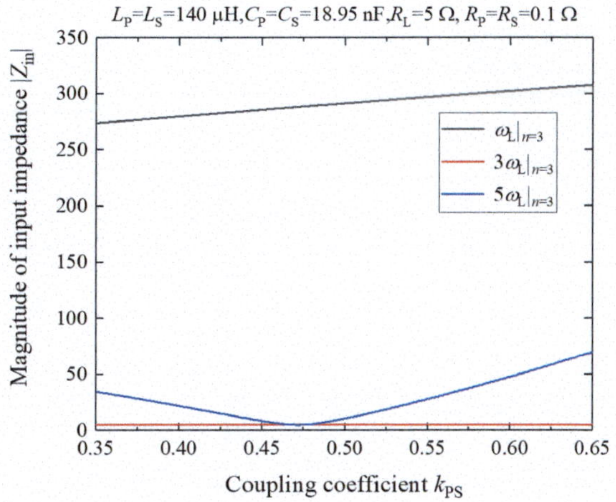

(a)

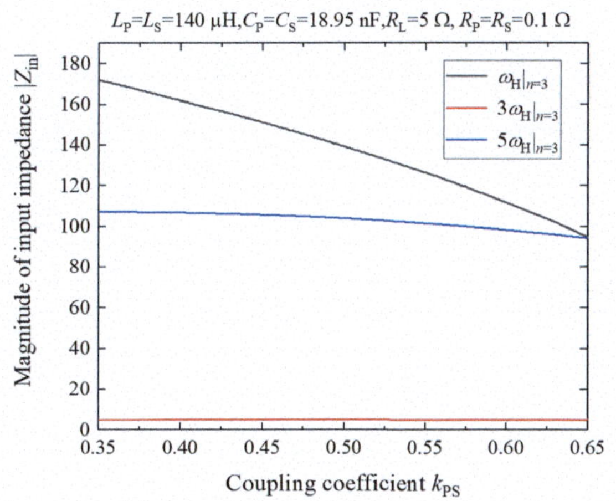

(b)

The generation mechanism of driver signals $U_{G1} \sim U_{G4}$ based on PSM is shown in Fig. 9.25. A capture module (CAP) and three pulse width modulation modules (PWM1, PWM2, and PWM3) in the digital signal processor (DSP) controller are used to produce $U_{G1} \sim U_{G4}$. Here, the CAP module is configured to be triggered by a rising edge. And the counters of the PWM1-3 (CTR1, CTR2, and CTR3) work in the down-count mode (that is, the CTR1-3 starts from the period value (PRD) and decrements until it equals zero. When it reaches zero, the CTR is reset to PRD and begins decrementing again). As shown in Fig. 9.25, a square wave (i'_P) in phase with i_P is first obtained by a zero-crossing detector. Then, the rising edge of the square wave is captured by the CAP module. When the third rising edge of the i'_P is captured, the i'_P is allowed as the synchronization signal of the PWM1 module (S_{n1}), and the

Fig. 9.22 Theoretical waveforms of i_P, i_{P3}, and u_P at $\omega = \omega_H$ **a** when $k_{PS} = 0.35$ and **b** $k_{PS} = 0.65$

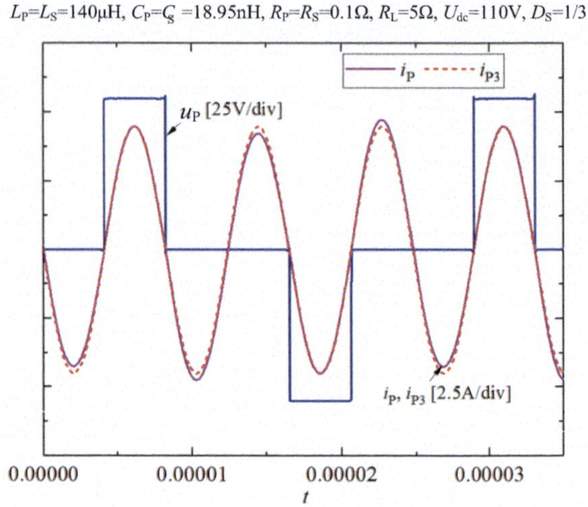

(a)

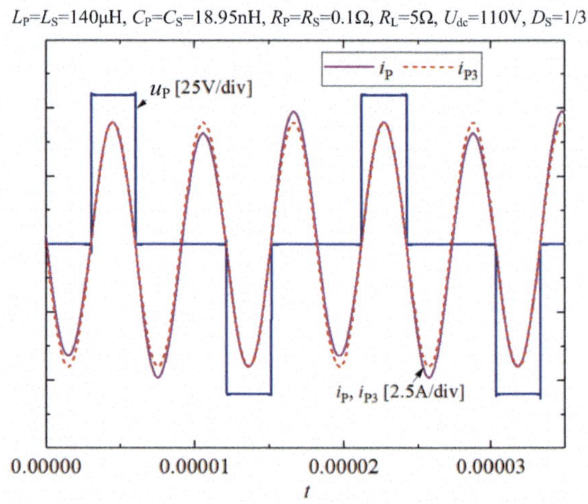

(b)

value of the counter of the PWM1 module (CTR1) is forced to the period value PRD when the fourth rising edge of the i'_P arrives. When CTR1 is equal to the compare value (CMP1), the synchronization signal S_{n2} is generated. At this point, the value of CTR2 is immediately updated to the initial value (Pha2). Once CTR2 equals zero, a synchronization signal S_{n3} is produced, and the value of CTR3 is updated to Pha3 immediately. The value of PRD, CMP1, Pha2, and Pha3 can be calculated by

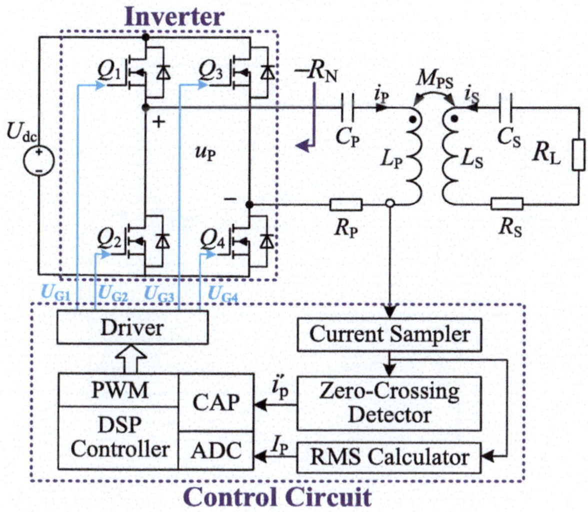

Fig. 9.23 Overall schematic diagram of the negative resistor

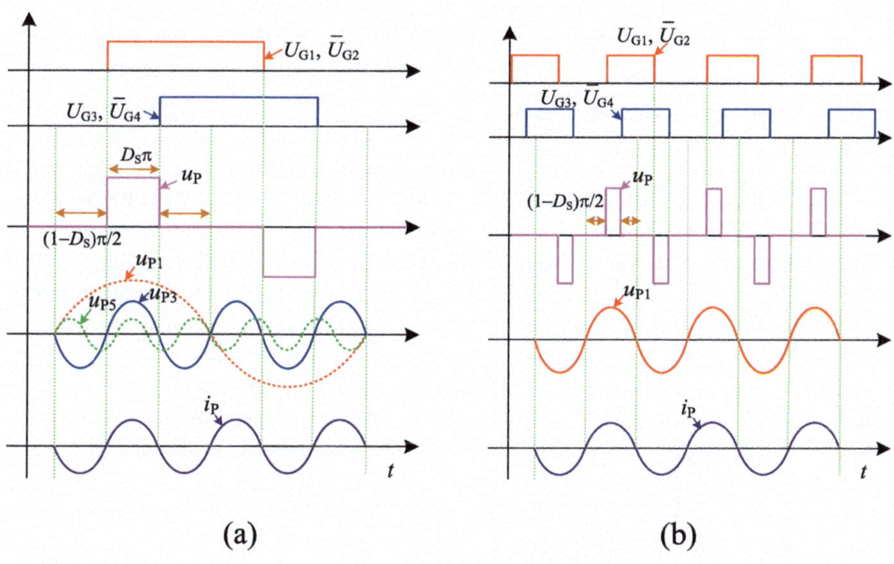

Fig. 9.24 Operating waveforms **a** in THO mode and **b** in FHO mode

$$\begin{cases} \text{PRD} = \dfrac{3T_{iP}}{T_{\text{CLK}}}, \text{CMP1} = \dfrac{5}{6}\text{PRD} \\ \text{Pha2} = \dfrac{1 - D_S}{4}\text{PRD}, \text{Pha3} = \dfrac{D_S}{2}\text{PRD} \end{cases}, \tag{9.70}$$

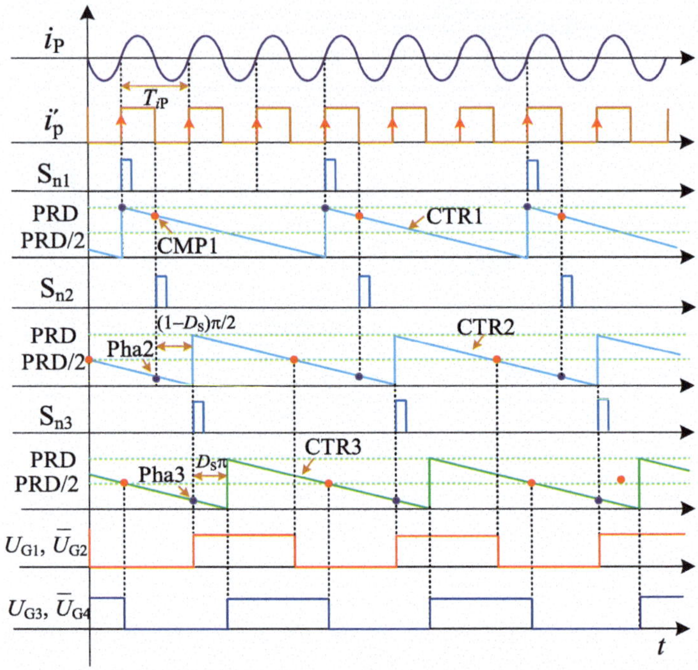

Fig. 9.25 Generation mechanism of the driver signals based on the phase synchronization method

where T_{iP} is the period of the current i_P, which is captured in real-time by the CAP module, and T_{CLK} represents the time required for the counter to decrease by 1. Next, according to Fig. 9.25, when CTR2 reduces to 0, U_{G1} is forced to output a high level, and when CTR2 is equal to half of the period (PRD/2), U_{G1} is forced to output a low level. U_{G2} is complementary to U_{G1}. Similarly, U_{G3} and U_{G4} can be generated based on CTR3. Controlling by $U_{G1} \sim U_{G4}$, the inverter outputs a desired ac square voltage to drive the resonant tank. To reliably start up the system, a drive signal with initial frequency ω_{int} will be given first. Then, according to the PSM, the switching frequency is constantly updated and becomes steady when the phase difference between the third harmonic u_{P3} and i_P is zero. Here, ω_{int} is set to larger than ω_0, and the value of PRD is limited to less than $PRD_0 = 3 \times 2\pi/(\omega_0 T_{CLK})$ to ensure that the system oscillates at $\omega = \omega_H$.

9.3 Reduction of Switching Losses

Recently, a robust MC-WPT system without the need for dual-side communication has been presented according to the PT-symmetric principle. It is shown that the PT-symmetric MC-WPT system can provide a consistent output power over a wide range of coupling coefficient variations. However, an operational amplifier circuit is employed to implement the negative resistor, which limits the power level of the PT-symmetric MC-WPT system and leads to extremely low system efficiency (only ~ 10%). By using a self-oscillating controlled inverter, a practical PT-symmetric MC-WPT system with high overall system efficiency is developed. However, the outputs of self-oscillating PT-symmetric MC-WPT systems cannot be adjusted according to demand unless a DC-DC converter is cascaded. In [10], phase-shift control is introduced to a PT-symmetric MC-WPT system for power control, and the merit is that only a single-stage inverter is required. However, soft switching cannot be achieved due to the varied duty cycle.

To solve the aforementioned problems, this work combines dual frequency modulation (DFM) and phase synchronization method (PSM) and applies them to the PT-symmetric MC-WPT system to realize controllable output power using a single-stage inverter while maintaining the soft switching or zero-voltage switching (ZVS) operation.

9.3.1 System Circuit Modeling

Figure 9.26 shows the equivalent circuit of the proposed PT-symmetric MC-WPT system, which is composed of a series compensated transmitter and receiver, and a load resistor. Different from traditional MC-WPT schemes, the transmitter contains a negative resistor $-R_N$. The power of the proposed system is instead injected by the negative resistor rather than a high-frequency source at a fixed frequency. R_P, L_P, and C_P (or R_S, L_S, and C_S) symbolize parasitic resistance, self-inductance, and compensation capacitance of the transmitter (or receiver), respectively. M_{PS} represents the mutual inductance between the transmitting and receiving coil. R_L is the load resistance. The natural frequencies of the transmitter and receiver are tuned to be equal, i.e., $f_0 = f_P = 1/(2\pi\sqrt{L_P C_P}) = f_S = 1/(2\pi\sqrt{L_S C_S})$, which is a necessary condition for PT-based systems. The series compensated resonant circuit presents a large impedance to harmonic components, which allows neglect of higher harmonics as they do not carry significant power. Thus, all the voltages and currents in the system are assumed to be sinusoidal. Using phasors, the voltage across the negative resistor (u_P), the voltage across the load (u_S), the transmitting current (i_P), and the receiving current (i_S) can be denoted as $\dot{U}_{P1}$, $\dot{U}_S$, $\dot{I}_P$, and $\dot{I}_S$, respectively. Based on the circuit theory, the PT-symmetric MC-WPT system can be expressed as

$$\begin{cases} (-R_N + R_P)\dot{I}_P + j\left(\omega L_P - \dfrac{1}{\omega C_P}\right)\dot{I}_P + j\omega M_{PS}\dot{I}_S = 0 \\ j\omega M_{PS}\dot{I}_P + (R_L + R_S)\dot{I}_S + j\left(\omega L_S - \dfrac{1}{\omega C_S}\right)\dot{I}_S = 0 \end{cases}, \tag{9.71}$$

where ω is the operating angular frequency of the system.

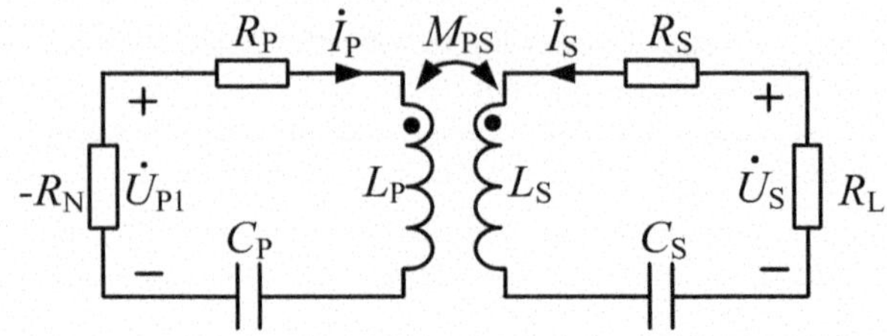

Fig. 9.26 First-harmonic equivalent circuit of the SS-based PT-symmetric MC-WPT system

Suppose that the transmitter and receiver have the matched resonance frequency, i.e., $\omega_P = 1/\sqrt{L_P C_P}$=$1/\sqrt{L_S C_S}$. Then, Eq. (9.71) can be rewritten into matrix form

$$\begin{bmatrix} \frac{-R_N+R_P}{L_P} + j\left(\omega - \frac{\omega_0^2}{\omega}\right) & j\omega k_{PS}\sqrt{\frac{L_S}{L_P}} \\ j\omega k_{PS}\sqrt{\frac{L_P}{L_S}} & \frac{R_L+R_S}{L_S} + j\left(\omega - \frac{\omega_0^2}{\omega}\right) \end{bmatrix} \begin{bmatrix} \dot{I}_P \\ \dot{I}_S \end{bmatrix} = 0, \tag{9.72}$$

where $k_{PS} = M_{PS}/\sqrt{L_P L_S}$ denotes the coupling coefficient between the transmitter and receiver.

To find the operating angular frequency ω, the imaginary and real parts of the coefficient matrix of (9.72) are separated, respectively

$$\begin{cases} \left(\omega - \frac{\omega_0^2}{\omega}\right)\left(\frac{R_L + R_S}{L_S} + \frac{-R_N + R_P}{L_P}\right) = 0 \\ \frac{-R_N + R_P}{L_P}\frac{R_L + R_S}{L_S} - \left(\omega - \frac{\omega_0^2}{\omega}\right)^2 + \omega^2 k_{PS}^2 = 0 \end{cases}. \tag{9.73}$$

According to (9.73), the operation frequency $f = \omega/2\pi$ can be derived as follows:

(1) In the exact PT-symmetric region.

$$f = f_{H,L} = f_0\sqrt{\frac{2 - \gamma^2 \pm \sqrt{\left[2 - \gamma^2\right]^2 - 4\left(1 - k_{PS}^2\right)}}{2\left(1 - k_{PS}^2\right)}}, k_C \le k_{PS} \le 1, \tag{9.74}$$

where $\gamma = (R_L + R_S)/(\omega_L L_S)$, $k_C = \sqrt{\gamma^2 - \gamma^4/4}$, and $f_0 = \omega_0/(2\pi)$. The + sign and—sign correspond to f_H and f_L, respectively.

(2) In the broken PT-symmetric region.

$$f = f_0,\ k_{PS} < k_C. \tag{9.75}$$

In the exact PT-symmetric range ($k_{PS} \ge k_C$), the system will automatically operate at $f = f_{H,L}$ to adapt to the variation of the coupling coefficients as long as the input power is provided by a negative resistor without the need to know the specific coupling coefficient [11].

In the exact PT-symmetric range, because of $\omega \neq \omega_0$, according to (9.72) and (9.73), the current gain is got

$$\frac{I_S}{I_P} = \left|\frac{\dot{I}_S}{\dot{I}_P}\right| = \sqrt{\frac{L_P}{L_S}}. \tag{9.76}$$

Combined with $U_{P1} = R_N I_P$ and $U_S = R_L I_S$, the voltage gain can be derived as

$$\frac{U_S}{U_{P1}} = \left|\frac{\dot{U}_S}{\dot{U}_{P1}}\right| = \frac{R_L\sqrt{L_P L_S}}{(R_L + R_S)L_P + R_P L_S} \tag{9.77}$$

Using (9.76) and (9.77), the coil-coil efficiency η_T and output power P_o can be expressed as

$$\begin{cases} \eta_T = \dfrac{I_S^2 R_L}{I_P^2 R_P + I_S^2 R_S + I_S^2 R_L} = \dfrac{R_L}{\frac{L_S}{L_P} R_P + R_S + R_L} \\ P_o = \dfrac{U_S^2}{R_L} = \dfrac{R_L V_{P1}^2}{\frac{L_P}{L_S}(R_L + R_S)^2 + 2R_P(R_L + R_S) + \frac{L_S}{L_P} R_P^2} \end{cases}. \tag{9.78}$$

From (9.78), it can be observed that the coil-coil efficiency and output power are determined only by the coil parameters, the input voltage U_{P1}, and the load resistor. Hence, the PT-symmetric system is inherently insensitive to the coupling coefficient. Besides, inspecting (9.78), the controllable output power that has nothing to the coupling coefficient can be easily obtained by controlling the value of U_{P1}. For the existing PT-symmetric MC-WPT systems, U_{P1} is adjusted by cascading dc-dc converters or changing the duty cycle of the inverter, which increases the system cost and size or destroys the soft switching operation. In this work, DFM and PSM are utilized to implement and control the PT-symmetric circuit, thereby achieving both adjustable U_{P1} and soft-switching operations based on a single-stage inverter. The proposed system will be given in the next section.

9.3.2 *Principle of DFM*

In the existing PT-symmetric MC-WPT systems, U_{P1} is adjusted by cascading DC-DC converters or changing the inverter's duty cycle [10, 12], which increases the system cost and size or destroys the soft switching operation. In this paper, DFP and PSM are utilized to implement and control the PT-symmetric circuit, thereby achieving both adjustable U_{P1} and soft-switching operations based on a single-stage inverter.

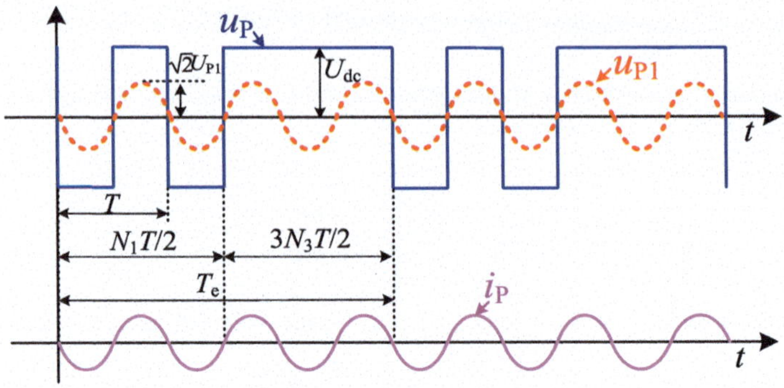

Fig. 9.27 Theoretical waveforms for DFM control scheme

The control scheme of the DFM technique is the operation at a constant duty cycle $D = 1$ and different switching frequencies. In this design, the proposed MC-WPT system is alternately modulated at two switching frequencies, namely, the frequency f and its subharmonic one $f/3$, as shown in Fig. 9.27 (it should be noted that, for PT-symmetric MC-WPT systems, the frequency f will automatically satisfy (9.74) in steady-state). By configuring the number of the two modulating pulses with frequency f and $f/3$, the root-mean-square (RMS) value of the equivalent fundamental component of the inverter output voltage (i.e., u_{P1}) will be adjusted. This guarantees a controllable output power while maintaining soft-switching or ZPA operation.

As presented in Fig. 9.27, it is assumed that the half-cycle number of the frequency-modulated square wave with frequency f is N_1, and the half-cycle number of the square wave with frequency $f/3$ is N_3. Using a Fourier series, the whole modulated-frequency square wave (u_P) can be expressed

$$u_P(t) = c_0 + \sum_{n=1}^{\infty} (a_n \cos 2\pi nft + b_n \sin 2\pi nft), \tag{9.79}$$

where

$$c_0 = \frac{1}{T_e}\int_0^{T_e} u_P(t)dt,\ a_n = \frac{2}{T_e}\int_0^{T_e} u_P(t)\cos 2\pi nft dt = 0$$
$$b_n = \frac{2}{T_e}\int_0^{T_e} u_P(t)\sin 2\pi nft dt,\ T_e = (N_1 + 3N_3)T/2, \tag{9.80}$$

and $T = 1/f$ is the period of the current i_P, and $T_e = N_1T/2 + 3N_3T/2$ is the period of the voltage u_P.

Due to the band-pass characteristics of SS-based MC-WPT systems, only the fundamental component of the square wave around the resonant frequency helps with energy transfer, and the harmonic components far away from the resonant frequency can be ignored. According to (9.79), the RMS of the fundamental component around the resonant frequency (i.e., u_{P1}), denoted as U_{P1}, can be deduced as

$$
\begin{aligned}
U_{\mathrm{P1}} = \frac{b_1}{\sqrt{2}} &= \frac{1}{\sqrt{2}} \frac{2}{N_1 \frac{T}{2} + N_3 \frac{3T}{2}} \left(\sum_{n=1}^{N_1} \left(\int_{\frac{(n-1)T}{2}}^{\frac{nT}{2}} (-1)^n U_{\mathrm{dc}} \sin 2\pi f t dt \right) \right. \\
&\left. + \sum_{n=N_1+1}^{N_1+N_3} \left(\int_{N_1 \frac{T}{2} + (n-N_1-1)\frac{3T}{2}}^{N_1 \frac{T}{2} + (n-N_1)\frac{3T}{2}} (-1)^n U_{\mathrm{dc}} \sin 2\pi f t dt \right) \right) \\
&= \frac{1}{\sqrt{2}} \frac{2}{N_1 \frac{T}{2} + N_3 \frac{3T}{2}} U_{\mathrm{dc}} \left(N_1 \frac{T}{\pi} + N_3 \frac{T}{\pi} \right) = \delta \frac{2\sqrt{2} U_{\mathrm{dc}}}{\pi},
\end{aligned}
\tag{9.81}
$$

where $\delta = (N_1 + N_3)/(N_1 + 3N_3)$. It can be found that the output voltage can be regulated by configuring N_1 and N_3 without changing the input DC voltage U_{dc} or duty cycle. The half-cycle numbers N_1 and N_3 are integers, and the solutions of N_1 and N_3 can be obtained according to the required output power.

9.3.3 Implementation of Negative Resistor Based on DFM

Different control schemes of the negative resistor have been studied in reported works, such as self-oscillating control and phase shift control. However, the former cannot adjust the voltage across the negative resistor unless cascaded with a DC-DC converter, and the latter suffers from a hard-switching problem. To achieve both adjustable voltage and soft-switching operations based on a single-stage H-bridge inverter, a new control scheme based on DFM and PSM is proposed in this work.

The configuration of the proposed PT-symmetric MC-WPT system is illustrated in Fig. 9.28, in which an H-bridge inverter controlled by DFM and PSM behaves like -R_{N}. From Fig. 9.28, the control circuit contains seven blocks. The current sampler is composed of a current transformer (CT), a differential amplifier circuit, and an *RC* time-leading compensator, which converts the i_{P} into a voltage signal. A zero-cross comparator and a 1/3 clock divider generate the square wave i'_{P} and $i'_{\mathrm{P\_1/3}}$, respectively. i'_{P} is in phase with i_{P} and has the same frequency as i_{P}, while the frequency of $i'_{\mathrm{P\_1/3}}$ is one-third of i_{P}, and its rising edge is consistent with i'_{P}. A digital signal processor (DSP) controller is used to capture the edges of $i'_{\mathrm{P\_1/3}}$ and output the control signal S_{C} based on DFM and PSM. The frequency selection circuit then determines the signal sent to the dead-time generation circuit (i'_{P} or $i'_{\mathrm{P\_1/3}}$) according to the control signal S_{C}. As a result, a new hybrid pulse sequence is generated, which involves two frequencies. Finally, the dead-time generation circuit

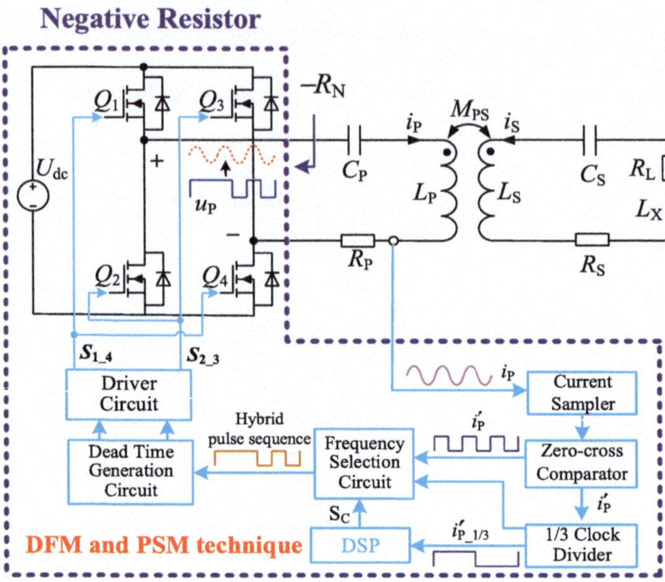

Fig. 9.28 Full schematic diagram of the proposed PT-symmetric WPT system

converts the hybrid pulse sequence into two complementary square waves and makes a proper dead-time, and the drive circuit generates the driving signal (S_{1_4} and S_{2_3}).

Here, taking the case of $N_1 = 3$ and $N_3 = 3$ as an example to illustrate the operating principle of realizing a negative resistor based on DFM. Figure 9.29 depicts the ideal waveforms of the inverter output voltage and current and the corresponding drive signals. From Fig. 9.29, it can be seen that the two drive signals are complementary square wave sequences with a constant duty cycle of 1 and two different frequencies (f and $f/3$). Besides, using the drive signals, the fundamental component of u_P can be controlled to be in phase with the current i_P, (that is, the inverter behaves like a resistor). Therefore, the core of control is to accurately obtain the required driving signals.

The generation process of the drive signals (S_{1_4} and S_{2_3}) is presented in Fig. 9.30. Firstly, the transmitting current i_P is sensed, and the square waves with a frequency of f (i'_P) and a frequency of $f/3$ ($i'_{P\_1/3}$) are generated by the zero-crossing comparator and a 1/3 clock divider, respectively. As shown in Fig. 9.30a, the rising edges of i'_P and $i'_{P\_1/3}$ correspond to the zero-crossings of the current i_P. Then, under the control of the signal S_C, i'_P and $i'_{P\_1/3}$ will be combined into a new pulse sequence by the frequency selection circuit. According to Fig. 9.30a, when the control signal S_C is at a high level, $i'_{P\_1/3}$ will be changed into two complementary square waves (S_{1_4} and S_{2_3}) to drive the inverter. When S_C becomes low level, i'_P is allowed to pass through the frequency selection circuit to generate the drive signal. Next, how to produce S_C will be analyzed in detail. From Fig. 9.30, the rising and falling edge of $i'_{P\_1/3}$ are first captured by the enhanced capture (eCAP) modules of DSP to count the captured

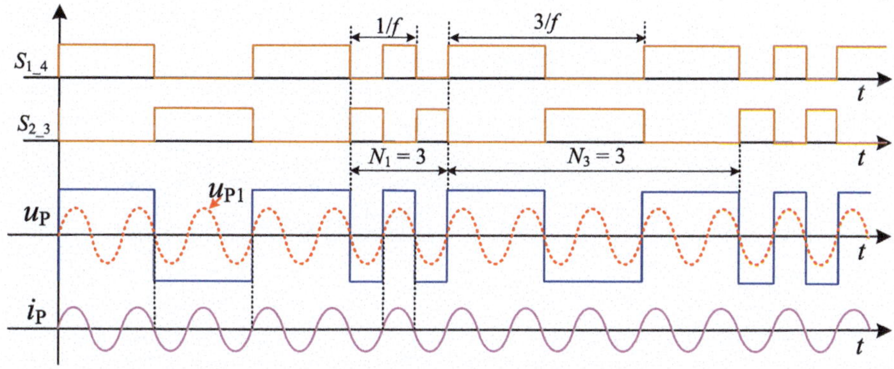

Fig. 9.29 Ideal voltage and current waveforms and corresponding drive signals under DFM when $N_1 = 3$ and $N_3 = 3$

half-cycle number of $i'_{P\_1/3}$ (variable $N_{\_Cap}$ is used to store the count value, that is, its value is incremented by 1 when a half-cycle of $i'_{P\_1/3}$ is captured), and then an enhanced pulse width modulator (ePWM) module of DSP is adopted to generate the control signal S_C according to the required N_1 and N_3. The counter of the ePWM module (CTR) works in the down-count mode as shown in Fig. 9.30a. When CTR = PRD, S_C will be forced to a high level, and S_C is set to be a low level when CTR is equal to the compare value (CMP). The PRD and CMP can be obtained by

$$\begin{cases} \text{PRD} = \dfrac{N_1 T/2 + 3N_3 T/2}{T_{\text{CLK}}} \\ \text{CMP} = \dfrac{N_1/3}{N_1/3 + N_3}\text{PRD} \end{cases}, \tag{9.82}$$

where $T = 1/f$ is the period of i_P, and T_{CLK} is the clock cycle of the ePWM modules. Besides, the moment when the level of S_C changes from low to high needs to be precisely controlled.

As shown in Fig. 9.30a, when the $(N_1/3 + N_3 - 1)th$ half-cycle number of $i'_{P\_1/3}$ is captured, the synchronization input of the ePWM module is enabled and $i'_{P\_1/3}$ is used as the external synchronization signal, which means that CTR is set to equal PRD when the next rising edge of $i'_{P\_1/3}$ arrives (Since only the rising edge of the synchronization signal is valid, if $(N_1/3 + N_3)$ is an odd number, the reverse signal $\overline{i'_{P\_1/3}}$ needs to replace $i'_{P\_1/3}$ as the synchronization signal of the ePWM module). After that, the synchronization input of the ePWM module is disabled, and the count value of the number of half-cycles is cleared to prepare for the next cycle (i.e., $N_{\_Cap} = 0$). Using the above synchronization control method, drive signals (S_{1_4} and S_{2_3}) would generate as expected, and the fundamental component of u_P will be in phase with the current i_P. As a result, the proposed PT-based system automatically oscillates at $f = f_{\text{H,L}}$ when $k_{\text{PS}} \geq k_C$ in a steady state (or $f = f_0$ when $k_{\text{PS}} < k_C$).

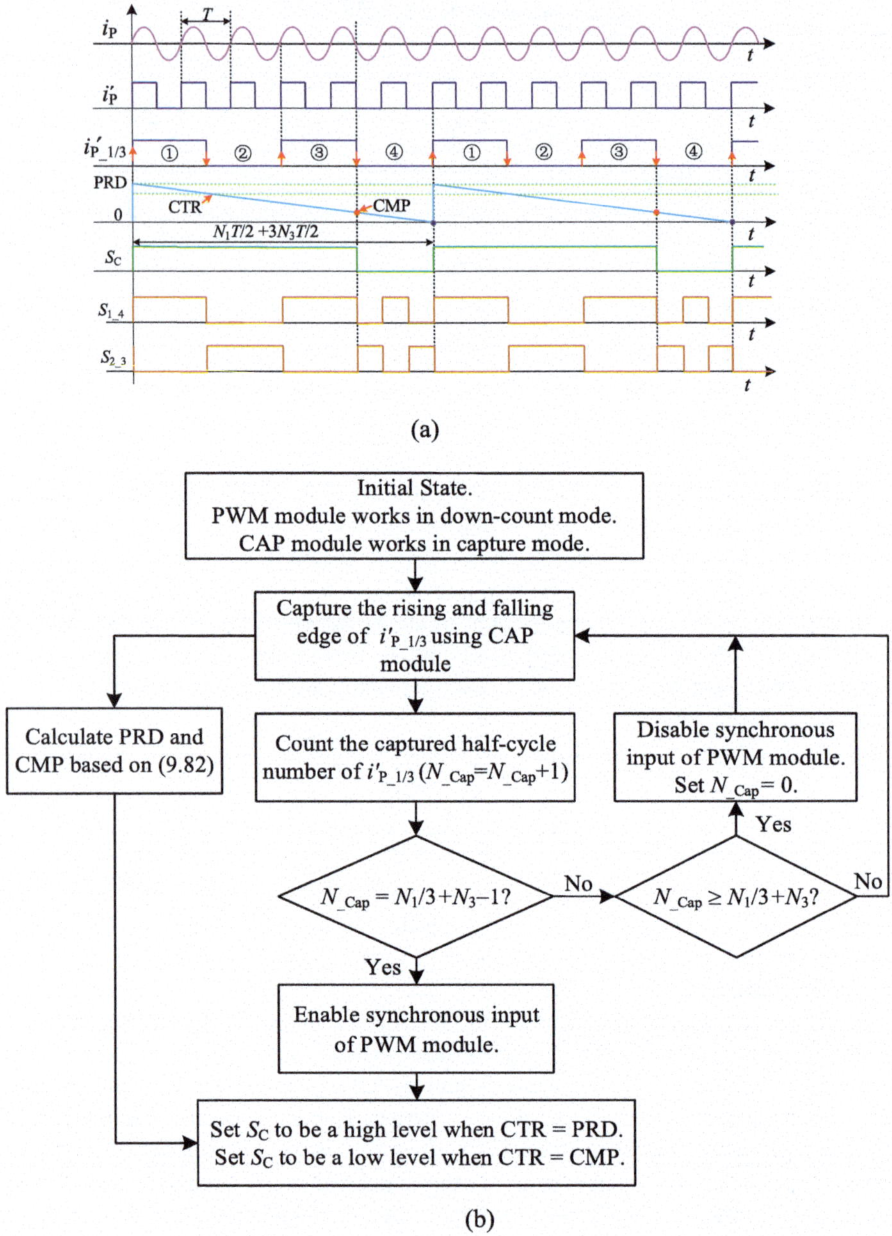

Fig. 9.30 Generation process of the drive signals based on DFM and PSM. **a** Detail waveforms. **b** Flowchart for generating control signal S_C

9.3.4 Soft-Switching Analysis and Discussion

To avoid the switching losses of MOSFETs, a soft-switching operation is commonly required. Figure 9.31 displays the theoretical waveforms of the inverter output voltage and current, voltage across the switches (u_{DS}), and current through switches (i_{DS}) under DFM and PSM for $N_1 = 3$ and $N_3 = 3$. It is shown that the turn-on/turn-off instant of Q_1 and Q_3 is always with zero current under DFM control, which means that soft-switching on and off are both realized. Besides, Q_4 and Q_1 (or Q_2 and Q_3) have similar behavior. Therefore, ideally, the proposed method can maintain full-range soft-switching against the variable output power for all switches. Furthermore, the average switching frequency of the inverter using DFM is lower compared with single frequency modulation, and thus it further reduces switching losses.

Because of the switching delay time of MOSFETs, a dead time t_{de} for the MOSFETs in the same bridge arm is required to prevent the shoot-through issue. Figure 9.32a displays the actual waveforms of the inverter output voltage and current, voltage across the switches, and current through switches, where the phase of inverter voltage leading to its current is expressed as $\Delta\varphi_L = \omega t_L$. From Fig. 9.32a, the ZVS can be obtained when $\Delta\varphi_L = \omega t_L \geq \omega t_{de}$ is satisfied. The detailed analysis is given as follows.

Taking the time period t_0-t_1 as an example. It can be found that the driving signals of the four MOSFETs during t_0-t_1 are all low-level. The equivalent circuit viewed from the transmitter during this time period is shown in Fig. 9.32b. To realize ZVS of Q_1 and Q_4, the current i_{DS} during t_0-t_1 should be sufficient to charge C_{oss1} and C_{oss4} from U_{dc} to 0 V and discharge C_{oss2} and C_{oss3} from 0 V to U_{dc}. Assumed that the parasitic output capacitance of the four MOSFETs is equal to C_{oss}, the integral of the transmitter coil current i_P (Q_{ZVS}) during t_{de} can be calculated by

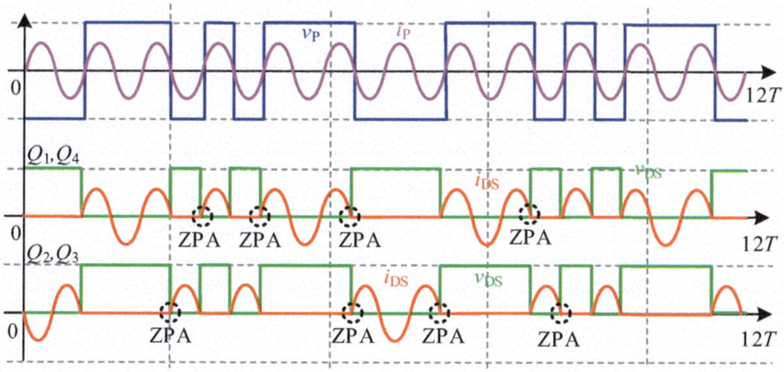

Fig. 9.31 Soft-switching analysis under DFM when $N_1 = 3$ and $N_3 = 3$

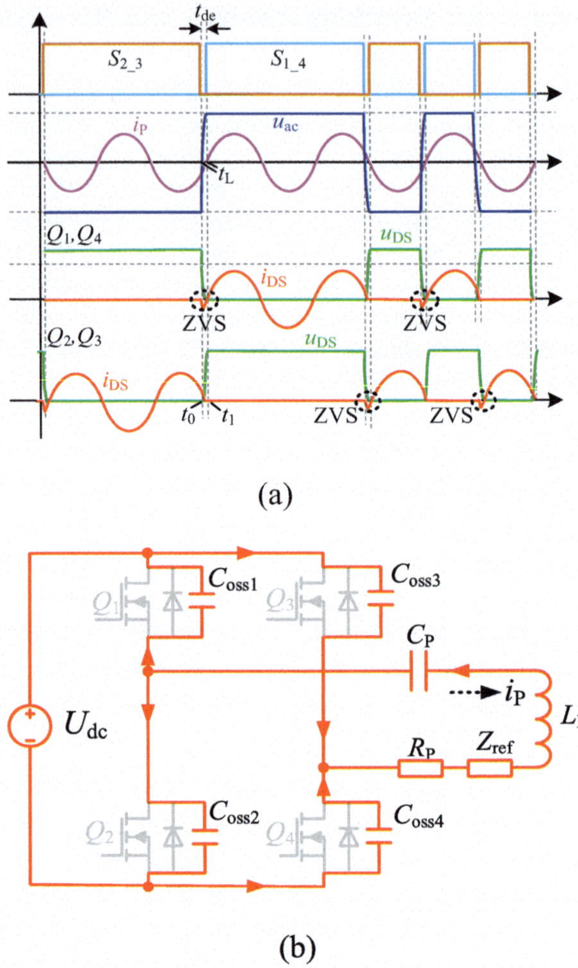

Fig. 9.32 Analysis of soft-switching technology considering the dead time. **a** Actual waveforms. **b** Equivalent circuit viewed from the transmitter during $t_0 - t_1$

$$Q_{ZVS} = \int_0^{t_{de}} i_P dt \geq \int_0^{U_{dc}} \frac{C_{oss}}{2} du. \tag{9.83}$$

Assuming that the waveform of i_P during t_{de} is a sine wave, based on (9.83), it can be obtained

$$\int_0^{t_{de}} \frac{-\sqrt{2}P_0 \sin(\omega t - \omega t_L)}{\eta_T U_P \cos(\omega t_L)} dt \geq \frac{U_{dc} C_{oss}}{2}. \tag{9.84}$$

From (9.84), to achieve ZVS, the dead time should be satisfied

$$t_{\mathrm{L}} \geq t_{\mathrm{de}} \geq t_{\mathrm{de\_min}} = \frac{1}{\omega} \arccos\left(\frac{1}{1 + \delta U_{\mathrm{dc}}^2 \omega C_{\mathrm{oss}} \eta_{\mathrm{T}} / (\pi P_{\mathrm{o}})}\right). \tag{9.85}$$

It can be found that $t_{\mathrm{de\_min}}$ is positively correlated with C_{oss} and η_{T} and negatively correlated with P_{o} and ω. In order to realize ZVS in the full range of operation conditions, the t_{L} and t_{de} should be greater than the maximum value of $t_{\mathrm{de\_min}}$, which can be obtained by substituting the maximum C_{oss} and η_{T}, and minimum P_{o} and ω into (9.85). It can be found that the larger the value of t_{L}, the more beneficial it is for soft-switching operation. However, a too-large t_{L} may destroy the negative-resistance output characteristics of the inverter. Since the C_{oss} of MOSFETs is usually very small, the required charging and discharging time is very short, that is, the required $t_{\mathrm{de\_min}}$ is very short. Therefore, it can be approximated that the fundamental component of the inverter output voltage and its output current still maintain a zero-phase angle relationship. Additionally, due to waveform symmetry, when $t_{\mathrm{L}} \geq t_{\mathrm{de}} \geq t_{\mathrm{de\_min}}$ is satisfied, the ZVS of Q_2 and Q_3 can also be achieved. It can be seen that when the required output power P_0 is small, the required t_{L} may need to be larger. In order to implement of negative resistor, the ZVS characteristic may disappear. In this case, a take-off needs to be made between the ZVS operation and the implementation of the negative resistor. However, it should be emphasized that although ZVS may not be achieved at this time, ZPA operation can still be achieved. When the system operates at ZPA, the switches always turn on/off at the moment of the current zero crossing. Therefore, the switching loss can also be kept very low at ZPA operation.

9.4 Three-Phase PT-Symmetric Wireless Power Transfer

At present, the research of MC-WPT technology mainly focuses on a single-phase MC-WPT system, which performs an uneven distribution of magnetic field, causing high sensitivity to the spatial of the coils. Starting from this perspective, multi-phase MC-WPT systems have been proposed successively, increasing the degree of freedom of receiving coils. The multiphase systems provide the possibility of realizing omnidirectional wireless power transmission, as well as a lower device requirement in high-power situations compared to single-phase systems. Therefore, with the introduction of PT symmetry, the three-phase PT-symmetric system has a higher distance-insensitive efficiency, as well as self-regulation of the phase sequence.

9.4.1 System Modeling

In the switching converters, the voltage-source inverter has the natural voltage limiting characteristic, which meets the requirement that the output equivalent resistance is negatively correlated with the oscillator current when the three-phase phase-locked loop (PLL) is used, so the three-phase negative resistor is constructed by a three-phase inverter with three-phase PLL controller. The system structure is shown in Fig. 9.33a, and the equivalent circuit of the three-phase PT-symmetric MC-WPT system is shown in Fig. 9.33b. Here, the inductances of the transmitting coils and receiving coils are L_i and L_j, each of which has a matching capacitor C_i and C_j. The intrinsic resistance of the transmitting coils is R_i, and the intrinsic resistance of the receiving coils is $R_{i'}$. R_{Lj} is the three-phase load. $-R_{Ni}$ is the equivalent three-phase negative resistance. Here, $i = 1, 2, 3$ and $j = 1', 2', 3'$.

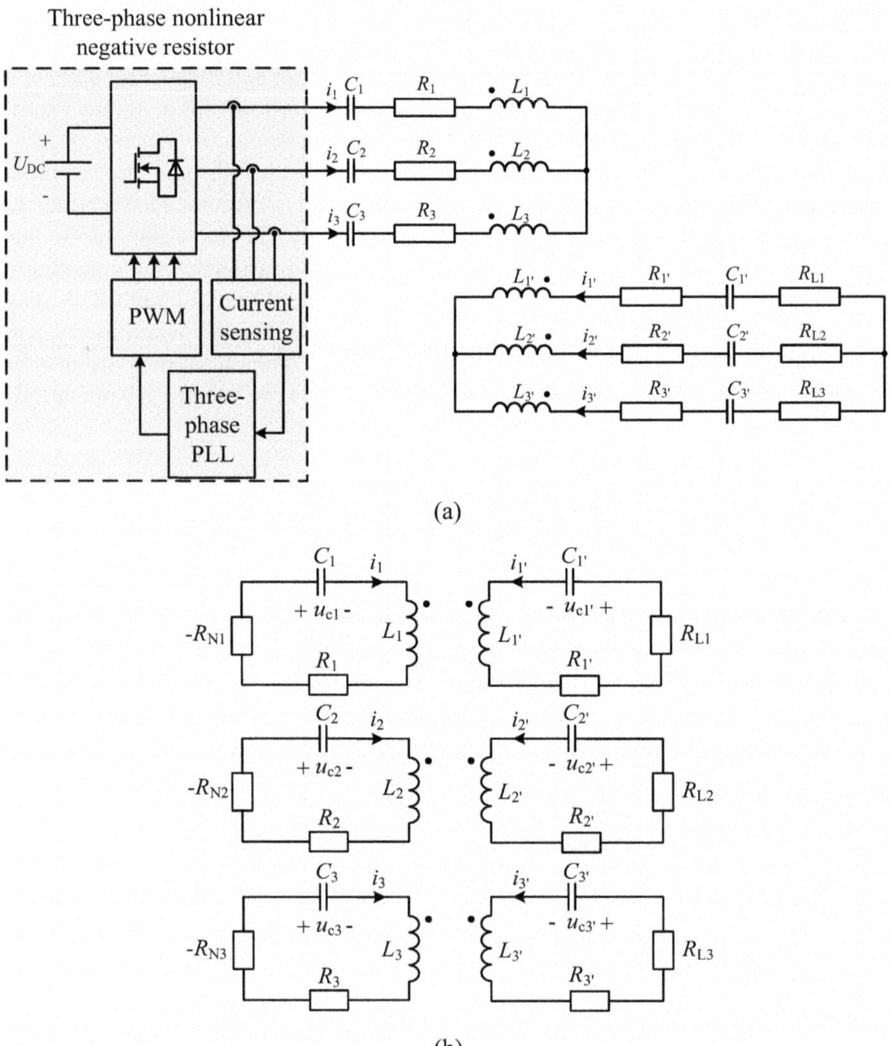

Fig. 9.33 The schematic diagram of the three-phase PT-symmetric MC-WPT system. **a** System structure. **b** Equivalent circuit of the system

According to the coupled-mode theory [2, 13], the coupled-mode equation of the system shown in Fig. 9.33 can be written as

$$\frac{d}{dt}\begin{bmatrix} \boldsymbol{a}_1 \\ \boldsymbol{a}_2 \\ \boldsymbol{a}_3 \\ \boldsymbol{a}_{1'} \\ \boldsymbol{a}_{2'} \\ \boldsymbol{a}_{3'} \end{bmatrix} = \begin{bmatrix} j\omega_1 + g_1 & 0 & 0 & -j\kappa_{11'} & -j\kappa_{12'} & -j\kappa_{13'} \\ 0 & j\omega_2 + g_2 & 0 & -j\kappa_{21'} & -j\kappa_{22'} & -j\kappa_{23'} \\ 0 & 0 & j\omega_3 + g_3 & -j\kappa_{31'} & -j\kappa_{32'} & -j\kappa_{33'} \\ -j\kappa_{1'1} & -j\kappa_{1'2} & -j\kappa_{1'3} & j\omega_{1'} - \tau_{1'} & 0 & 0 \\ -j\kappa_{2'1} & -j\kappa_{2'2} & -j\kappa_{2'3} & 0 & j\omega_{2'} - \tau_{2'} & 0 \\ -j\kappa_{3'1} & -j\kappa_{3'2} & -j\kappa_{3'3} & 0 & 0 & j\omega_{3'} - \tau_{3'} \end{bmatrix} \begin{bmatrix} \boldsymbol{a}_1 \\ \boldsymbol{a}_2 \\ \boldsymbol{a}_3 \\ \boldsymbol{a}_{1'} \\ \boldsymbol{a}_{2'} \\ \boldsymbol{a}_{3'} \end{bmatrix}, \tag{9.86}$$

where $\boldsymbol{a}_i$ and $\boldsymbol{a}_j$ represent the energy modes of the transmitter and receiver. κ_{ij} is the coupling rate between the transmitter and receiver, which is listed in Table 9.3. Besides, $g_i = g_{Ni}$-τ_{i0} is the whole gain rate of the transmitters, $g_{Ni} = G_0/|\boldsymbol{a}_i|$ is the gain rate of the negative resistor which merely depends on the amplitude $|\boldsymbol{a}_i|$ of the modes, τ_{i0} is the intrinsic loss rate of transmitter internal resistance. $\tau_j = \tau_{j0} + \tau_{Lj}$ is the whole loss rate of the receivers, τ_{j0} is the intrinsic loss rate of receiver internal resistance, τ_{Lj} is the loss rate of the load resistance. Especially, the parameters M_{tra} and M_{rec} can be neglected by constructing the structure of coils.

In the field of electronics, the parity operation is equivalent to the interchange of labels corresponding to pairs of associated circuit components, while the time reversal merely represents the reversal of the origin energy flow [14]. Performing combined parity and time transformations on (9.86), the PT-symmetric conditions can be obtained as

$$\begin{cases} \omega_1 = \omega_{1'},\ \omega_2 = \omega_{2'},\ \omega_3 = \omega_{3'} \\ \kappa_{11'} = \kappa_{1'1},\ \kappa_{22'} = \kappa_{2'2},\ \kappa_{33'} = \kappa_{3'3}\,. \\ \kappa_{12'} = \kappa_{21'},\ \kappa_{13'} = \kappa_{31'},\ \kappa_{23'} = \kappa_{32'} \end{cases} \tag{9.87}$$

Table 9.3 The definition of the mutual inductances and coupling rates

Categories	Symbols
Mutual inductance between the transmitting coils M_{tr}	M_{12}, M_{13}, M_{23}
Mutual inductance between the receiving coils M_{rec}	$M_{1'2'}, M_{1'3'}, M_{2'3'}$
Mutual inductance between the main coupling coils of transmitter and receiver M_{main}	$M_{11'}, M_{22'}, M_{33'}$
Mutual inductance between the cross-coupling ones M_{cross}	$M_{12'}, M_{13'}, M_{21'}, M_{23'}, M_{31'}, M_{32'}$
Coupling rate between the transmitting coils κ_{tr}	$\kappa_{12}, \kappa_{13}, \kappa_{23}$
Coupling rate between the receiving coils κ_{rec}	$\kappa_{1'2'}, \kappa_{1'3'}, \kappa_{2'3'}$
Coupling rate between the main coupling coils of transmitter and receiver κ_{main}	$\kappa_{11'}, \kappa_{22'}, \kappa_{33'}$
Coupling rate between the cross-coupling ones κ_{cross}	$\kappa_{12'}, \kappa_{13'}, \kappa_{21'}, \kappa_{23'}, \kappa_{31'}, \kappa_{32'}$

It can be seen that (9.87) is not much different from the single-phase system described in Chap. 3, but the cross-coupling rate of any two phases is required to be equal, which indicates the spatial symmetry requirement of the coils to a certain extent.

Assuming that the natural resonant frequencies of the coils, gain rates of the transmitting coils, and the loss rate of the receiving coils are identical, that is $\omega_1 = \omega_{1'} = \omega_2 = \omega_{2'} = \omega_3 = \omega_{3'} = \omega_0$, $g_1 = g_2 = g_3 = g$ and $\tau_{1'} = \tau_{2'} = \tau_{3'} = \tau$, (9.86) can be rewritten as

$$\frac{d}{dt}\begin{bmatrix} a_1 \\ a_2 \\ a_3 \\ a_{1'} \\ a_{2'} \\ a_{3'} \end{bmatrix} = \begin{bmatrix} j\omega_0 + g & 0 & 0 & -j\kappa_{11'} & -j\kappa_{12'} & -j\kappa_{13'} \\ 0 & j\omega_0 + g & 0 & -j\kappa_{21'} & -j\kappa_{22'} & -j\kappa_{23'} \\ 0 & 0 & j\omega_0 + g & -j\kappa_{31'} & -j\kappa_{32'} & -j\kappa_{33'} \\ -j\kappa_{11'} & -j\kappa_{21'} & -j\kappa_{31'} & j\omega_0 - \tau & 0 & 0 \\ -j\kappa_{12'} & -j\kappa_{22'} & -j\kappa_{32'} & 0 & j\omega_0 - \tau & 0 \\ -j\kappa_{13'} & -j\kappa_{23'} & -j\kappa_{33'} & 0 & 0 & j\omega_0 - \tau \end{bmatrix} \begin{bmatrix} a_1 \\ a_2 \\ a_3 \\ a_{1'} \\ a_{2'} \\ a_{3'} \end{bmatrix}, \tag{9.88}$$

9.4.2 Characteristics Analysis

According to the circuit model of the three-phase PT-symmetric MC-WPT system, the characteristic equation can be derived as

$$\begin{vmatrix} j(\omega-\omega_0) - g & 0 & 0 & j\kappa_{11'} & j\kappa_{12'} & j\kappa_{13'} \\ 0 & j(\omega-\omega_0) - g & 0 & j\kappa_{21'} & j\kappa_{22'} & j\kappa_{23'} \\ 0 & 0 & j(\omega-\omega_0) - g & j\kappa_{31'} & j\kappa_{32'} & j\kappa_{33'} \\ j\kappa_{11'} & j\kappa_{21'} & j\kappa_{31'} & j(\omega-\omega_0) + \tau & 0 & 0 \\ j\kappa_{12'} & j\kappa_{22'} & j\kappa_{32'} & 0 & j(\omega-\omega_0) + \tau & 0 \\ j\kappa_{13'} & j\kappa_{23'} & j\kappa_{33'} & 0 & 0 & j(\omega-\omega_0) + \tau \end{vmatrix} = 0 \tag{9.89}$$

By expanding the determinant, there is

$$ax^3 + bx^2 + cx + d = 0, \tag{9.90}$$

where $a = 1, b = \kappa_{11'}^2 + \kappa_{12'}^2 + \kappa_{13'}^2 + \kappa_{21'}^2 + \kappa_{22'}^2 + \kappa_{23'}^2 + \kappa_{31'}^2 + \kappa_{32'}^2 + \kappa_{33'}^2$, $c = (\kappa_{11'}\kappa_{22'} - \kappa_{12'}\kappa_{21'})^2 + (\kappa_{11'}\kappa_{23'} - \kappa_{13'}\kappa_{21'})^2 + (\kappa_{11'}\kappa_{32'} - \kappa_{12'}\kappa_{31'})^2 + (\kappa_{11'}\kappa_{33'} - \kappa_{13'}\kappa_{31'})^2 + (\kappa_{12'}\kappa_{23'} - \kappa_{13'}\kappa_{21'})^2 + (\kappa_{11'}\kappa_{33'} - \kappa_{13'}\kappa_{32'})^2 + (\kappa_{21'}\kappa_{32'} - \kappa_{22'}\kappa_{31'})^2 + (\kappa_{21'}\kappa_{33'} - \kappa_{23'}\kappa_{31'})^2 + (\kappa_{22'}\kappa_{33'} - \kappa_{23'}\kappa_{31'})^2$, $d = [\kappa_{11'}(\kappa_{22'}\kappa_{33'} - \kappa_{23'}\kappa_{32'}) + \kappa_{12'}(\kappa_{23'}\kappa_{31'} - \kappa_{21'}\kappa_{33'}) + \kappa_{13'}(\kappa_{21'}\kappa_{32'} - \kappa_{22'}\kappa_{31'})]^2$, $x = [j(\omega - \omega_0) - g][j(\omega - \omega_0) + \tau]$.

Then the characteristic equation can be transformed into a cubic equation of variable x, whose solution could be discussed by adopting the Van Shengjin formula.

(1) Ignoring the cross-coupling between transmitting and receiving circuits.

The situation represents $\kappa_{11'}$, $\kappa_{22'}$, $\kappa_{33'} \neq 0$ and $\kappa_{12'} = \kappa_{13'} = \kappa_{21'} = \kappa_{23'} = \kappa_{31'} = \kappa_{32'} \neq 0$, and applying these conditions to (9.90) the coefficients of a, b, c, d turns to

$$\begin{cases} a = 1 \\ b = \kappa_{11'}^2 + \kappa_{22'}^2 + \kappa_{33'}^2 \\ c = \kappa_{11'}\kappa_{22'}^2 + \kappa_{11'}\kappa_{33'}^2 + \kappa_{22'}\kappa_{33'}^2 \\ d = \kappa_{11'}^2\kappa_{22'}^2\kappa_{33'}^2 \end{cases}. \tag{9.91}$$

According to the Van Shengjin formula, currently the discriminant and the correlation discriminant coefficient of the root can be given by

$$\begin{cases} A = \kappa_{11'}^4 + \kappa_{22'}^2 + \kappa_{33'}^2 - \kappa_{11'}^2\kappa_{22'}^2 - \kappa_{11'}^2\kappa_{33'}^2 - \kappa_{22'}^2\kappa_{33'}^2 \\ B = \left(\kappa_{11'}^2 + \kappa_{22'}^2 + \kappa_{33'}^2\right)\left(\kappa_{11'}^2\kappa_{22'}^2 + \kappa_{11'}^2\kappa_{33'}^2 + \kappa_{22'}^2\kappa_{33'}^2\right) - 9\kappa_{11'}^2\kappa_{22'}^2\kappa_{33'}^2 \\ C = \kappa_{11'}^4\kappa_{22'}^4 + \kappa_{11'}^4\kappa_{33'}^4 + \kappa_{22'}^4\kappa_{33'}^4 - \left(\kappa_{11'}^2 + \kappa_{22'}^2 + \kappa_{33'}^2\right)\kappa_{11'}^2\kappa_{22'}^2\kappa_{33'}^2 \\ \Delta = -3\left(\kappa_{11'}^2 - \kappa_{22'}^2\right)^2\left(\kappa_{11'}^2 - \kappa_{33'}^2\right)^2\left(\kappa_{22'}^2 - \kappa_{33'}^2\right)^2 \end{cases}. \tag{9.92}$$

Then the specific situation of the root of (9.90) can be obtained explicitly.

(a) $\kappa_{11'} = \kappa_{22'} = \kappa_{33'}$.

In this case, $A = B = \Delta = 0$, which means that the solution of (9.90) is triple real roots, that is

$$x_1 = x_2 = x_3 = -\kappa_{11'}^2. \tag{9.93}$$

Combining the detailed expression of variable x, the characteristic Eq. (9.89) can be written as

$$[j(\omega - \omega_0) - g][j(\omega - \omega_0) + \tau] + \kappa_{11'}^2 = 0. \tag{9.94}$$

Consequently, the eigenfrequency of the three-phase PT-symmetric MC-WPT system is given by

$$\omega = \omega_0 \pm \sqrt{\kappa_{11'}^2 - \tau^2}, \tag{9.95}$$

which is consistent with that of the characteristic equation of single-phase PT-symmetric MC-WPT system [3, 6, 15], as well as the subsequent analysis, so it will not be further discussed here.

(b) $\kappa_{11'} = \kappa_{22'} \neq \kappa_{33'}$.

In this case, $\Delta = 0$ but $A \neq B \neq 0$, implying the solution of (9.90) has three real roots, one of which is a double root, that is

$$\begin{cases} x_1 = -\dfrac{b}{a} + K = -\kappa_{33'}^2 \\ x_2 = x_3 = -\dfrac{1}{2}K = -\kappa_{11'}^2 \end{cases}, \tag{9.96}$$

where $K = B/A$.

The characteristic Eq. (9.89) can be described by

$$\begin{cases} \left[j(\omega - \omega_0) - g\right]\left[j(\omega - \omega_0) + \tau\right] + \kappa_{11'}^2 = 0 \\ \left[j(\omega - \omega_0) - g\right]\left[j(\omega - \omega_0) + \tau\right] + \kappa_{33'}^2 = 0 \end{cases}. \tag{9.97}$$

Accordingly, the eigenfrequency of the three-phase PT-symmetric MC-WPT system is given by

$$\begin{cases} \omega = \omega_0 \pm \sqrt{\kappa_{11'}^2 - \tau^2} \\ \omega = \omega_0 \pm \sqrt{\kappa_{33'}^2 - \tau^2} \end{cases}. \tag{9.98}$$

(c) $\kappa_{11'} \neq \kappa_{22'} \neq \kappa_{32'}$.

In this case, $\Delta < 0$, which means that the solution of (9.90) are three unique real roots, that is

$$\begin{cases} x_1 = \dfrac{-b - 2\sqrt{A}\cos\frac{\theta}{3}}{3a} \\ x_{2,3} = \dfrac{-b + \sqrt{A}\left(cos\frac{\theta}{3} \pm \sqrt{3}sin\frac{\theta}{3}\right)}{3a} \end{cases}, \tag{9.99}$$

where $\theta = \arccos T$, $T = (2Ab\text{-}3aB)/(2\sqrt{A^3})$, $A > 0$ and $|T|<1$.

Then the characteristic Eq. (9.89) can be described by

$$\begin{cases} \left[j(\omega - \omega_0) - g\right]\left[j(\omega - \omega_0) + \tau\right] + \kappa_{11'}^2 = 0 \\ \left[j(\omega - \omega_0) - g\right]\left[j(\omega - \omega_0) + \tau\right] + \kappa_{22'}^2 = 0. \\ \left[j(\omega - \omega_0) - g\right]\left[j(\omega - \omega_0) + \tau\right] + \kappa_{33'}^2 = 0 \end{cases} \tag{9.100}$$

Therefore, the eigenfrequency of the three-phase PT-symmetric MC-WPT system can be obtained as

$$\begin{cases} \omega = \omega_0 \pm \sqrt{\kappa_{11'}^2 - \tau^2} \\ \omega = \omega_0 \pm \sqrt{\kappa_{22'}^2 - \tau^2} \\ \omega = \omega_0 \pm \sqrt{\kappa_{33'}^2 - \tau^2} \end{cases}. \tag{9.101}$$

It can be seen from the above analysis that the three-phase PT-symmetric MC-WPT system degenerates into three independent single-phase systems when the cross-coupling between transmitters and receivers is ignored, which is consistent with the facts. In other words, the eigenfrequency of each individual single-phase system is determined by its own main coupling coefficient.

(2) Considering the cross-coupling between transmitting and receiving circuits.

Since the receiving coils of the three-phase PT-symmetric MC-WPT system are always designed as one unit in practice, it can be assumed that the main coupling coils of transmitter and receiver have the same mutual inductance, i.e., $\kappa_{11'} = \kappa_{22'} = \kappa_{33'}$. Meanwhile, combined with the three-phase PT-symmetric conditions derived before, we can reconsider Eq. (9.90) and obtain the expression of the Van Shengjin formula discriminant as follows:

$$\Delta = -48t\left(4\kappa_{11'}^3 - \kappa_{11'}\kappa_{12'}^2 - \kappa_{11'}\kappa_{13'}^2 - \kappa_{11'}\kappa_{23'}^2 - \kappa_{12'}\kappa_{13'}\kappa_{23'}\right)^2 \tag{9.102}$$

where $t = \kappa_{12'}^6 + \kappa_{13'}^6 + \kappa_{23'}^6 + 3\kappa_{12'}^4\kappa_{13'}^2 + 3\kappa_{12'}^4\kappa_{23'}^2 + 3\kappa_{12'}^4\kappa_{23'}^2 + 3\kappa_{12'}^2\kappa_{13'}^4 + 3\kappa_{12'}^2\kappa_{23'}^4 + 3\kappa_{13'}^2\kappa_{23'}^4 - 21\kappa_{12'}^2\kappa_{13'}^2\kappa_{23'}^2$.

It is worth noting that the discriminant obtained by using Kaldan's formula, another solution to a cubic equation, also has a similar form as

$$\Delta_K = -\frac{4}{27}t\left(4\kappa_{11'}^3 - \kappa_{11'}\kappa_{12'}^2 - \kappa_{11'}\kappa_{13'}^2 - \kappa_{11'}\kappa_{23'}^2 - \kappa_{12'}\kappa_{13'}\kappa_{23'}\right)^2 \tag{9.103}$$

The symbolic nature of expression t determines that of the discriminant, which further indicates the situation of the roots. It is observed that when the cross-coupling coefficient between transmitter and receiver are equal, i.e., $\kappa_{12'} = \kappa_{13'} = \kappa_{23'}$, then $t = 0$ and $\Delta = 0$ can be obtained. The coefficients of the Van Shengjin formula can be transformed to

$$\begin{cases} A = 9\kappa_{12'}^4(2\kappa_{11'} + \kappa_{12'})^2 \\ B = 18\kappa_{12'}^4(2\kappa_{11'} + \kappa_{12'})^2(\kappa_{11'} - \kappa_{12'})^2 \end{cases}, \tag{9.104}$$

Since $A > 0$, (9.90) has two unequal real roots, one of which is a double root, that is

$$\begin{cases} x_1 = -\dfrac{b}{a} + K = -(\kappa_{11'} + 2\kappa_{12'})^2 \\ x_2 = x_3 = -\dfrac{1}{2}K = -(\kappa_{11'} - \kappa_{12'})^2 \end{cases}. \tag{9.105}$$

Combining the detailed expression of variable x, the characteristic Eq. (9.89) can be written as

$$\begin{cases} [j(\omega - \omega_0) - g][j(\omega - \omega_0) + \tau] + (\kappa_{11'} + 2\kappa_{12'})^2 = 0 \\ [j(\omega - \omega_0) - g][j(\omega - \omega_0) + \tau] + (\kappa_{11'} - \kappa_{12'})^2 = 0 \end{cases}. \tag{9.106}$$

Then the eigenfrequency of the three-phase PT-symmetric MC-WPT system can be written as

$$\begin{cases} \omega = \omega_0 \pm \sqrt{(\kappa_{11'} - \kappa_{12'})^2 - \tau^2} \\ \omega = \omega_0 \pm \sqrt{(\kappa_{11'} + 2\kappa_{12'})^2 - \tau^2} \end{cases}. \tag{9.107}$$

Unlike single-phase systems, three-phase systems have two cases of frequency splitting situations, exhibiting similar but different characteristic frequency characteristics than single-phase systems. According to (9.107), the system performance can be divided into two cases: in the strong coupling region ($\kappa_{11'}$-$\kappa_{12'} > \tau$) of case 1, the system supports two modes with operating frequencies $\omega = \omega_0 \pm \sqrt{(\kappa_{11'} - \kappa_{12'})^2 - \tau^2}$, which have the same gain rate, exactly balancing the loss, i.e., $g = \tau$. Besides, equal amplitude distribution has been realized, i.e., $|a_1/a_{1'}|=1$. Meanwhile, in the weak coupling region ($\kappa_{11'}$-$\kappa_{12'} < \tau$) of case 1, the symmetry of energy distribution would be destroyed, which means $|a_1/a_{1'}| \neq 1$. At this time, the system only has a mode with frequency $\omega = \omega_0$, where the balance of gain and loss rates no longer exists. Case 2 has the same performance, but changes the critical condition to $\kappa_{11'} + 2\kappa_{12'} = \tau$. Detailed system features are listed in Table 9.4.

Figure 9.34 shows the frequency bifurcation phenomenon of the three-phase PT-symmetric MC-WPT system, it can be seen that the strong coupling region of case 1 decreases with the increase of cross-coupling coefficient, while the opposite is true for case 2. In the extreme situation of case 2, no matter what $\kappa_{11'}$ is, the system would stay in the strong coupling region. Since the coupling coefficient is always greater than the cross-coupling coefficient, this does not exist in practical applications. Therefore, it is desired to achieve that the system always exhibits the characteristics of case 2, which indicates that the strong coupling region is broadened compared to the single-phase system. Figure 9.35 shows the transfer characteristics of the three-phase PT-symmetric MC-WPT system, it can be seen that in the strong coupling region, both two cases have the same output power and transfer efficiency, independent with coupling rates, while in the weak coupling region, constant transfer power and efficiency is destroyed, and the concrete expressions of each case are related to the critical situation respectively.

Table 9.4 Detailed system characteristics

		Case 1	Case 2
Critical condition		$\kappa_{11'}\text{-}\kappa_{12'} = \tau$	$\kappa_{11'} + 2\kappa_{12'} = \tau$
Strong coupling region	Frequency	$\omega = \omega_0 \pm \sqrt{(\kappa_{11'} - \kappa_{12'})^2 - \tau^2}$	$\omega = \omega_0 \pm \sqrt{(\kappa_{11'} + 2\kappa_{12'})^2 - \tau^2}$
	Gain rate	$g = \tau$	$g = \tau$
	Ration of amplitude	$\lvert a_1/a_{1'}\rvert$=1	$\lvert a_1/a_{1'}\rvert$=1
	Output power	$2\tau_{L1}\ G_0^2/(\tau_{10} + \tau_{1'})^2$	$2\tau_{La1}\ G_0^2/(\tau_{10} + \tau_{1'})^2$
	Transfer efficiency	$\tau_{L1}/(\tau_{10} + \tau_{1'})$	$\tau_{La1}/(\tau_{10} + \tau_{1'})$
Weak coupling region	Frequency	$\omega = \omega_0$	$\omega = \omega_0$
	Gain rate	$g = (\kappa_{11'}\text{-}\kappa_{12'})^2/\tau$	$g = (\kappa_{11'} + 2\kappa_{12'})^2/\tau$
	Ration of amplitude	$\lvert a_1/a_{1'}\rvert$=$\tau/(\kappa_{11'}\text{-}\kappa_{12'})$	$\lvert a_1/a_{1'}\rvert$=$\tau/(\kappa_{11'} + 2\kappa_{12'})$
	Output power	$2\tau_{La1}\ G_0^2/ (\kappa_{11'} - \kappa_{12'} + \frac{\tau_{10}\tau_{1'}}{\kappa_{11'}-\kappa_{12'}})^2$	$2\tau_{La1}\ G_0^2/ (\kappa_{11'} + 2\kappa_{12'} + \frac{\tau_{10}\tau_{1'}}{\kappa_{11'}+2\kappa_{12'}})^2$
	Transfer efficiency	$\tau_{La1}/[\tau_{10}(\frac{\tau_{10}\tau_{1'}}{\kappa_{11'}-\kappa_{12'}})^2 + \tau_{1'}]$	$\tau_{La1}/[\tau_{10}(\frac{\tau_{10}\tau_{1'}}{\kappa_{11'}+2\kappa_{12'}})^2 + \tau_{1'}]$

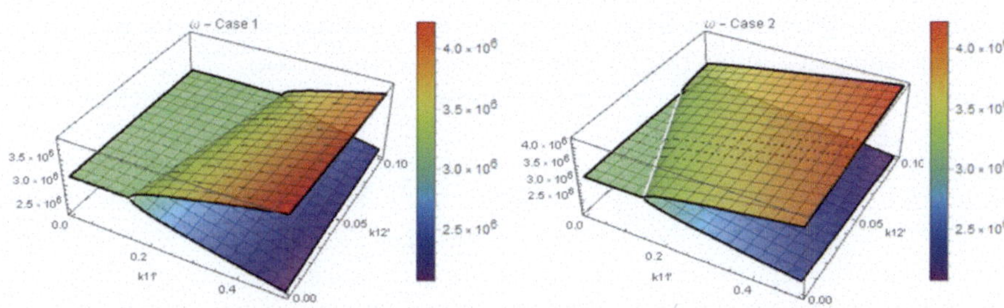

(a) (b)

Fig. 9.34 Frequency splitting phenomenon of three-phase PT-symmetric MC-WPT system. **a** Frequency splitting phenomenon in case 1. **b** Frequency splitting phenomenon in case 2

9.5 Quasi-PT-Symmetric Wireless Power Transfer

In the field of optics, when an exact PT-symmetric system is in a symmetric state, the gain and loss of the system are exactly equal, which also means that the gain and loss of the system are balanced, but the gain and loss of the system can exist in an unbalanced state [16]. This gain and loss imbalance is not a PT broken state, but an

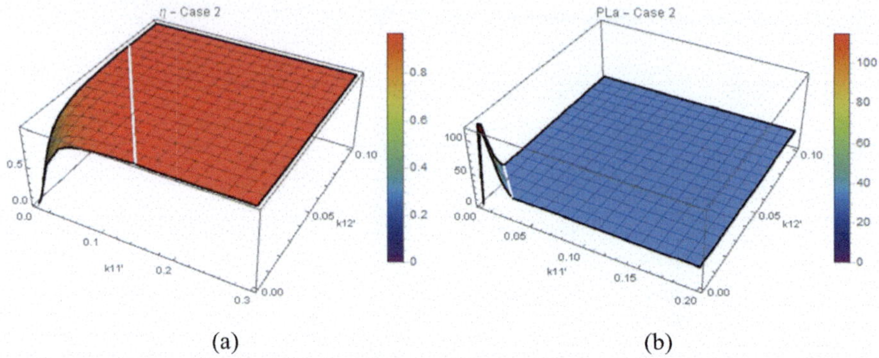

Fig. 9.35 Output power and transfer efficiency of case 2. **a** Transfer efficiency. **b** Output power

approximate quasi-PT symmetric state [17]. Taking a second-order coupled resonant circuit as an example, its coupled-mode equation can be written as

$$j\frac{d}{dt}\begin{bmatrix} \boldsymbol{a}_1 \\ \boldsymbol{a}_2 \end{bmatrix} = \underbrace{\begin{bmatrix} \omega_1 - j\gamma_1 & \kappa_{12} \\ \kappa_{21} & \omega_2 - j\gamma_2 \end{bmatrix}}_{\text{Hamiltonian:}H}\begin{bmatrix} \boldsymbol{a}_1 \\ \boldsymbol{a}_2 \end{bmatrix}, \tag{9.108}$$

where γ_1, γ_2 are the gain /loss rates in the corresponding resonator ($\gamma_{1,2} < 0$ represents the gain, $\gamma_{1,2} > 0$ represents the loss, $\gamma_{1,2} = 0$ indicates neither loss nor gain), κ_{12}, κ_{21} are the energy coupling rates between the resonators, and $\kappa_{12} = \kappa_{21}$.

Hamiltonian matrix $\boldsymbol{H}$ of the second-order coupled resonant circuit can be written as

$$\boldsymbol{H} = \begin{bmatrix} \omega_1 - j\gamma_1 & \kappa \\ \kappa & \omega_2 - j\gamma_2 \end{bmatrix}. \tag{9.109}$$

If the gain and loss rate in matrix $\boldsymbol{H}$ are not equal, i.e., $|\gamma_1| \neq |\gamma_2|$, defining $\tau = (\gamma_1 + \gamma_2)/2$, $\upsilon = (\gamma_1 - \gamma_2)/2$, $\omega_0 = (\omega_1 + \omega_2)/2$, $\varpi = (\omega_1 - \omega_2)/2$, and setting $[\boldsymbol{a}_1, \boldsymbol{a}_2]^{\mathrm{T}} = e^{\tau t}[\boldsymbol{a}_1', \boldsymbol{a}_2']^{\mathrm{T}}$, by using Gauge transformation, (9.108) can be converter to the following form:

$$\begin{aligned} j\frac{\mathrm{d}}{\mathrm{d}t}\begin{bmatrix} \boldsymbol{a}_1' \\ \boldsymbol{a}_2' \end{bmatrix} &= \begin{bmatrix} \omega_1 - j\gamma_1 & \kappa \\ \kappa & \omega_2 - j\gamma_2 \end{bmatrix}\begin{bmatrix} \boldsymbol{a}_1' \\ \boldsymbol{a}_2' \end{bmatrix} + j\tau\begin{bmatrix} \boldsymbol{a}_1' \\ \boldsymbol{a}_2' \end{bmatrix} \\ &= \begin{bmatrix} \omega_0 + \varpi + j\upsilon & \kappa \\ \kappa & \omega_0 - \varpi + j\upsilon \end{bmatrix}\begin{bmatrix} \boldsymbol{a}_1' \\ \boldsymbol{a}_2' \end{bmatrix}. \end{aligned} \tag{9.110}$$

From (9.110), we can get a new Hamiltonian matrix $\boldsymbol{H}'$, which $\boldsymbol{H}'$ satisfies the following mathematical relationship with the original Hamiltonian matrix $\boldsymbol{H}$:

$$\underbrace{\begin{bmatrix} \omega_1 - \mathrm{j}\gamma_1 & \kappa \\ \kappa & \omega_2 - \mathrm{j}\gamma_2 \end{bmatrix}}_{\text{Original Hamiltonian: } H} = \underbrace{\begin{bmatrix} \omega_0 - \mathrm{j}\nu & \kappa \\ \kappa & \omega_0 - \mathrm{j}\nu \end{bmatrix}}_{\text{New Hamiltonian: } H'} + \underbrace{\begin{bmatrix} \bar{\omega} - \mathrm{j}\tau & 0 \\ 0 & -\bar{\omega} - \mathrm{j}\tau \end{bmatrix}}_{\text{Uncoupledpart: } \Delta H} \tag{9.111}$$

where the new Hamiltonian $\boldsymbol{H}'$ still satisfies the exact PT-symmetric condition, that is

$$\begin{aligned} [PT,\ \boldsymbol{H}'] =& PT\boldsymbol{H}' - \boldsymbol{H}'PT \\ =& \begin{bmatrix} 0 & 1 \\ 1 & 0 \end{bmatrix}\left(\begin{bmatrix} 1 & 0 \\ 0 & 1 \end{bmatrix}\mathbf{K}\right)\begin{bmatrix} \omega_0 + \mathrm{j}\upsilon & \kappa \\ \kappa & \omega_0 - \mathrm{j}\upsilon \end{bmatrix} \\ & - \begin{bmatrix} \omega_0 + \mathrm{j}\upsilon & \kappa \\ \kappa & \omega_0 - \mathrm{j}\upsilon \end{bmatrix}\begin{bmatrix} 0 & 1 \\ 1 & 0 \end{bmatrix}\left(\begin{bmatrix} 1 & 0 \\ 0 & 1 \end{bmatrix}\mathbf{K}\right) \\ =& \begin{bmatrix} \kappa & \omega_0 - \mathrm{j}\upsilon \\ \omega_0 + \mathrm{j}\upsilon & \kappa \end{bmatrix} - \begin{bmatrix} \kappa & \omega_0 - \mathrm{j}\upsilon \\ \omega_0 + \mathrm{j}\upsilon & \kappa \end{bmatrix} \\ =& 0 \end{aligned} \tag{9.112}$$

where $\mathbf{K}$ stands for the complex conjugate operation.

The Gauge transformation reveals the hidden PT symmetry of the original Hamiltonian matrix $\boldsymbol{H}$ by changing the gain/loss reference benchmark: Before the transformation, the gain/loss in the original Hamiltonian matrix $\boldsymbol{H}$ takes zero as the reference benchmark, i.e., $\gamma_{1,2} < 0$ represents gain, $\gamma_{1,2} > 0$ represents loss, and $\gamma_{1,2} = 0$ is neutral, meaning no gain/loss. After the transformation, the average value τ of the gain/loss in the new Hamiltonian matrix $\boldsymbol{H}'$ is taken as the reference benchmark, i.e., $\gamma_{1,2} < \tau$ represents gain, $\gamma_{1,2} > \tau$ represents loss, and $\gamma_{1,2} = \tau$ is neutral, meaning no gain/loss [18]. It is worth noting that the gain/loss in the original Hamiltonian matrix $\boldsymbol{H}$ can be unbalanced (i.e., $|\gamma_1| \neq |\gamma_2|$), so when the system has unbalanced gain/loss (i.e., $\gamma_1 \neq -\gamma_2$), it can be viewed as being in an approximate "quasi-PT-symmetric" state. While when the gain/loss is balanced (i.e., $\gamma_1 = -\gamma_2$), it can be viewed as being in a strict "exact PT-symmetric" state, and "broken PT-symmetric state" indicates that the gain/loss of the system changes with the coupling rate.

9.5.1 System Modeling

According to the above analysis, the approximate "quasi-PT-symmetric" state can be regarded as an "exact PT-symmetric" state with gain/loss bias, and the mathematical conditions for forming a "quasi-PT-symmetric system" are broader and more general. As for (9.111), from the perspective of matrix transformation, the approximate "quasi-PT symmetric system" represented by the original matrix $\boldsymbol{H}$ can be equivalent to the sum of the exact PT-symmetric system represented by the new

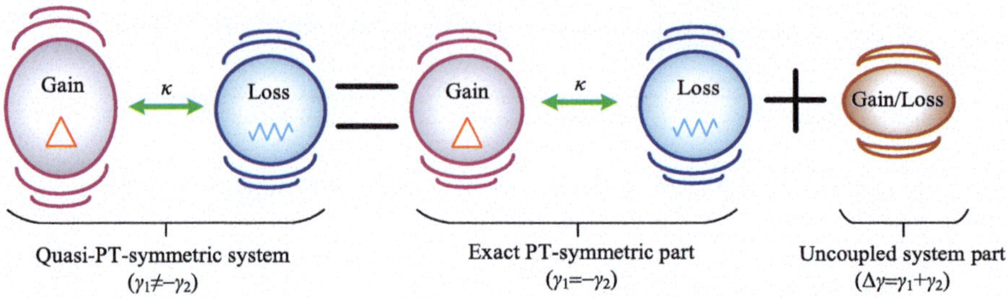

Fig. 9.36 Relationship between quasi-PT-symmetric state and exact PT-symmetric state

matrix $\boldsymbol{H}'$ and the uncoupled system represented by the matrix $\Delta\boldsymbol{H}$, as shown in Fig. 9.36 [19].

Therefore, a quasi-PT-symmetric MC-WPT system can be described in Fig. 9.37, the transmitter consists of the resonant capacitor C_P, transmitting coil L_P, internal resistance of transmitter R_P, equivalent virtual impedance of inverter -Z_N, the receiver consists of the resonant capacitor C_S, receiving coil L_S, internal resistance of receiver R_S, equivalent load resistance R_L. According to the coupled-mode theory [13], the energy modes of the transmitter and receiver resonators can be defined as $\boldsymbol{a}_T$ and $\boldsymbol{a}_R$ and $\boldsymbol{a} = [\boldsymbol{a}_T, \boldsymbol{a}_R]^T$. The energy stored in the transmitter and receiver are $|\boldsymbol{a}_T|^2$ and $|\boldsymbol{a}_R|^2$, respectively. The dynamics of the PT-symmetric MC-WPT system shown in Fig. 9.37 are governed by the differential equation $j\partial_t\boldsymbol{a}(t) = \boldsymbol{H}\boldsymbol{a}(t)$ [3, 6]:

$$j\frac{d}{dt}\begin{bmatrix}\boldsymbol{a}_T\\ \boldsymbol{a}_R\end{bmatrix} = \begin{bmatrix}\omega_T + jg_T & \kappa\\ \kappa & \omega_R - j\gamma_R\end{bmatrix}\begin{bmatrix}\boldsymbol{a}_T\\ \boldsymbol{a}_R\end{bmatrix}, \tag{9.113}$$

where $\boldsymbol{H}$ is a 2 × 2 Hamiltonian matrix [18, 20], ω_T and ω_R denote the natural resonant frequencies of the transmitter and receiver resonators, respectively.

In order to ensure complete energy transmission, the circuit parameters of the coupled resonator need to meet the following relationship [3]:

$$\omega_0 = \frac{1}{\sqrt{L_P C_P}} = \frac{1}{\sqrt{L_S C_S}}. \tag{9.114}$$

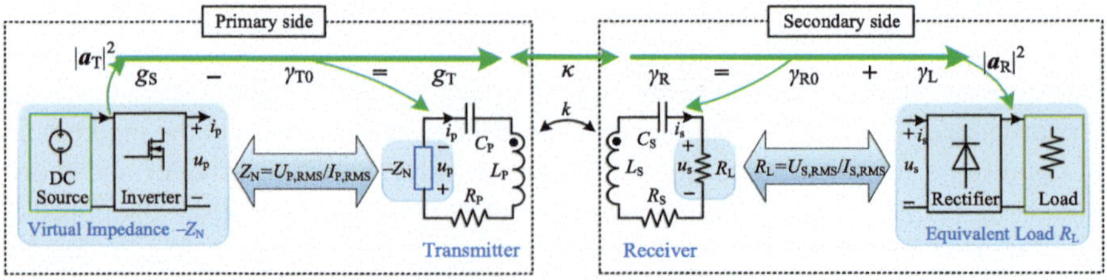

Fig. 9.37 The quasi-PT-symmetric MC-WPT system

It is noted that the resonant frequency ω_R of the receiver always meets $\omega_R = \omega_0$ and does not change with the load resistance, while the resonant frequency ω_T of the transmitter is tunable by manipulating the angle θ of the virtual impedance $-Z_N$, and there is $\omega_T = \omega_R = \omega_0$ as $\theta = 0°$. The relationship between the equivalent resonant frequency ω_T, the absolute value of the virtual impedance $|Z_N|$ and the impedance angle θ of the transmitter resonator is as follows:

$$\omega_T = \sqrt{\left(\frac{|Z_N|\sin\theta}{2L_P}\right)^2 + \frac{1}{L_P C_P}} - \frac{|Z_N|\sin\theta}{2L_P}. \tag{9.115}$$

Meanwhile, $g_T = g_S\ \gamma_{T0}$ is the net gain rate of the transmitter, g_S is the gain rate causing by the power supply in the transmitter, γ_{T0} is the intrinsic loss rate causing by the transmitting coil resistance R_P, which can be expressed as

$$g_T = g_S - \gamma_{T0} = \frac{|Z_N|\cos\theta}{2L_P} - \frac{R_P}{2L_P}. \tag{9.116}$$

where $\gamma_R = \gamma_L + \gamma_{R0}$ is the net loss rate of the receiver, γ_L is the loss rate causing by the load, and γ_{R0} the intrinsic loss rate causing by the receiving coil resistance R_S, which can be expressed as

$$\gamma_R = \gamma_L + \gamma_{R0} = \frac{R_L}{2L_S} + \frac{R_S}{2L_S}. \tag{9.117}$$

where κ is the coupling rate between the two resonators, which is determined by the mutual coupling coefficient k $(0 < k < 1)$ and the resonant frequencies ω_T and ω_R. The coupling rate κ is defined as

$$\kappa = \frac{k}{2}\sqrt{\omega_T\omega_R}. \tag{9.118}$$

Assuming that the operating angular frequency of the coupled resonators at the steady state is ω, the energy modes of the resonators satisfy $\boldsymbol{a} = [\boldsymbol{a}_T, \boldsymbol{a}_R]^T \propto e^{j\omega t}$, and it is governed by $(\boldsymbol{H}\text{-}\omega\boldsymbol{I})\boldsymbol{a} = 0$, where $\boldsymbol{I}$ is a 2×2 identity matrix. The characteristic equation of (9.113) satisfies determinate $\mathbf{det}(\boldsymbol{H}\text{-}\omega\boldsymbol{I}) = 0$ if (9.113) has real eigen-frequencies, which means

$$(\omega - \omega_T + jg_T)(\omega - \omega_R - j\gamma_R) - \kappa^2 = 0. \tag{9.119}$$

Separating the real and imaginary parts of (9.119), the operating angular frequency ω at steady-state is derived as

$$\begin{cases} (\omega - \omega_T)(\omega - \omega_R) + g_T\gamma_R - \kappa^2 = 0 \\ (\omega - \omega_R)g_T = (\omega - \omega_T)\gamma_R \end{cases}. \tag{9.120}$$

According to the CMT, the net power injected into the transmitter $\boldsymbol{a}_{\mathrm{T}}$ is $P_{\mathrm{in}} = 2g_{\mathrm{T}}|\boldsymbol{a}_{\mathrm{T}}|^2$, and the net power dissipated by the receiver $\boldsymbol{a}_{\mathrm{R}}$ is $P_{\mathrm{out}} = 2\gamma_{\mathrm{R}}|\boldsymbol{a}_{\mathrm{R}}|^2$. Based on the energy conservation of the WPT system, there is $P_{\mathrm{in}} = P_{\mathrm{out}}$, then the gain–loss-ratio ε can be defined as

$$\varepsilon = \frac{\gamma_{\mathrm{R}}}{g_{\mathrm{T}}} = \frac{|\boldsymbol{a}_{\mathrm{T}}|^2}{|\boldsymbol{a}_{\mathrm{R}}|^2}. \tag{9.121}$$

Meanwhile, substituting (9.121) and (9.118) into (9.120), and simplifying it with $\omega_R = \omega_0$, the solutions of the steady-state operating angular frequency can be finally solved. The analytic expression of the root locus can be divided into two cases:

(1) Exact PT-symmetry (balanced gain–loss condition with $\varepsilon = \gamma_{\mathrm{R}}/g_{\mathrm{T}} = 1$).

$$\omega = \begin{cases} \left(1 \pm \sqrt{\frac{k^2}{4} - \frac{\gamma_{\mathrm{R}}^2}{\omega_0^2}}\right)\omega_0 & , 1 \ge k \ge k_{\mathrm{C}} \\ \omega_0 & , 0 < k < k_{\mathrm{C}} \end{cases}. \tag{9.122}$$

(2) Quasi PT-symmetry (imbalanced gain–loss condition with $\varepsilon = \gamma_{\mathrm{R}}/g_{\mathrm{T}} \ne 1$).

$$\omega = \begin{cases} \left[1 + \frac{k^2}{8}(\varepsilon - 1) \pm \sqrt{\frac{(\varepsilon-1)^2}{64}k^4 + \frac{\varepsilon k^2}{4} - \frac{\gamma_{\mathrm{R}}^2}{\omega_0^2}}\right]\omega_0 & , 1 \ge k \ge k_{\mathrm{C}} \\ \omega_0 & , 0 < k < k_{\mathrm{C}} \end{cases}, \tag{9.123}$$

where k_{C} is defined as the critical coupling coefficient, that is

$$k_{\mathrm{C}} = \begin{cases} \frac{2\gamma_{\mathrm{R}}}{\omega_0} & , \varepsilon = 1 \\ \sqrt{\sqrt{\frac{64\gamma_{\mathrm{R}}^2}{(\varepsilon-1)^2\omega_0^2} + \left[\frac{8\varepsilon}{(\varepsilon-1)^2}\right]^2} - \frac{8\varepsilon}{(\varepsilon-1)^2}} & , \varepsilon \ne 1 \end{cases}. \tag{9.124}$$

When the system is in different states, the nonlinear gain g_{S} at the transmitter can be expressed as a function of the absolute value of the virtual impedance $|Z_{\mathrm{N}}|$ and its impedance angle θ. The specific expression is as follows:

$$g_{\mathrm{S}}(\theta) = \frac{|Z_{\mathrm{N}}|\cos\theta}{2L_{\mathrm{P}}} = \frac{R_{\mathrm{N}}}{2L_{\mathrm{P}}} = \begin{cases} \frac{\gamma_{\mathrm{R}}}{\varepsilon} + \gamma_{\mathrm{TO}} & , 1 \ge k \ge k_{\mathrm{C}} \\ \frac{k^2\omega_0^2}{4\gamma_{\mathrm{R}}} + \gamma_{\mathrm{TO}} & , 0 < k < k_{\mathrm{C}} \end{cases}. \tag{9.125}$$

It should be noted that the analytical expression of the operating angular frequency under the balanced condition ($\varepsilon = 1$) is the same as the conventional PT-symmetric MC-WPT system (as shown in Sect. 3.3.1). Specifically, there are other analytical solutions of the working angular frequency of the differential Eq. (9.113) that also meet the requirements of Eq. (9.120), that is, in the case of an approximate quasi-PT symmetry with an unbalanced condition ($\varepsilon \ne 1$), there are other root locus represented by Eq. (9.123) that meets the solving conditions of the differential Eq. (9.113). Therefore, the exact PT symmetry and quasi-PT symmetry can be divided by defining the concept of gain–loss ratio, and the frequency analytic expression of the MC-WPT system reveals mathematically that the conventional PT-symmetric MC-WPT system is only one special solution of the quasi PT-symmetric MC-WPT system.

The normalized operating angular frequency ω/ω_0 obtained by changing the gain–loss ratio ε and coupling coefficient k is shown in Fig. 9.38. When the coupling coefficient between coils $k > k_C$, the quasi PT-symmetric MC-WPT system operates at the symmetric phase, and the operating angular frequency has two different branches: ω_H and ω_L. When $k \le k_C$, there is only one root locus, that is $\omega = \omega_0$, and the system enters the broken PT-symmetric state. Importantly, Fig. 9.38c implies that the critical coupling coefficient k_C varies with the gain–loss ratio ε.

Therefore, if the MC-WPT system satisfies the condition of quasi PT-symmetry, its circuit parameters and operating frequency should meet the following conditions:

(1) The passive circuit element parameters of the transmitter and receiver satisfy (9.114).
(2) The nonlinear source gain g_s satisfies (9.125).
(3) The steady-state operating angular frequency of the system satisfies (9.122) or (9.123).

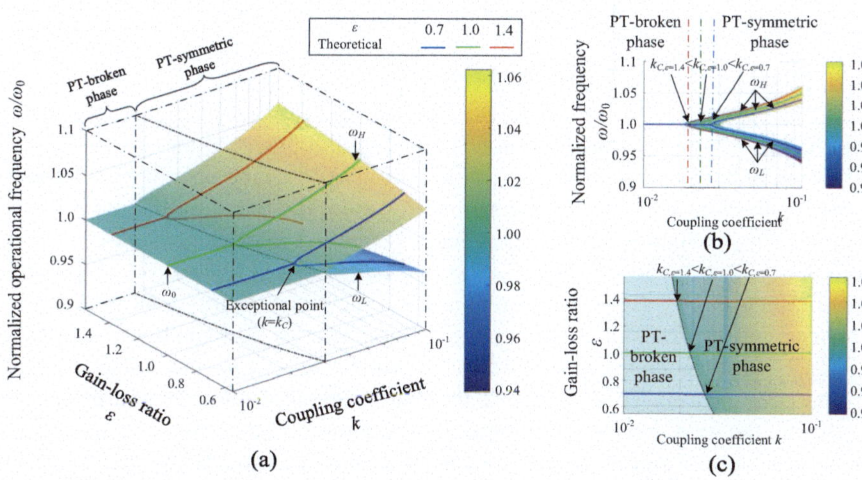

Fig. 9.38 Normalized operational angular frequency ω/ω_0 versus different coupling coefficient k, wherein solid lines (yellow, red, and green) represent the desired frequency loci with specific gain–loss ratio ε. **a** 3-D surface of the root locus ω/ω_0. **b** ε-axis projection of (**a**). **c** ω/ω_0-axis projection of (**a**)

9.5.2 Transfer Performance and Analysis

According to CMT, the input power of the system is $P_{\rm N,ac} = 2g_{\rm S}|\boldsymbol{a}_{\rm T}|^2 = 2\gamma_{\rm TO}|\boldsymbol{a}_{\rm R}|^2 + 2\gamma_{\rm R}|\boldsymbol{a}_{\rm R}|^2$, and the received power of the load is $P_{\rm L,ac} = 2\gamma_{\rm L}|\boldsymbol{a}_{\rm R}|^2$. Hence, the transfer efficiency η can be deduced as

$$\eta = \frac{P_{\rm L,ac}}{P_{\rm N,ac}} = \frac{2\gamma_{\rm L}|\boldsymbol{a}_{\rm R}|^2}{2\gamma_{\rm TO}|\boldsymbol{a}_{\rm T}|^2 + 2\gamma_{\rm R}|\boldsymbol{a}_{\rm R}|^2} = \begin{cases} \frac{\gamma_{\rm L}}{\gamma_{\rm R}} \frac{1}{1+\varepsilon\frac{\gamma_{\rm TO}}{\gamma_{\rm R}}} &, 1 \geq k \geq k_{\rm C} \\ \frac{\gamma_{\rm L}}{\gamma_{\rm R}} \frac{1}{1+\frac{4\gamma_{\rm TO}\gamma_{\rm R}}{k^2\omega_0^2}} &, 0 < k < k_{\rm C} \end{cases}. \quad (9.126)$$

The total real power $P_{\rm N,ac}$ generated by the real part $R_{\rm N}$ of the virtual impedance $Z_{\rm N}$ is $P_{\rm N,ac} = V_{\rm P,RMS}{}^2/|Z_{\rm N}|\cos\theta = I_{\rm P,RMS}{}^2|Z_{\rm N}|\cos\theta = I_{\rm P,RMS}{}^2 R_{\rm N}$, where $V_{\rm P,RMS}$ denotes the fundamental RMS value of the output voltage of the inverter. Define the source gain coefficient as $G_{\rm S} = V_{\rm P,RMS}\sqrt{\cos\theta/(2L_{\rm P})}$, and combine the transfer efficiency η described in (9.126) with the AC output power $P_{\rm L,ac} = \eta P_{\rm N,ac}$, the output power $P_{\rm L,ac}$ received by the equivalent load $R_{\rm L}$ can be derived as:

$$P_{\rm L,ac} = \begin{cases} \frac{G_{\rm S}^2}{(\gamma_{\rm R}/\varepsilon+\gamma_{\rm TO})} \frac{\gamma_{\rm L}}{(\gamma_{\rm R}+\varepsilon\gamma_{\rm TO})} &, 1 \geq k \geq k_{\rm C} \\ \frac{G_{\rm S}^2}{\left(\frac{k^2\omega_0^2}{4\gamma_{\rm R}}+\gamma_{\rm TO}\right)} \frac{\gamma_{\rm L}}{\left(\gamma_{\rm R}+\frac{4\gamma_{\rm TO}\gamma_{\rm R}^2}{k^2\omega_{\rm R}^2}\right)} &, 0 < k < k_{\rm C} \end{cases}. \quad (9.127)$$

Figure 9.39 shows the transfer efficiency and output power of the system versus coupling coefficient k with different gain–loss ratio ε. When the coupling coefficient $k \geq k_{\rm C}$, the system works in the PT-symmetric state, the transfer efficiency η and load power $P_{\rm L,ac}$ remains constant regardless of coupling coefficient k, which means the efficiency η and load power $P_{\rm L,ac}$ are insensitive to the fluctuation of transfer distance. When the coupling coefficient $0<k<k_{\rm C}$, the system works in the broken PT-symmetric state, the transfer efficiency η and load power $P_{\rm L,ac}$ are no longer constant and change with the coupling coefficient k, which means that the stable transfer performance is broken. This is because the gain of the transmitter is nonlinear and controllable, the gain and loss of the quasi PT-symmetric MC-WPT system remain stable with a constant gain–loss ratio ε in the PT-symmetric state. While in the broken PT-symmetric state, the gain–loss ratio ε of the quasi PT-symmetric MC-WPT system will change with the coupling coefficient k.

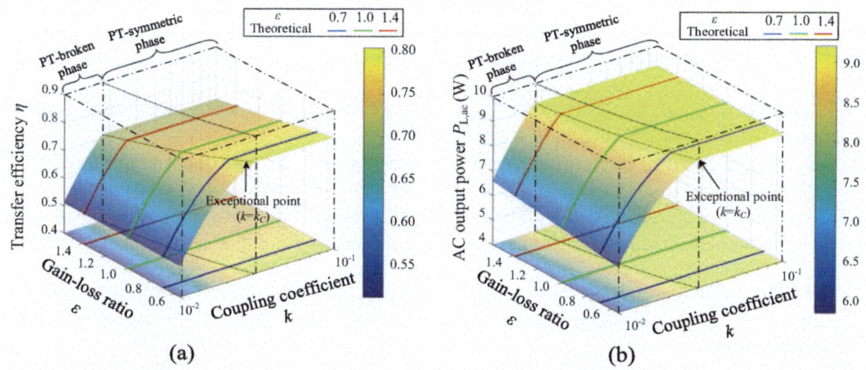

Fig. 9.39 Theoretical transfer performance versus coupling coefficient k with different gain–loss ratio ε, where solid lines (yellow, red and green) represent the loci with specific gain–loss ratio ε. **a** Transfer efficiency η. **b** Output power $P_{\mathrm{L,ac}}$

9.6 Summary

This chapter mainly introduces several techniques to improve the performance of the PT-symmetric MC-WPT system. Firstly, the methods of adding passive elements to the receiver to expand the transfer distance are introduced, the results show that adding an LC element to the receiver can improve the transfer distance of the system to a greater extent than that of a single L or C element. Then, the methods of reducing switching frequency and switching loss are described successively. Meanwhile, the single-phase PT-symmetric MC-WPT system is extended to the three-phase PT-symmetric MC-WPT system, the basic characteristics of the three-phase PT-symmetric MC-WPT system are analyzed, and the application range of PT-symmetric MC-WPT system is further expanded. Finally, a quasi PT-symmetric MC-WPT system is introduced, which generalizes the original balanced gain–loss ratio to the unbalanced gain–loss ratio. The results demonstrate that the conventional PT-symmetric MC-WPT system is only one special case of the quasi PT-symmetric MC-WPT system.

References

1. Wei Z, Zhang B (2021) Transmission range extension of PT-symmetry-based wireless power transfer system. IEEE Trans Power Electron 36(10):11135–11147
2. Li H, Wang K, Huang L, Chen W, Yang X (2015) Dynamic modeling based on coupled modes for wireless power transfer systems. IEEE Trans Power Electron 30(11):6245–6253
3. Zhou J, Zhang B, Xiao W, Qiu D, Chen Y (2019) Nonlinear parity-time-symmetric model for constant efficiency wireless power transfer: application to a drone-in-flight wireless charging platform. IEEE Trans Industr Electron 66(5):4097–4107

4. Qu Y, Zhang B, Gu W, Li J, Shu X (2023) Distance extension of S-PS wireless power transfer system based on parity-time symmetry. IEEE Trans Circuits Syst II Express Briefs 70(8):2954–2958
5. Wei Z, Zhang B, Lin S, Wang C (2024) A self-oscillation WPT system with high misalignment tolerance. IEEE Trans Power Electron 39(1):1870–1887
6. Assawaworrarit S, Yu X, Fan S (2017) Robust wireless power transfer using a nonlinear parity–time-symmetric circuit. Nature 546(7658):387–390
7. Assawaworrarit S, Fan S (2020) Robust and efficient wireless power transfer using a switch-mode implementation of a nonlinear parity–time symmetric circuit. Nature Electron 3(5):273–279
8. Zeng H, González-Santini NS, Yu Y, Yang S, Peng FZ (2017) Harmonic burst control strategy for full-bridge series-resonant converter-based EV charging. IEEE Trans Power Electron 32(5):4064–4073
9. Wu L, Zhang B, Jiang Y (2022) Position-independent constant current or constant voltage wireless electric vehicles charging system without dual-side communication and DC-DC converter. IEEE Trans Industr Electron 69(8):7930–7939
10. Liu Z, Su M, Zhu Q, Zhao L, Hu AP (2021) A dual frequency tuning method for improved coupling tolerance of wireless power transfer system. IEEE Trans Power Electron 36(7):7360–7365
11. Li Y et al (2019) Reconfigurable intermediate resonant circuit based WPT system with load-independent constant output current and voltage for charging battery. IEEE Trans Power Electron 34(3):1988–1992
12. Yang L, Li X, Liu S, Xu Z, Cai C (2021) Analysis and design of an LCCC/S-compensated WPT system with constant output characteristics for battery charging applications. IEEE J Emerg Sel Topics Power Electron 9(1):1169–1180
13. L. W. H (1960) Coupled mode and parametric electronics. New York, NY, USA: Wiley
14. Schindler J, Lin Z, Lee J, Ramezani H, Ellis F, Kottos T (2012) PT-symmetric electronics. J Phys A: Math Theor 45(444029):2077–2082
15. Wu L, Zhang B, Zhou J (2020) Efficiency improvement of the parity-time-symmetric wireless power transfer system for electric vehicle charging. IEEE Trans Power Electron 35(11):12497–12508
16. Dmitriev SV, Sukhorukov AA, Kivshar YS (2010) Binary parity-time-symmetric nonlinear lattices with balanced gain and loss. Opt Lett 35(17):2976–2978
17. Ornigotti M, Szameit A JJO (2014) Quasi PT-symmetry in passive photonic lattices. J Opt 16(065501):1–6
18. Özdemir ŞK, Rotter S, Nori F, Yang L (2019) Parity-time symmetry and exceptional points in photonics. Nat Mater 18(8):783–798
19. Li J, Zhang B (2023) A wireless power transfer system based on quasi-parity-time symmetry with gain-loss ratio modulation. Int J Circuit Theory Appl 51(3):1039–1056
20. Zyablovsky AA, Vinogradov AP, Pukhov AA, Dorofeenko AV, Lisyansky AA (2014) PT-symmetry in optics. Phys Usp 57(11):1063

Chapter 10
PT-Symmetric Electric-Field Coupled Wireless Power Transfer System

Electric-field coupled wireless power transfer (EC-WPT) system, also known as a capacitive wireless power transfer (CPT) system, can achieve contactless power transfer through electric-field coupling between transmitting and receiving metal plates. EC-WPT system utilizes the displacement current between the transmitting and receiving metal plates to transfer power, which has several remarkable advantages, such as light weight, low cost, low eddy current loss, small electromagnetic interference, insensitive to metal obstacles, etc. Currently, EC-WPT technology has been gradually applied to underwater equipment, electric vehicles, implantable biomedical equipment, and industrial robots. However, it is a significant challenge for the existing EC-WPT system to achieve constant output power and invariable transfer efficiency over a wide transfer distance. Since the PT-symmetric principle was applied to the MC-WPT system, the constant transfer against the coupling coefficient can be achieved without introducing communication and additional control circuits. Based on the duality between the EC-WPT system and MC-WPT system, this chapter attempts to extend the concept of PT symmetry to the EC-WPT system, establish the mathematical model of the PT-symmetric EC-WPT system, and analyze the PT-symmetric conditions and performance of the system.

10.1 System Structure

The PT-symmetric EC-WPT system is mainly composed of a negative resistance component, the capacitive coupler, transmitting and receiving compensation networks, and the load equipment [1–3], which can be divided into energy transmitter and energy receiver, as shown in Fig. 10.1. The transmitter and receiver are coupled with each other through a pair of metal plates, which is called the capacitive coupler. The negative resistance component, as the power supply of the PT-symmetric EC-WPT system, can generate a high-frequency alternating electric field between

B. Zhang et al., *Parity-Time Symmetric Wireless Power Transfer*, CPSS Power Electronics Series, https://doi.org/10.1007/978-981-96-3149-0_10

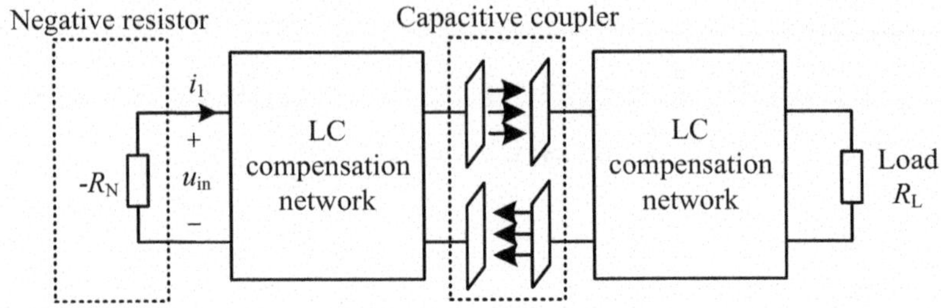

Fig. 10.1 System structure of the PT-symmetric EC-WPT system

the transmitting and receiving metal plates. The displacement current is generated between the coupling metal plates under the action of an electric field, so as to realize the energy transfer between coupling metal plates.

A common design of the capacitive coupler consists of parallel metal plats forming two fair of coupling capacitances, as shown in Fig. 10.2a, in which the plates P1 and P2 are denoted as the transmitting plates, and the plates P3 and P4 are denoted as the receiving plates. Due to proximity, cross-coupling between metal plates cannot be neglected, so the four electrode plates of the capacitive coupler are coupled to form six equivalent capacitances (C_{12}, C_{13}, C_{14}, C_{23}, C_{24}, and C_{34}), which can be further equivalent to the π-type circuit model, as shown in Fig. 10.2b [4, 5]. Here, C_P and C_S are the self-capacitances, and C_M is the mutual (coupling) capacitance, which can be represented by [6, 7]

$$\begin{cases} C_M = \dfrac{C_{13}C_{24} - C_{14}C_{23}}{C_{13} + C_{14} + C_{23} + C_{24}} \\ C_P = C_{12} + \dfrac{(C_{13} + C_{14})(C_{23} + C_{24})}{C_{13} + C_{14} + C_{23} + C_{24}} \\ C_S = C_{34} + \dfrac{(C_{13} + C_{23})(C_{14} + C_{24})}{C_{13} + C_{14} + C_{23} + C_{24}} \end{cases}. \tag{10.1}$$

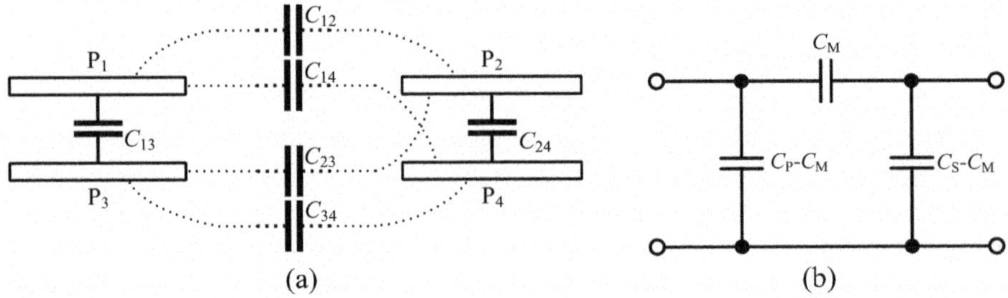

Fig. 10.2 Capacitive coupler. **a** electrode plates and their coupling capacitors, **b** Equivalent π-type circuit model

In a completely symmetric capacitive coupler, it can be considered that $C_{12} = C_{34}$, $C_{14} = C_{23}$, and $C_{13} = C_{24}$, so C_p is equal to C_s. It should be pointed out that this chapter mainly studies the capacitive coupler with a symmetric structure, that is, the self-capacitances of transmitting and receiving plates always remain equal when the coupling between transmitting and receiving plates changes.

Because the mutual capacitance of the capacitive coupler is generally small, the EC-WPT system usually needs to adopt compensation networks to improve the performance of the capacitive coupler. The simplest compensation network for the EC-WPT system is a double-side LC compensation network, which can adjust the equivalent self-capacitance by paralleling compensation capacitors [4]. According to the connection of compensation inductors and capacitors, the PT-symmetric EC-WPT system has four fundamental topologies, including SS, PP, SP, and PS compensation topologies, as shown in Fig. 10.3. Here, L_1, L_2, C_1, C_2 are the components of LC compensation networks, where C_1 and C_2 are composed of self-capacitances and additional compensation capacitances in parallel. R_1 and R_2 are the internal resistances of the transmitting and receiving sides, respectively. i_1 and i_2 are the currents flowing through compensation inductors L_1 and L_2, respectively. u_1 and u_2 are the voltages across the capacitors C_1 and C_2, respectively. $-R_N$ is the negative resistor, u_{in} is the voltage across the negative resistor $-R_N$, which can also be regarded as the input voltage of the PT-symmetric EC-WPT system, and the phase difference between the voltage u_{in} and current i_1 is equal to 180°. u_o is the voltage across the load resistor R_L, which can also be regarded as the output voltage of the PT-symmetric EC-WPT system.

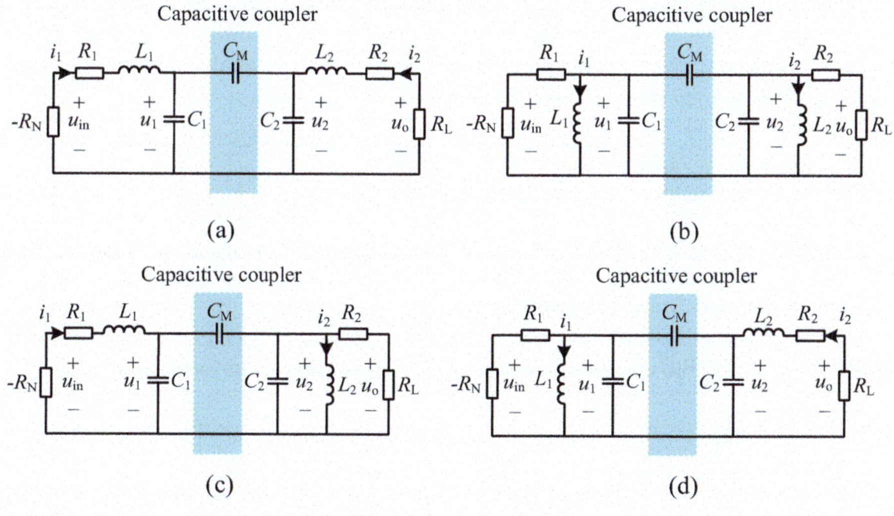

Fig. 10.3 System structure of the PT-symmetric EC-WPT system. **a** SS compensation topology, **b** PP compensation topology, **c** SP compensation topology, **d** PS compensation topology

Without loss of generality, this chapter mainly analyzes the PT-symmetric EC-WPT system with SS compensation topology shown in Fig. 10.3a, and other compensation topologies can be referred to the following analysis.

10.2 Modeling

10.2.1 Circuit Model

According to Kirchhoff's laws, the capacitor voltages u_1, u_2 and the inductor currents i_1, i_2 are selected as the state variables of the EC-WPT system, the state equation of the PT-symmetric EC-WPT system can be obtained as

$$\begin{cases} \dfrac{di_1}{dt} = \dfrac{i_1}{L_1}(-R_N + R_1) - \dfrac{u_1}{L_1} \\ \dfrac{du_1}{dt} = \dfrac{C_2 + C_M}{C_M(C_1 + C_2) + C_1C_2} i_1 + \dfrac{C_M}{C_M(C_1 + C_2) + C_1C_2} i_2 \\ \dfrac{di_2}{dt} = \dfrac{i_2}{L_2}(R_2 + R_L) - \dfrac{u_2}{L_2} \\ \dfrac{du_2}{dt} = \dfrac{C_M}{C_M(C_1 + C_2) + C_1C_2} i_1 + \dfrac{C_1 + C_M}{C_M(C_1 + C_2) + C_1C_2} i_2 \end{cases}. \tag{10.2}$$

Given that the quality factor of LC compensation network is generally high, the current can be considered as a sine wave, so that the steady-state circuit analysis of the system can be carried out by using the fundamental wave analysis method. Based on the KVL law, the circuit phasor equation can be derived as

$$\begin{bmatrix} -R_N + R_1 + j\omega L_1 + \frac{C_2+C_M}{j\omega\Delta} & \frac{C_M}{j\omega\Delta} \\ \frac{C_M}{j\omega\Delta} & R_e + R_2 + j\omega L_2 + \frac{C_1+C_M}{j\omega\Delta} \end{bmatrix} \begin{bmatrix} \dot{I}_1 \\ \dot{I}_2 \end{bmatrix} = 0, \tag{10.3}$$

where ω is the operating angular frequency, $\Delta = C_M(C_1 + C_2) + C_1C_2$, $\dot{I}_1$ and $\dot{I}_2$ are phasor forms of i_1 and i_2, respectively.

When symmetric circuit parameters are selected, i.e., $L_1 = L_2 = L$, $R_1 = R_2 = R$ and $C_1 = C_2 = C$, the circuit equations of (10.3) can be further written as

$$\begin{bmatrix} jQ_T^{-1} - \frac{\omega}{\omega_0} + \frac{\omega_0(k_C+1)}{\omega(2\kappa_C+1)} & \frac{\omega_0 k_C}{\omega(2k_C+1)} \\ \frac{\omega_0 k_C}{\omega(2k_C+1)} & jQ_R^{-1} - \frac{\omega}{\omega_0} + \frac{\omega_0(k_C+1)}{\omega(2k_C+1)} \end{bmatrix} \begin{bmatrix} \dot{I}_1 \\ \dot{I}_2 \end{bmatrix} = 0, \tag{10.4}$$

where $\omega_0 = \sqrt{\mathrm{LC}}$ is the natural resonant angular frequency, $k_C = C_M/C$ is defined as the capacitive coupling coefficient corresponding to the transfer distance, $Q_T = \omega_0 L/$

$(-R_N + R)$ and $Q_R = \omega_0 L/(R_L + R)$ are defined as the quality factor of transmitting and receiving sides, respectively.

10.2.2 Coupled-Mode Model

According to the coupled-mode theory, the relationship between coupled modes and the state variables in (10.2) can be represented as

$$a_n = \sqrt{\frac{L_n}{2}} i_n + j\sqrt{\frac{C_n}{2}} u_n = A_n e^{j(\omega t + \theta_n)},\ n = 1, 2. \tag{10.5}$$

By representing i_n and u_n with A_n and θ_n according to (10.5), we have

$$\begin{cases} u_n = \sqrt{\dfrac{2}{C_n}} A_n \cos(\omega t + \theta_n) \\ i_n = \sqrt{\dfrac{2}{L_n}} A_n \sin(\omega t + \theta_n) \\ \dfrac{du_n}{dt} = \sqrt{\dfrac{2}{C_n}} \dfrac{dA_n}{dt} \cos(\omega t + \theta_n) - \sqrt{\dfrac{2}{C_n}} A_n \left(\omega + \dfrac{d\theta_n}{dt}\right) \sin(\omega t + \theta_n) \\ \dfrac{di_n}{dt} = \sqrt{\dfrac{2}{L_n}} \dfrac{dA_n}{dt} \sin(\omega t + \theta_n) + \sqrt{\dfrac{2}{L_n}} A_n \left(\omega + \dfrac{d\theta_n}{dt}\right) \cos(\omega t + \theta_n) \end{cases}. \tag{10.6}$$

By substituting (10.6) into (10.2), the dynamics of the coupled modes can be derived. Assuming that the slowly varying variables A_n and θ_n are constant during a self-oscillating period, the high-frequency terms which vary at several times of operating frequency f can be eliminated by the averaging method [8, 9]. Then, only considering low-frequency characteristics and ignoring small high-frequency ripples, the approximate model related to A_n and θ_n is derived as follows [10, 11]:

$$\frac{dA_1}{dt} = -\frac{(-R_N + R_1)}{2L_1} A_1 + \frac{C_M\sqrt{C_1 C_2}}{2\Delta} \omega_2 A_2 \sin(\theta_2 - \theta_1), \tag{10.7}$$

$$A_1\left(\omega + \frac{d\theta_1}{dt}\right) = -\frac{1}{2}\omega_1 A_1 - \frac{(C_M + C_2)C_1}{2\Delta}\omega_1 A_1 - \frac{C_M\sqrt{C_1 C_2}}{2\Delta} \omega_2 A_2 \cos(\theta_2 - \theta_1), \tag{10.8}$$

$$\frac{dA_2}{dt} = -\frac{(R_2 + R_L)}{2L_2} A_2 - \frac{C_M\sqrt{C_1 C_2}}{2\Delta} \omega_1 A_1 \sin(\theta_2 - \theta_1), \tag{10.9}$$

$$A_2\left(\omega + \frac{d\theta_2}{dt}\right) = -\frac{1}{2}\omega_2 A_2 - \frac{(C_M + C_1)C_2}{2\Delta}\omega_2 A_2 - \frac{C_M\sqrt{C_1C_2}}{2\Delta}\omega_1 A_1 \cos(\theta_2 - \theta_1). \quad (10.10)$$

Selecting complex variables $\boldsymbol{a}_n$ as the new state variables, the state space can be described by

$$\frac{d\boldsymbol{a}_n}{dt} = \frac{dA_n}{dt}e^{j(\omega t + \theta_n)} + jA_n\left(\omega + \frac{d\theta_n}{dt}\right)e^{j(\omega t + \theta_n)}\ n = 1, 2. \quad (10.11)$$

By substituting (10.7)–(10.10) into (10.11), the coupled-mode model can be described as follows:

$$\frac{d\boldsymbol{a}_1}{dt} = \left[j\frac{\omega_1}{2}\left(1 + \frac{(C_M + C_2)C_1}{\Delta}\right) - \frac{(R_1 - R_N)}{2L_1}\right]\boldsymbol{a}_1 - j\frac{\omega_2 C_M\sqrt{C_1C_2}}{2\Delta}\boldsymbol{a}_2 \quad (10.12)$$

$$\frac{d\boldsymbol{a}_2}{dt} = \left[j\frac{\omega_2}{2}\left(1 + \frac{(C_M + C_1)C_2}{\Delta}\right) - \frac{(R_2 + R_L)}{2L_2}\right]\boldsymbol{a}_2 - j\frac{\omega_1 C_M\sqrt{C_1C_2}}{2\Delta}\boldsymbol{a}_1 \quad (10.13)$$

where natural frequencies are defined as $\omega_1 = 1/\sqrt{L_1C_1}$ and $\omega_2 = 1/\sqrt{L_2C_2}$.

For the EC-WPT system, considering symmetric circuit parameters, i.e., $L_1 = L_2 = L$, $R_1 = R_2 = R$ and $C_1 = C_2 = C$, (10.12) and (10.13) can be rewritten as

$$\frac{d\boldsymbol{a}_1}{dt} = \left[j\omega_0\left(2 - \frac{k_C}{2k_C + 2}\right) + \frac{R_N - R_1}{2L_1}\right]\boldsymbol{a}_1 - j\frac{\omega_0 k_C}{2(2k_C + 1)}\boldsymbol{a}_2, \quad (10.14)$$

$$\frac{d\boldsymbol{a}_2}{dt} = -j\frac{\omega_0 k_C}{2(2k_C + 1)}\boldsymbol{a}_1 + \left[j\left(2 - \frac{k_C}{2k_C + 2}\right) - \frac{R_2 + R_L}{2L_2}\right]\boldsymbol{a}_2, \quad (10.15)$$

where $k_C = C_M/\sqrt{C_1C_2}$ is the capacitive coupling coefficient.

Considering that the mutual capacitance in EC-WPT systems is generally much smaller than the self-capacitances, i.e., $k_C << 1$, the coupled-mode model can be further simplified as

$$\frac{d}{dt}\begin{bmatrix}\boldsymbol{a}_1\\ \boldsymbol{a}_2\end{bmatrix} = \begin{bmatrix} j\omega_0 + g_N - \tau_{10} & -j\kappa_C \\ -j\kappa_C & j\omega_0 - (\tau_2 + \tau_L)\end{bmatrix}\begin{bmatrix}\boldsymbol{a}_1\\ \boldsymbol{a}_2\end{bmatrix} \quad (10.16)$$

where $g_N = R_N/(2L_1)$ is the gain coefficient, $\tau_{10} = R_1/(2L_1)$ and $\tau_{20} = R_2/(2L_2)$ are intrinsic loss rates, $\tau_L = R_L/(2L_2)$ is the loss rate due to the load resistance, $\kappa_C = \omega_0 k_C/2$ is the energy coupling rate.

10.3 Fundamental Characteristic Analysis

Based on the equivalence of the circuit model and coupled-mode model, see Sect. 10.3.3, the fundamental characteristics of the PT-symmetric EC-WPT system will be analyzed based on the circuit model shown in (10.4).

10.3.1 Frequency Characteristic

The circuit equation of (10.4) has a non-zero solution only if the coefficient matrix determinant of the binary homogeneous linear equation system equals to zero. Thus, the characteristic equation of the PT-symmetric EC-WPT system can be deduced as

$$\left[jQ_T^{-1} - \frac{\omega}{\omega_0} + \frac{\omega_0(k_C+1)}{\omega(2k_C+1)}\right]\left[jQ_R^{-1} - \frac{\omega}{\omega_0} + \frac{\omega_0(k_C+1)}{\omega(2k_C+1)}\right] - \frac{\omega_0^2 k_C^2}{\omega^2(2k_C+1)^2} = 0. \tag{10.17}$$

Separating the real and imaginary parts of (10.17), we have

$$\text{Re}: \left[-\frac{\omega}{\omega_0} + \frac{\omega_0(k_C+1)}{\omega(2k_C+1)}\right]^2 - \frac{\omega_0^2 k_C^2}{\omega^2(2k_C+1)^2} - Q_T^{-1}Q_R^{-1} = 0, \tag{10.18}$$

$$\text{Im}: \left(Q_T^{-1} + Q_R^{-1}\right)\left[-\frac{\omega}{\omega_0} + \frac{\omega_0(k_C+1)}{\omega(2k_C+1)}\right] = 0. \tag{10.19}$$

According to (10.19), the characteristic equation has two solutions as follows:

(1) When $Q_T^{-1} + Q_R^{-1} = 0$ is satisfied.

Substituting $Q_T^{-1}+Q_R^{-1} = 0$ into (10.18), the operating angular frequency can be derived as

$$\omega_{o1} = \omega_0\sqrt{\frac{k_C+1}{2k_C+1} - \frac{1}{2Q_R^2} + \sqrt{\left(\frac{k_C}{2k_C+1} + \frac{1}{2Q_R^2}\right)^2 - \frac{1}{Q_R^2}}}, \tag{10.20}$$

$$\omega_{o2} = \omega_0\sqrt{\frac{k_C+1}{2k_C+1} - \frac{1}{2Q_R^2} - \sqrt{\left(\frac{k_C}{2k_C+1} + \frac{1}{2Q_R^2}\right)^2 - \frac{1}{Q_R^2}}}. \tag{10.21}$$

In order to ensure that the operating angular frequency is positive real number solution, the capacitive coupling coefficient k_C has to satisfy

$$k_C \geq k_{CR} = \frac{2Q_R - 1}{2(Q_R - 1)^2}, \tag{10.22}$$

where k_{CR} is defined as the critical coupling coefficient of the PT-symmetric EC-WPT system. Due to $0 < k_{CR} < 1$, $Q_R > (3 + \sqrt{3})/2$ is required, which means that the load resistance should satisfy $R_L < 2\omega_0 L_2/(3 + \sqrt{3})\text{-}R_2$.

(2) When $-\frac{\omega}{\omega_0} + \frac{\omega_0(k_C+1)}{\omega(2k_C+1)} = 0$ is satisfied.

There is only one solution for the operating angular frequency, which is

$$\omega_{o3} = \omega_0 \sqrt{\frac{k_C + 1}{2k_C + 1}}. \tag{10.23}$$

From (10.20)–(10.23), it can be seen that the PT-symmetric EC-WPT system has only one angular frequency solution in the weak coupling region of $0 < k_C < k_{CR}$ and three angular frequency solutions in the strong coupling region of $1 > k_C \geq k_{CR}$. In fact, the PT-symmetric EC-WPT system cannot work stably at the angular frequency defined by (10.23) when $1 > k_C \geq k_{CR}$ [9, 12]. Therefore, the frequency characteristics of the PT-symmetric EC-WPT system can be summarized as

$$\omega = \begin{cases} \omega_0 \sqrt{\frac{k_C+1}{2k_C+1}}, & 0 < k_C < k_{CR} \\ \omega_0 \sqrt{\frac{k_C+1}{2k_C+1} - \frac{1}{2Q_R^2} \pm \sqrt{\left(\frac{k_C}{2k_C+1} + \frac{1}{2Q_R^2}\right)^2 - \frac{1}{Q_R^2}}}, & 1 > k_C \geq k_{CR} \end{cases} \tag{10.24}$$

In particular, the coupling region of $1 > k_C \geq k_{CR}$ is also called the exact PT-symmetric region, while the coupling region of $0 < k_C < k_{CR}$ is called the broken PT-symmetric region. As can be seen from (10.24), the operating angular frequency ω changes with the capacitive coupling coefficient k_C. When the PT-symmetric EC-WPT system works in the exact PT-symmetric region, the operating angular frequency appears bifurcation phenomenon, and there are two frequency branches, which can be defined as the high-frequency branch and the low-frequency branch, respectively. As the capacitive coupling coefficient k_C decreases, the high and low-frequency branches coalesce at the critical coupling coefficient k_{CR}. In the broken PT-symmetric region, the operating angular frequency does not bifurcate and will decrease with k_C.

Figure 10.4 shows the curves of the system operating frequency f versus the capacitive coupling coefficient k_C under different load resistance R_L, and the circuit parameters are listed in Table 10.1. From Fig. 10.4, it can be seen that with the increase of capacitive coupling coefficient k_C, the system operating frequency appears bifurcation phenomenon, that is, in the exact PT-symmetric region $k_C \geq k_{CR}$, there are two steady-state operating frequencies defined as high-frequency and low-frequency branches, and their difference becomes larger when the capacitive coupling coefficient k_C increases or the load resistance R_L decreases. While in the broken PT-symmetric region $k_C < k_{CR}$, the system only has one operating frequency, which is only dependent on the capacitive coupling coefficient k_C and independent of the load resistance R_L. In addition, the critical capacitive coupling coefficient k_{CR}

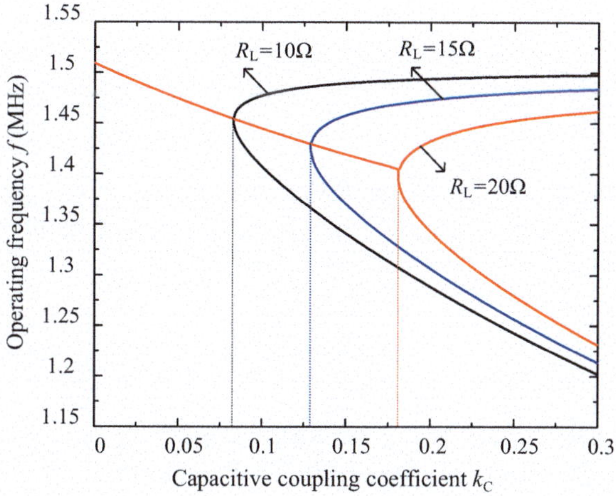

Fig. 10.4 The operating frequency f versus the capacitive coupling coefficient k_C

Table 10.1 Main circuit parameters

Description	Symbol	Values
Compensation inductance of transmitter	L_1	15.24 μH
Compensation inductance of receiver	L_2	15.24 μH
Compensation capacitance of transmitter	C_1	730 pF
Compensation capacitance of receiver	C_2	730 pF
Internal resistance of transmitter	R_1	0.66 Ω
Internal resistance of receiver	R_2	0.66 Ω
Load resistance	R_L	10 Ω
Natural resonant frequency of transmitter	f_1	1.509 MHz
Natural resonant frequency of receiver	f_2	1.509 MHz
DC input voltage	U_{DC}	20 V

decreases with the decrease of the load resistance, which indicates that the exact PT-symmetric region (i.e., strong coupled region) is extended with the decrease of the load resistance. The circuit parameters are listed in Table 10.1.

10.3.2 Output Power and Transfer Efficiency

Based on the above frequency characteristic of the PT-symmetric EC-WPT system, when the system works in the exact PT-symmetric region, i.e., $1 > k_C \geq k_{CR}$, the negative resistance -R_N can be deduced from $Q_T^{-1}+Q_R^{-1} = 0$ as follows

$$R_N = R_1 + R_2 + R_L. \tag{10.25}$$

Submitting (10.25) and (10.24) into (10.4), the current gain G_I can be obtained as

$$G_I = \frac{I_2}{I_1} = \frac{|\dot{I}_2|}{|\dot{I}_1|} = 1, \tag{10.26}$$

where I_1 and I_2 are the RMS values of i_1 and i_2, respectively.

Then, according to $U_{in} = R_N I_1$ and $U_o = R_L I_2$, the voltage gain G_U in this strong coupling region can be derived as

$$G_U = \frac{U_o}{U_{in}} = \frac{|\dot{U}_o|}{|\dot{U}_{in}|} = \frac{R_L}{R_1 + R_2 + R_L}, \tag{10.27}$$

where U_{in} and U_o are the RMS values of u_{in} and u_o, respectively.

Based on (10.26) and (10.27), the output power and transfer efficiency in the exact PT-symmetric region can be derived as

$$P_O = U_o I_2 = \frac{U_o^2}{R_L} = \frac{U_{in}^2 G_U^2}{R_L} = \frac{R_L U_{in}^2}{(R_1 + R_2 + R_L)^2}, \tag{10.28}$$

$$\eta = \frac{U_o I_2}{U_{in} I_1} = \frac{I_2^2 R_L}{I_1^2 R_N} = \frac{R_L}{R_N} G_I^2 = \frac{R_L}{R_1 + R_2 + R_L}. \tag{10.29}$$

Similarly, when the system works in the broken PT-symmetric region, i.e., $0 < k_C < k_{CR}$, by substituting (10.23) into (10.18), the negative resistance can be derived as

$$R_N = \frac{\omega_0^2 L_1 L_2 k_C^2}{(k_C + 1)(2k_C + 1)(R_2 + R_L)} + R_1. \tag{10.30}$$

By substituting (10.30) into, the current gain G_I and voltage gain G_U in the broken PT-symmetric region can be deduced as

$$G_I = \frac{k_C Q_R}{\sqrt{(k_C + 1)(2k_C + 1)}}, \tag{10.31}$$

$$G_U = \frac{R_L \omega_0 L_2 k_C \sqrt{(k_C + 1)(2k_C + 1)}}{\omega_0^2 L_1 L_2 k_C^2 + R_1 (R_2 + R_L)(k_C + 1)(2k_C + 1)}. \tag{10.32}$$

Then, the output power and transfer efficiency in the broken PT-symmetric region can be presented as

$$P_O = \frac{U_{in}^2 R_L \omega_0^2 L_2^2 k_C^2 (k_C + 1)(2k_C + 1)}{\left[\omega_0^2 L_1 L_2 k_C^2 + R_1 (R_2 + R_L)(k_C + 1)(2k_C + 1)\right]^2}, \tag{10.33}$$

$$\eta = \frac{R_L \omega_0^2 L_2^2 k_C^2}{(R_2 + R_L)\left[\omega_0^2 L_1 L_2 k_C^2 + R_1 (R_2 + R_L)(k_C + 1)(2k_C + 1)\right]}. \tag{10.34}$$

Therefore, the output power characteristics of the PT-symmetric EC-WPT system can be summarized as

$$P_O = \begin{cases} \dfrac{U_{in}^2 R_L \omega_0^2 L_2^2 k_C^2 (k_C + 1)(2k_C + 1)}{\left[\omega_0^2 L_1 L_2 k_C^2 + R_1 (R_2 + R_L)(k_C + 1)(2k_C + 1)\right]^2}, & k_C < k_C \\ \dfrac{R_L U_{in}^2}{(R_1 + R_2 + R_L)^2}, & k_C \geq k_{CR} \end{cases}. \tag{10.35}$$

The transfer efficiency characteristics of the PT-symmetric EC-WPT system can be summarized as

$$\eta = \begin{cases} \frac{R_L \omega_0^2 L_2^2 k_C^2}{(R_2+R_L)\left[\omega_0^2 L_1 L_2 k_C^2 + R_1 (R_2+R_L)(k_C+1)(2k_C+1)\right]}, & k_C < k_C \\ \frac{R_L}{R_1+R_2+R_L}, & k_C \geq k_{CR} \end{cases}. \tag{10.36}$$

From (10.35) and (10.36), it can be seen that in the PT-symmetric region $k_C \geq k_{CR}$, the output power P_O and transfer efficiency η only depend on the load resistance and internal resistances, but are independent of the capacitive coupling coefficient k_C, which indicates that the PT-symmetric EC-WPT system is robust against the variation of the coupling capacitance. Furthermore, when the capacitive coupling coefficient is small enough to meet $k_C < k_{CR}$, the output power P_O and transfer efficiency η will no longer maintain a constant value, but change with the variation of capacitive coupling coefficient k_C, which is not desirable in practical applications.

Figure 10.5 gives the curves of output power and transfer efficiency versus the capacitive coupling coefficient, in which the negative resistance is constructed by the full-bridge inverter, i.e., $U_{in} = 2\sqrt{2}\, U_{DC}/\pi$. As can be depicted in Fig. 10.5, the PT-symmetric EC-WPT system can achieve constant output power and transfer efficiency that is independent of the capacitive coupling coefficient in the PT-symmetric region, and the coupling region of constant output power and transfer efficiency increases with the decrease of the load resistance, which demonstrates that the PT-symmetric region can be broadened by reducing the load resistance.

10.3.3 Comparison Between PT-Symmetric EC-WPT and MC-WPT Systems

Although the MC-WPT system and the EC-WPT system are dual circuits, there are still differences between the MC-WPT system and the EC-WPT system when applying the PT-symmetric principle. By comparing the circuit state Eqs. (3.1) and (10.2), it can be seen that the coupling coils in the MC-WPT system are modeled

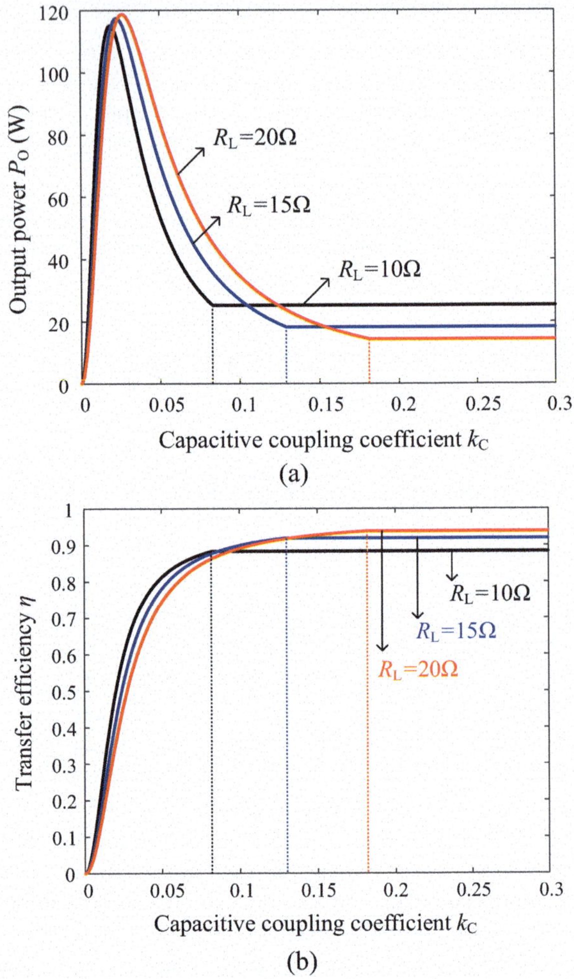

Fig. 10.5 The output power and transfer efficiency characteristics versus the capacitive coupling coefficient k_C. **a** Output power P_O, **b** Transfer efficiency η

as a loosely coupled transformer, and the change of its mutual inductance M only affects the inductance currents i_1 and i_2 of state variables, but does not affect the state variables of capacitance voltages u_1 and u_2. For the PT-symmetric EC-WPT system, its coupling plates are modeled as a discrete capacitor, and the variation of mutual capacitance C_M has s direct effect on all state variables.

Comparing the steady-state circuit Eqs. (3.2) and (10.3), it can also be seen that the mutual inductance M, as an equivalent element that cannot exist independently of self-inductance, changes only affect the energy transmission in the loosely coupled transformer, and does not change the original circuit characteristics of the transmitter and receiver. While the mutual capacitance C_M, as an independent element, can directly participate in the resonance of the dual-side LC compensation network, and its change can directly affect the system circuit characteristics.

Comparing the frequency characteristic of the PT-symmetric MC-WPT system and PT-symmetric EC-WPT system, see (3.41)–(3.44) and (10.20)–10.23, it can be found that both systems exhibit bifurcation characteristics in the PT-symmetric state. However, in the broken PT-symmetric state, the PT-symmetric MC-WPT system operates at the natural resonant frequency ω_0, and the frequency does not change with the coupling coefficient, while the operating frequency of the PT-symmetric EC-WPT system decreases with the capacitive coupling coefficient, and only when $k_C = 0$, there is $\omega = \omega_0$. In other words, the mutual inductance M in the PT-symmetric MC-WPT system only participates in the transfer of energy, but does not participate in the resonance of the circuit. While in the PT-symmetric EC-WPT system, the mutual capacitance C_M directly participates in the resonance of the LC compensation network, and the change of the capacitive coupling coefficient k_C also directly changes the actual resonance frequency of the system.

Comparing the transfer characteristics of PT-symmetric MC-WPT and EC-WPT systems in the PT-symmetric state, i.e., the output power characteristic (3.66) and (10.28), the transfer efficiency characteristic (3.67) and (10.29), it can be observed that PT-symmetric MC-WPT and EC-WPT systems have similar transfer characteristics, and both can achieve constant output power and transfer efficiency characteristics, which are independent of the coupling coefficient in the PT-symmetric state.

Moreover, when the transfer distance between the coupling coils in the PT-symmetric MC-WPT system is increased, the mutual inductance M decreases, but it does not cause the self-inductance L_1 and L_2 to change at the same time. On the other hand, for the PT-symmetric EC-WPT system, as can be seen from (10.1), when the distance between the coupling capacitance plates increases, the mutual capacitance C_M will decrease rapidly, and the self-capacitances C_1 and C_2 will slightly increase. While when the coupling capacitance plates are laterally shifted, the mutual capacitance C_M will linearly decrease, and the self-capacitance C_1 and C_2 will increase rapidly. Thus, the change of self-capacitance with the displacement of the coupling capacitance plates is a major characteristic to distinguish PT-symmetric EC-WPT system from PT-symmetric MC-WPT system, and should be taken into account in circuit modeling and coupling mechanism design.

10.4 System Design and Verification

10.4.1 Negative Resistor

To fully utilize the benefits of the PT-symmetric EC-WPT system mentioned in the above analysis, it is of great significance to implement an efficient and robust negative resistor. The negative resistor serves as the power source while ensuring that the system automatically operates at the eigen frequencies determined by (10.24). According to the description of the negative resistor, see Sect. 2.4, Fig. 10.6 gives the circuit structure of a full-bridge inverter to construct the negative resistor, which consists of a full-bridge inverter circuit and a self-oscillating controlled feedback circuit. The input DC voltage of the prototype is 20 V, the maximum input current is 2 A, and the maximum operating frequency is 1.5 MHz.

In Fig. 10.6, the full-bridge inverter is built based on two integrated power half-bridge chips SiC532. The chip integrates two half-bridge switches and a bootstrap gate drive circuit, with an adaptive dead-time insertion function, maximum output current of 30 A, maximum input voltage of 24 V, and maximum operating frequency of 2 MHz, which is suitable for the high-frequency and low-voltage full-bridge inverter circuit. The schematic diagram of the full-bridge inverter is depicted in Fig. 10.7.

Figure 10.8 shows the circuit diagram of the self-oscillating controlled circuit, in which the transmitting current i_1 is sampled by a current transformer CU8965 and a differential amplifier OPA690, and then passes through an RC compensator and a zero-crossing comparator TL3016 to counteract the effect of circuit delay and generate two complementary driving pluses for the switches S_1 (S_4) and S_2 (S_3), which ensure that the output voltage u_{in} and current i_1 of the full-bridge inverter maintain a phase difference of 180° for the negative resistance. The schematic diagram

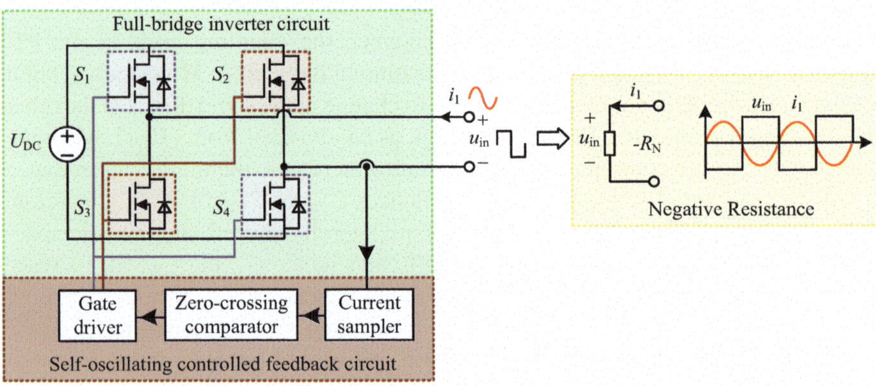

Fig. 10.6 The circuit structure diagram of the negative resistor based on the full-bridge inverter

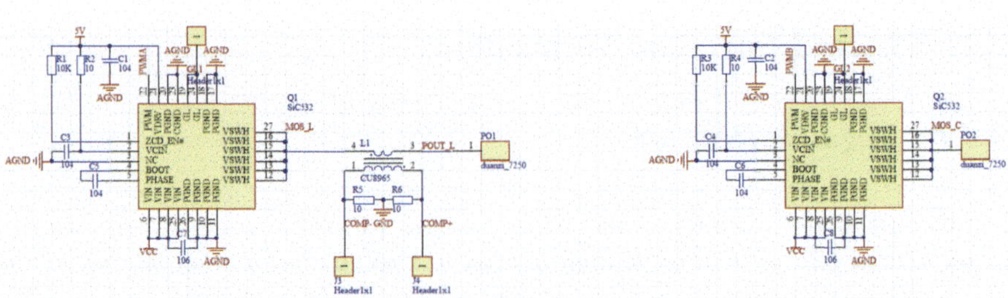

Fig. 10.7 The schematic diagram of the full-bridge inverter

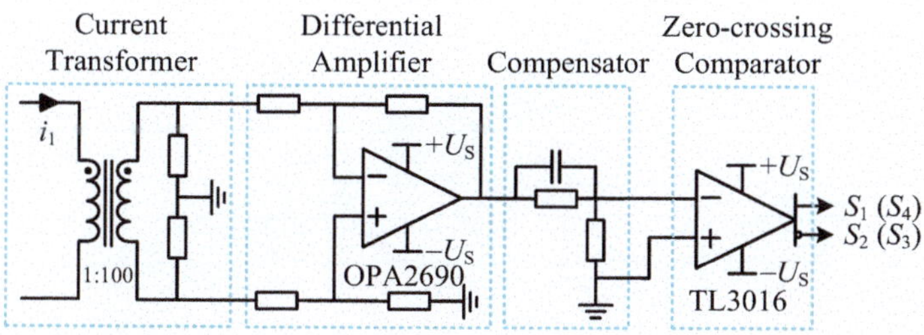

Fig. 10.8 The circuit diagram of the self-oscillating controller

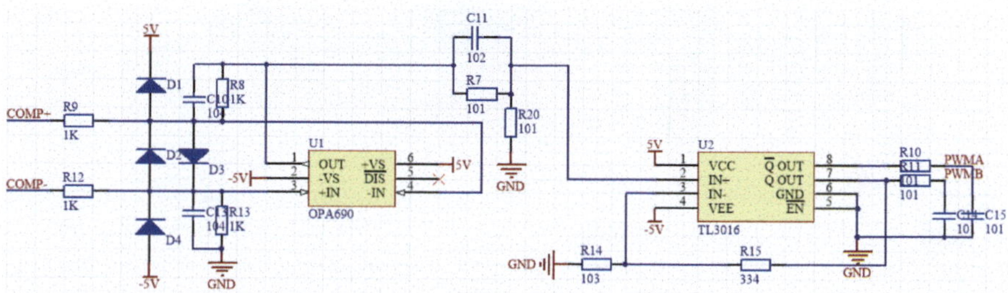

Fig. 10.9 The schematic diagram of the self-oscillating controlled circuit

of the self-oscillating controlled feedback circuit is shown in Fig. 10.9. It can be found that the self-oscillating controlled full-bridge inverter is only controlled on the transmitter without any communication.

10.4.2 *Capacitive Coupler*

For the parallel plate coupler shown in Fig. 10.2, the coupling capacitance can be written as

$$C = \frac{\varepsilon_r \varepsilon_0 S}{d}, \tag{10.37}$$

where ε_0 is the vacuum dielectric constant, ε_r is the relative permittivity of the medium, S is the relative coupling area of the parallel plates, and d is the distance between the parallel plates.

Since the relative dielectric constant of air is only 1, the larger plate area S or closer transfer distance d is required for the traditional parallel plate coupler to obtain a larger coupling capacitance, which obviously limits the flexibility and transfer distance of the EC-WPT system. A stacked array coupler can achieve high coupling capacitance through double-sided coupling of plates [1, 13, 14], as shown in Fig. 10.10. The plates P1 and P2 are the transmitting plates, the plates P3 and P4

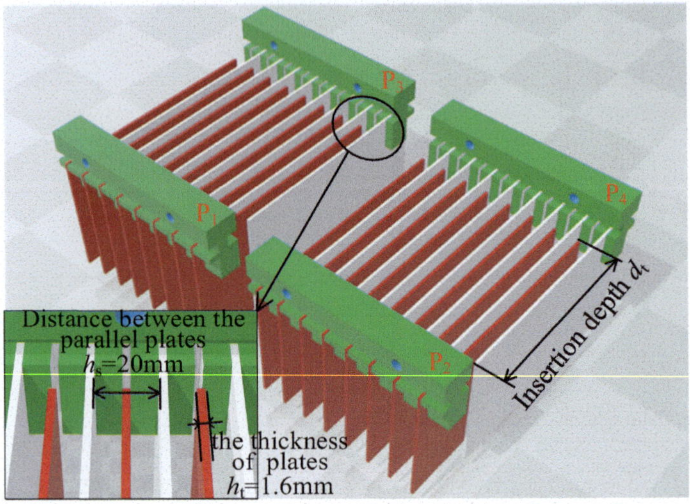

Fig. 10.10 Structure of stack array coupler

are the receiving plates, and each plate uses multiple PCB plates with the same size, which are connected in parallel at a fixed spacing. The transmitting plates and the receiving plates are inserted cross-over. Particularly, both sides of the PCB plates in the middle layer can be coupled to produce coupling capacitance, so a larger coupling area can be obtained in a smaller volume.

When the capacitive coupler is inserted in front of each other, that is, the spacing between adjacent PCB plates is equal, the coupling capacitance C_{13} (or C_{24}) of the stack array coupler can be defined by

$$C_{13} = (2n - 1)\frac{2\varepsilon_{\text{air}}\varepsilon_0 a d_t}{h_s - h_t}, \tag{10.38}$$

where n is the number of parallel electrode plates, a is the width of the plate, h_s is the distance between the parallel plates, h_t is the thickness of the electrode plate, d_t is the insertion depth between the transmitting and receiving plates, and the maximum value of d_t is the length of the plates b.

When the capacitive coupler has a lateral shift, the spacing between the adjacent PCB plates is inconsistent, the coupling capacitance C_{13} (or C_{24}) will become

$$C_{13} = n\frac{2\varepsilon_{\text{air}}\varepsilon_0 a d_t}{h_s - h_t - 2h_X} + (n - 1)\frac{2\varepsilon_{\text{air}}\varepsilon_0 a d_t}{h_s - h_t + 2h_X}, \tag{10.39}$$

where h_X is the lateral offset distance between the transmitting and receiving plates.

From Fig. 10.10 and (10.38), it can be seen that compared with the traditional four-plate capacitive coupler, the stack array coupler effectively increases the coupling area under the same volume and improves the volumetric capacitance density of the coupling mechanism. If eight 120 × 200 × 1.6 mm^3 PCB plates are connected in parallel at 20 mm intervals, that is, $n = 8$, $a = 120$, $b = 200$, $h_t = 1.6$ mm, and $h_s = 20$ mm, the coupling capacitance can be deduced as $C_{13} = 1732d_t$ pF, then C_{13} varies linearly with d_t, as shown in Fig. 10.11a, where the simulation value can be estimated by Maxwell and the measured value is obtained by the impedance analyzer at 1.5 MHz. Moreover, the external surface area of the stack array coupler is reduced due to the parallel connection of the plates, thereby the self-capacitance of the stack array coupler is too small and can be negligible. Thus, according to (10.1), the mutual-capacitance can be simplified by $C_M = C_{13}C_{24}/(C_{13} + C_{24})$, or $C_M = C_{13}/2$, which can be obtained by the value of C_{13} in Fig. 10.11a. Due to $k_C = C_M/C$, the capacitive coupling coefficient k_C of EC-WPT system varies linearly with insertion depth d_t.

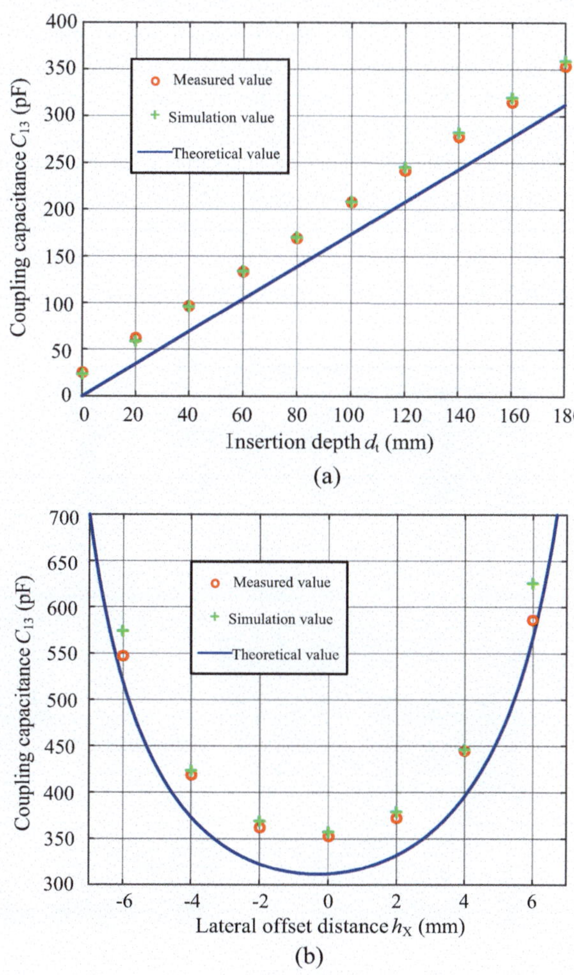

Fig. 10.11 Coupling capacitance C_{13} versus the relative position of the coupling plates. **a** C_{13} versus d_t, **b** C_{13} versus h_X

When the insertion depth is fixed at $d_t = 180$ mm, the transmitting plates and receiving plates are offset laterally, and the relationship between the coupling capacitance C_{13} and the lateral offset distance h_X is shown in Fig. 10.11b, where the theoretical value is calculated by (10.39). it can be seen that the coupling capacitance C_{13} increases with the increase of $|h_X|$, which means that the stack array coupler has high misalignment tolerance.

10.4.3 Experimental Verification

To practically evaluate the performance of the PT-symmetric EC-WPT system, a prototype shown in Fig. 10.12 is built, where circuit parameters are listed in Table 10.2. By substituting the parameters into (10.22), the critical coupling coefficient is $k_{CR} = 0.0823$, and the corresponding range of the exact PT-symmetric region is $d_t \geq 60$ mm. Figures 10.13 and 10.14 show the theoretical curves, simulation, and experimental results of operating frequency, output power, and transfer efficiency of the PT-symmetric EC-WPT system.

From Fig. 10.13, it can be seen that with the increase of insertion depth d_t, the operating frequency f decreases, and follows the low-frequency branch of the theoretical curves in the strong coupled region of $k_C > k_{CR}$ (i.e., the exact PT-symmetric region). The variation trend of the measured frequency is basically consistent with the theoretical analysis and simulation values.

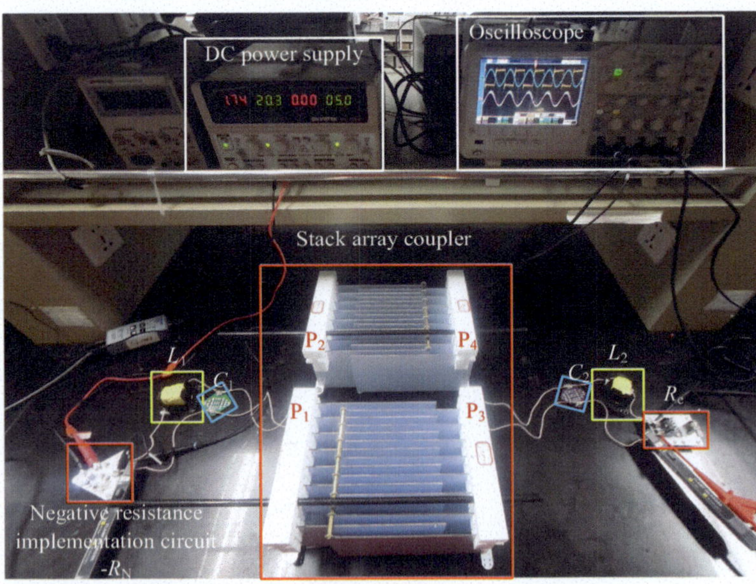

Fig. 10.12 Experimental prototype

Table 10.2 The circuit parameters of the experimental prototype

Description	Symbol	Values
Compensation inductance of transmitter	L_1	15.070 μH
Compensation inductance of receiver	L_2	15.247 μH
Compensation capacitance of transmitter	C_1	738.086 pF
Compensation capacitance of receiver	C_2	730.054 pF
Internal resistance of transmitter	R_1	647.746 mΩ
Internal resistance of receiver	R_2	660.262 mΩ
Load resistance	R_L	10 Ω
Natural resonant frequency of transmitter	f_1	1.5090 MHz
Natural resonant frequency of receiver	f_2	1.5085 MHz
DC input voltage	U_{DC}	20 V

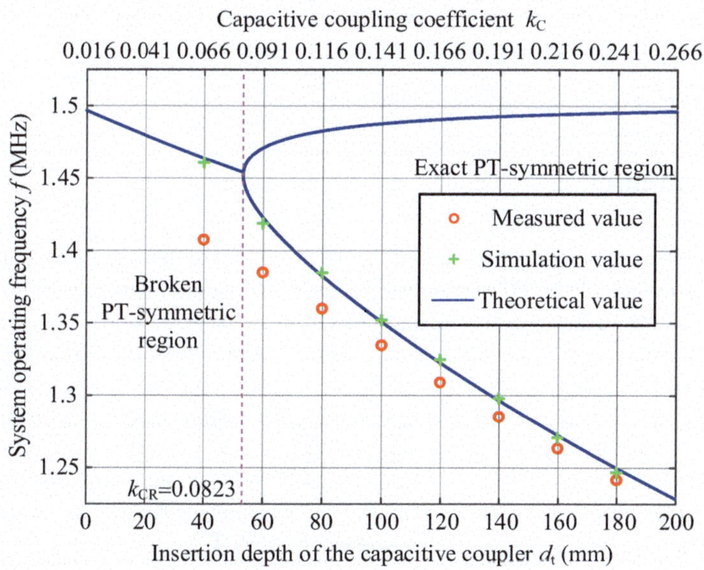

Fig. 10.13 Operating frequency f versus k_C and d_t

From Fig. 10.14, it can be observed that both output power P_O and transfer efficiency η keep constant in the exact PT-symmetric region of $d_t > 60$ mm, where the output power P_O is stable near 25 W and the transfer efficiency η is basically constant at 88%. When the system works in the broken PT-symmetric region, both output power and transfer efficiency are no longer constant.

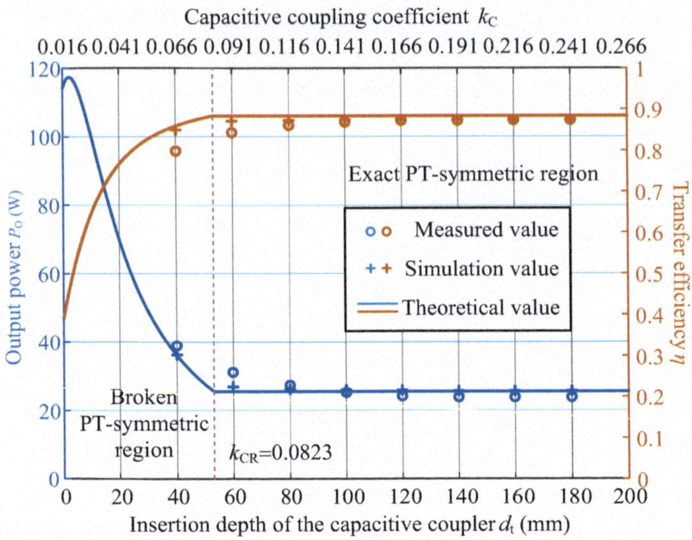

Fig. 10.14 Output power P_O and transfer efficiency η versus k_C and d_t

10.5 Summary

This chapter first introduces the equivalent circuit model of the capacitive coupler and the basic structure of the PT-symmetric EC-WPT system. Then, the PT-symmetric EC-WPT system with a double LC compensation network is modeled and analyzed based on circuit theory and coupled-mode theory. Meanwhile, the fundamental characteristics of the PT-symmetric EC-WPT system are analyzed in detail, including frequency characteristics, current and voltage gain, output power, and transfer efficiency. The analytical results show that the current and voltage gain, output power, and transfer efficiency of the PT-symmetric EC-WPT system are independent of the capacitive coupling coefficient in the PT-symmetric region. In addition, the system characteristics of the PT-symmetric EC-WPT system and PT-symmetric MC-WPT system are compared, which lays a foundation for the hybrid electric–magnetic coupled WPT system based on PT-symmetric theory. Finally, the experimental prototype of the PT-symmetric EC-WPT system is built, and the correctness of the theoretical analysis is verified.

References

1. Gu W, Qiu D, Shu X, Zhang B, Xiao W, Chen Y (2023) A constant output capacitive wireless power transfer system based on parity-time symmetric. IEEE Trans Circuits Syst II Express Briefs 70(7):2585–2589

2. Zhang H, Lu F, Hofmann H, Liu W, Mi CC (2016) A four-plate compact capacitive coupler design and LCL-compensated topology for capacitive power transfer in electric vehicle charging application. IEEE Trans Power Electron 31(12):8541–8551
3. Wu X, Su Y, Hu AP, Qing X, Hou X (2021) Multi-objective parameter optimization of a four-plate capacitive power transfer system. IEEE J Emerg Sel Topics Power Electron 9(2):2328–2342
4. Lu F, Zhang H, Hofmann H, Mi CC (2018) A double-sided LC-compensation circuit for loosely coupled capacitive power transfer. IEEE Trans Power Electron 33(2):1633–1643
5. Pratik U, Pantic Z (2023) Design of a capacitive wireless power transfer system with a vertical four-plate coupler for minimum stray electric field. IEEE J Emerg Sel Topics Power Electron 11(5):5486–5499
6. Yang ZL (1992) Mutual capacitance-duality principle evolved from planar network. IEEE Trans Circuit Syst I: Fundamental Theory Appl 39(12):1005–1006
7. Luo B, Mai R, Guo L, Wu D, He Z (2019) LC-CLC compensation topology for capacitive power transfer system to improve misalignment performance. IET Power Electron 12(10):2626–2633
8. Sanders JA, Verhulst F, Murdock (1985) Averaging methods in nonlinear dynamical systems. Springer, New York
9. Zhou J, Zhang B, Xiao W, Qiu D, Chen Y (2019) Nonlinear parity-time-symmetric model for constant efficiency wireless power transfer: application to a drone-in-flight wireless charging platform. IEEE Trans Industr Electron 66(5):4097–4107
10. Shu X, Zhang B (2018) Single-wire electric-field coupling power transmission using nonlinear parity-time-symmetric model with coupled-mode theory. Energies 11(3):532
11. Liu G, Zhang B (2019) Analytical model of a 25–50 m robust single-wire electric-field coupling power transfer system using a limiter. IEEE Trans Circuits Syst II Express Briefs 66(6):978–982
12. Assawaworrarit S, Yu X, Fan S (2017) Robust wireless power transfer using a nonlinear parity-time-symmetric circuit. Nature 546(7658):387–390
13. Dai J, Ludois DC (2015) A survey of wireless power transfer and a critical comparison of inductive and capacitive coupling for small gap applications. IEEE Trans Power Electron 30(11):6017–6029
14. Ludois DC, Erickson MJ, Reed JK (2014) Aerodynamic fluid bearings for translational and rotating capacitors in noncontact capacitive power transfer systems. IEEE Trans Ind Appl 50(2):1025–1033

Chapter 11
PT-Symmetric Electric–Magnetic Coupled Hybrid Wireless Power Transfer System

Due to the technical advantages and application limitations of EC-WPT and MC-WPT technologies, a hybrid wireless power transfer (HWPT) system is formed by combining the magnetic-coupled mechanism (i.e., coupling coils) of MC-WPT system with the electric-coupled mechanism (i.e., capacitive plates) of EC-WPT system. Compared with EC-WPT and MC-WPT systems, the HWPT system can make full use of high-frequency electric and magnetic fields, and has stronger transmission performance and power density of the coupler under the same circuit structure and parameters, but also introduces higher parameter sensitivity. Hence, achieving stable output power and constant transmission efficiency over a wide range of coupling coefficient variations is still a major challenge. Inspired by PT-symmetric MC-WPT and EC-WPT systems, the PT-symmetric theory can be utilized to realize the robustness of the HWPT system. This chapter first introduces the basic structure of the PT-symmetric HWPT system, and establishes the mathematical model of the PT-symmetric HWPT system by using mutual capacitance and mutual inductance. Then, the fundamental characteristics of operating frequency, output power, and transfer efficiency are analyzed. Finally, the effect of connecting the capacitive plates to the homonymous end or heteronymous end of the coupling coils on the PT-symmetry range of the HWPT system is discussed, and the HWPT system is compared with EC-WPT and MC-WPT systems from the aspects of coupling range, frequency characteristic, output power, and transfer efficiency.

11.1 System Structure

Generally, the MC-WPT system utilizes the inductive coils to transfer power via the magnetic field, and the EC-WPT system makes use of the capacitive plates to transfer power via electric field. The difference is that the HWPT system utilizes two kinds of

B. Zhang et al., *Parity-Time Symmetric Wireless Power Transfer*, CPSS Power Electronics Series, https://doi.org/10.1007/978-981-96-3149-0_11

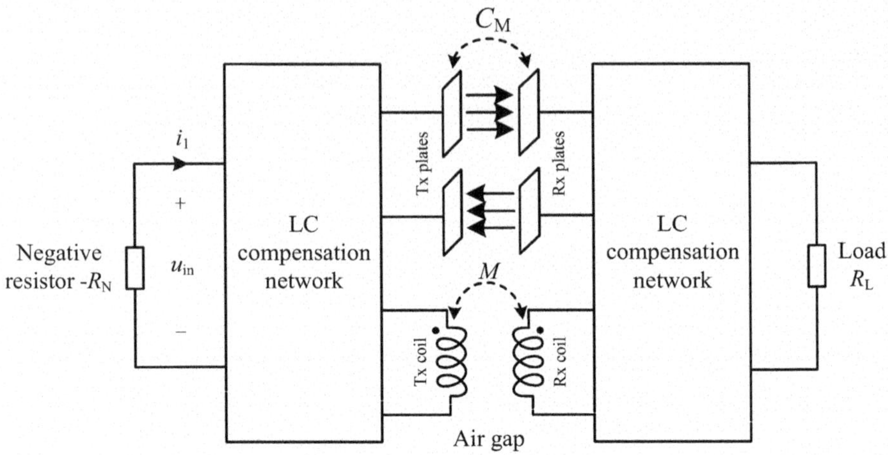

Fig. 11.1 System structure of the PT-symmetric HWPT system

coupling mechanisms simultaneously [1]. Figure 11.1 shows the structure of the PT-symmetric HWPT system, which consists of a negative resistance, LC compensation networks of transmitter (Tx) and receiver (Rx), capacitive coupling plates (including Tx and Rx metal plates), and inductive coupling coils (including Tx and Rx coils).

The compensation networks of the HWPT system can be divided into the independent-branch compensation network with different resonant frequency and the shared-branch compensation network with uniform resonant frequency. Because the magnetic-field coupling of the HWPT system needs to be compensated by capacitance, and the electric-field coupling needs to be compensated by inductance, different coupling mechanisms can be reused. Although the HWPT system of the independent-branch compensation network achieves flexible power distribution between the electric-field and magnetic-field coupling branches, the frequency bifurcation curves of the two coupling branches are not identical when the PT-symmetric principle is applied, so the system cannot operate at a stable frequency. For the HWPT system of the shared-branch compensation network, its different coupling branches act as resonant elements with each other, so that the compensation network has a uniform resonant frequency, and a common frequency bifurcation curve of PT-symmetric system can be derived. Therefore, a PT-symmetric HWPT system with shared-branch compensation network is selected, as shown in Fig. 11.2, where the electric-field and magnetic-field coupling branches are connected in series, -R_N is the negative resistor, u_{in} and i_1 are the voltage and current of the negative resistor, respectively. R_L is the equivalent resistance of the load, u_o and i_2 are the load voltage and load current, respectively. C_M is the mutual-capacitance of coupling plates, C_1 and C_2 are composed of self-capacitances C_P and C_S of capacitive coupler and additional compensation capacitance in parallel, both as the compensation capacitance of the coupling plates and as the series resonant capacitance of the coupling coils. M is the mutual inductance of coupling coils, L_1 and L_2 are the self-inductances

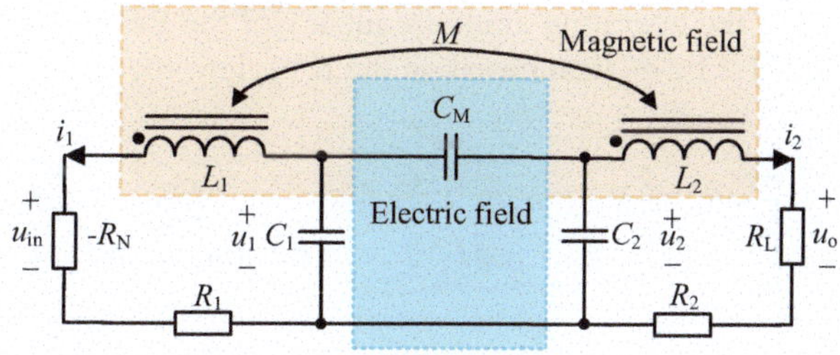

Fig. 11.2 Equivalent circuit of the PT-symmetric HWPT system

of transmitter and receiver, and also serve as the series resonant inductance of the coupling plates. It is worth noting that the π-type circuit model of coupling plates should be connected with the heteronymous end of the loosely coupled transformer model of coupling coils. R_1 and R_2 are the total internal resistances of the transmitter and receiver, respectively.

11.2 Modeling

According to Kirchhoff's law, the capacitor voltages u_1, u_2 of C_1, C_2 and the inductor currents i_1, i_2 of L_1, L_2 are selected as the state variables of the HWPT system, Then the state equation of the PT-symmetric HWPT system can be obtained as

$$\begin{cases} \dfrac{di_1}{dt} = \dfrac{(-R_N + R_1)L_2}{M^2 - L_1L_2}i_1 + \dfrac{L_2}{M^2 - L_1L_2}u_1 + \dfrac{(-R_N + R_1)M}{M^2 - L_1L_2}i_2 + \dfrac{M}{M^2 - L_1L_2}u_2 \\ \dfrac{di_2}{dt} = \dfrac{(-R_N + R_1)M}{M^2 - L_1L_2}i_1 + \dfrac{M}{M^2 - L_1L_2}u_1 + \dfrac{(-R_N + R_1)L_1}{M^2 - L_1L_2}i_2 + \dfrac{L_1}{M^2 - L_1L_2}u_2 \\ \dfrac{du_1}{dt} = \dfrac{C_2 + C_M}{\Delta}i_1 + \dfrac{C_M}{\Delta}i_2 \\ \dfrac{du_2}{dt} = \dfrac{C_M}{\Delta}i_1 + \dfrac{C_1 + C_M}{\Delta}i_2 \end{cases}, \tag{11.1}$$

where $\Delta = C_1C_M + C_2C_M + C_1C_2$.

11.2.1 Circuit Model

Due to the high-quality factor of LC compensation networks, i_1 and i_2 can be regarded as sine waves, so the steady-state analysis of the PT-symmetric HWPT system can be performed by using the fundamental wave analysis method. $\dot{U}_{in}$, $\dot{U}_o$, $\dot{I}_1$, and $\dot{I}_2$ are phasor forms of u_{in}, u_o, i_1, and i_2, respectively.

Let the operating angle frequency of the PT-symmetric HWPT system shown in Fig. 11.2 be ω, then based on the KVL law, the steady-state equation can be written as

$$\begin{bmatrix} -R_N + R_1 + j\omega L_1 + \frac{C_2+C_M}{j\omega\Delta} & \frac{C_M}{j\omega\Delta} - j\omega M \\ \frac{C_M}{j\omega\Delta} - j\omega M & R_L + R_2 + j\omega L_2 + \frac{C_1+C_M}{j\omega\Delta} \end{bmatrix} \begin{bmatrix} \dot{I}_1 \\ \dot{I}_2 \end{bmatrix} = 0. \tag{11.2}$$

Considering the symmetric circuit parameters are generally selected, i.e., $L_1 = L_2 = L$, $R_1 = R_2 = R$ and $C_1 = C_2 = C$, the characteristic equation of (11.2) can be further simplified as

$$\begin{bmatrix} jQ_T^{-1} - \frac{\omega}{\omega_0} + \frac{\omega_0(k_C+1)}{\omega(2k_C+1)} & \frac{\omega_0 k_C}{\omega(2k_C+1)} + \frac{\omega}{\omega_0}k_L \\ \frac{\omega_0 k_C}{\omega(2k_C+1)} + \frac{\omega}{\omega_0}k_L & jQ_R^{-1} - \frac{\omega}{\omega_0} + \frac{\omega_0(k_C+1)}{\omega(2k_C+1)} \end{bmatrix} \begin{bmatrix} \dot{I}_1 \\ \dot{I}_2 \end{bmatrix} = 0, \tag{11.3}$$

where $\omega_0 = 1/\sqrt{L_1C_1} = 1/\sqrt{L_2C_2}$ is the natural resonant angular frequency, $k_C = C_M/C_1 = C_M/C_2$ is the capacitive coupling coefficient, $k_L = M/L_1 = M/L_2$ is the inductive coupling coefficient, $Q_T = \omega_0 L_1/(-R_N + R_1)$ and $Q_R = \omega_0 L_2/(R_L + R_2)$ are defined as the quality factors of transmitter and receiver, respectively.

11.2.2 Coupled-Mode Model

According to the derivation process in Sect. 10.2.2, the simplified coupled-mode equation of the PT-symmetric HWPT system can be obtained as

$$\frac{d}{dt}\begin{bmatrix} \boldsymbol{a}_1 \\ \boldsymbol{a}_2 \end{bmatrix} = \begin{bmatrix} j\omega_0 + g_N - \tau_{10} & -j\kappa_H \\ -j\kappa_H & j\omega_0 - \tau_{20} - \tau_L \end{bmatrix} \begin{bmatrix} \boldsymbol{a}_1 \\ \boldsymbol{a}_2 \end{bmatrix}, \tag{11.4}$$

where $g_N = R_N/(2L_1)$ is the gain coefficient, $\tau_{10} = R_1/(2L_1)$ and $\tau_{20} = R_2/(2L_2)$ are intrinsic loss rates, $\tau_L = R_L/(2L_2)$ is the loss rate due to the load resistance, $\kappa_H = \omega_0(k_C + k_L)/2$ is the energy coupling rate.

11.3 Fundamental Characteristic Analysis

Based on the equivalence of the circuit model and coupled-mode model, see Sect. 3.3, the fundamental characteristics of the PT-symmetric HWPT system will be analyzed based on the circuit model shown in (11.3).

11.3.1 Frequency Characteristic

If the coefficient matrix determinant of the binary homogeneous linear Eq. (11.3) equals to zero, there exists a non-zero solution. Thus, according to (11.3), the characteristic frequency equation can be deduced as

$$\left[jQ_T^{-1} - \frac{\omega}{\omega_0} + \frac{\omega_0(k_C+1)}{\omega(2k_C+1)}\right]\left[jQ_R^{-1} - \frac{\omega}{\omega_0} + \frac{\omega_0(k_C+1)}{\omega(2k_C+1)}\right] - \left[\frac{\omega_0 k_C}{\omega(2k_C+1)} + \frac{\omega}{\omega_0}k_L\right]^2 = 0. \tag{11.5}$$

Separating the real and imaginary parts of (11.5), there is

$$\text{Re}\ \left[-\frac{\omega}{\omega_0} + \frac{\omega_0(k_C+1)}{\omega(2k_C+1)}\right]^2 - \left[\frac{\omega_0 k_C}{\omega(2k_C+1)} + \frac{\omega}{\omega_0}k_L\right]^2 - Q_T^{-1}Q_R^{-1} = 0, \tag{11.6}$$

$$\text{Im}\ \left(Q_T^{-1} + Q_R^{-1}\right)\left[-\frac{\omega}{\omega_0} + \frac{\omega_0(k_C+1)}{\omega(2k_C+1)}\right] = 0. \tag{11.7}$$

By solving (11.7), the characteristic frequency equation of (11.5) has two solutions as follows:

(1) $Q_T^{-1} + Q_R^{-1} = 0$ is satisfied.

The operating angular frequency can be derived as

$$\omega_H = \omega_0\sqrt{\frac{A \pm \sqrt{A^2 - 4(2k_C+1)\left(1-k_L^2\right)}}{2(2k_C+1)\left(1-k_L^2\right)}}, \tag{11.8}$$

where $A = 2k_Ck_L + 1 - \left(Q_R^{-2} - 1\right)(2k_C+1)$.

In order to ensure that (11.8) has a positive real number solution, the PT-symmetric HWPT system has to satisfy

$$A^2 - 4(2k_C+1)\left(1-k_L^2\right) \geq 0. \tag{11.9}$$

It can be seen that (11.9) is a binary inequality of the capacitive coupling coefficient k_C and the inductive coupling coefficient k_L, the analytical solution cannot be directly evaluated. Thus, let the hybrid coupling coefficient k_H be used to describe the overall coupling state of the PT-symmetric HWPT system, that is, $k_H = y(k_C, k_L)$, then (11.9) can be solved as

$$k_H \geq k_{HR} = y(Q_R), \tag{11.10}$$

where k_{HR} is defined as the critical hybrid coupling coefficient of the PT-symmetric HWPT system. The hybrid coupling coefficient k_H is correlated with k_C and k_L, and the relationship between k_H, k_C, and k_L in the PT-symmetric HWPT system will be analyzed in detail later.

(2) $-\frac{\omega}{\omega_0}+\frac{\omega_0(k_C+1)}{\omega(2k_C+1)}=0$ is satisfied.

There is only one solution for the operating angular frequency, which is

$$\omega_{H0}=\omega_0\sqrt{\frac{k_C+1}{2k_C+1}} \tag{11.11}$$

It can be seen from the above analysis that the PT-symmetric HWPT system has three operating angular frequency solutions in the region of $k_H \geq k_{HR}$ and only one operating angular frequency solution in the region of $0 < k_H < k_{HR}$. In fact, the PT-symmetric HWPT system cannot work stably at the operating angular frequency defined by (11.11) when $k_H \geq k_{HR}$ [2, 3]. Therefore, the frequency characteristics of the PT-symmetric HWPT system shown in Fig. 11.2 are summarized as follows:

$$\omega=\begin{cases}\omega_0\sqrt{\dfrac{k_C+1}{2k_C+1}}, & 0<k_H<k_{HR}\\ \omega_0\sqrt{\dfrac{A\pm\sqrt{A^2-4(2k_C+1)\left(1-k_L^2\right)}}{2(2k_C+1)\left(1-k_L^2\right)}}, & k_H\geq k_{HR}\end{cases}, \tag{11.12}$$

where $\omega = 2\pi f$.

Generally, the region of $k_H \geq k_{HR}$ is called the exact PT-symmetric region, and the region of $0 < k_H < k_{HR}$ is called the broken PT-symmetric region. From (11.12), it can be seen that the operating angular frequency ω changes with the capacitive coupling coefficient k_C and the inductive coupling coefficient k_L. When the PT-symmetric HWPT system works in the exact PT-symmetric region, the operating angular frequency has a bifurcation phenomenon, and there are two frequency branches, which are defined as the high-frequency branch and low-frequency branch, respectively. However, in the broken PT-symmetric region, the operating angular frequency does not bifurcate and will decrease with the capacitive coupling coefficient k_C.

11.3.2 Output Power and Transfer Efficiency

When the PT-symmetric HWPT system operates in the exact PT-symmetric region, i.e., $k_H \geq k_{HR}$, according to $Q_T^{-1}+Q_R^{-1}=0$, the negative resistance can be presented by

$$R_N = R_1 + R_2 + R_L. \tag{11.13}$$

Submitting (11.13) and (11.8) into (11.3), the current gain can be derived as

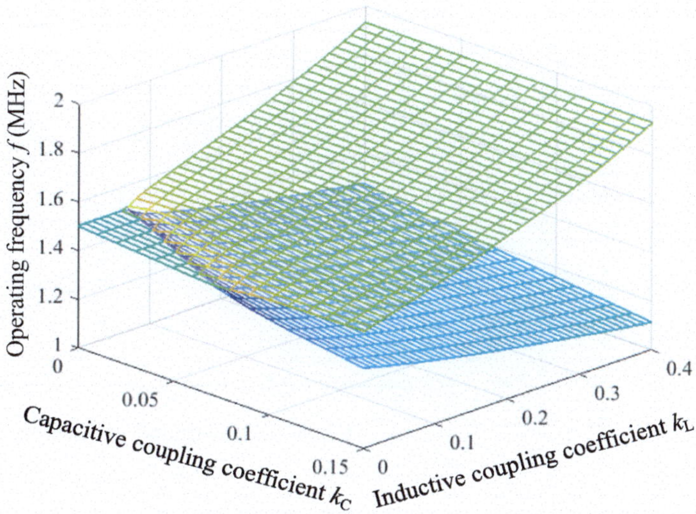

Fig. 11.3 Three-dimensional graph of the operating frequency f as a function of the capacitive coupling coefficient k_C and inductive coupling coefficient k_L. Figure 11.3 shows the relationship between the operating frequency f, capacitive and inductive coupling coefficients k_C and k_L, and Fig. 11.4 depicts the variation of operating frequency with the other coupling coefficient when the capacitive or inductive coupling coefficient is fixed. It can be seen that the operating frequency is a function of the capacitive and inductive coupling coefficients. When the inductive coupling coefficient k_L is fixed, the variation law of the operating frequency with the capacitive coupling coefficient k_C is consistent with that of the PT-symmetric EC-WPT system. And when the capacitive coupling coefficient k_C is fixed, the variation law of the operating frequency with the inductive coupling coefficient k_L is also consistent with that of the PT-symmetric MC-WPT system. In addition, when the capacitive (inductive) coupling coefficient k_C (k_L) is fixed, the critical value of inductive (capacitive) coupling coefficient k_L (k_C) will decrease with the increase of the capacitive (inductive) coupling coefficient k_C (k_L)

$$G_{\mathrm{I}} = \frac{I_2}{I_1} = \frac{|\dot{I}_2|}{|\dot{I}_1|} = 1, \tag{11.14}$$

where I_1 and I_2 are the RMS values of the transmitter and receiver currents i_1 and i_2, respectively.

Then, according to $U_{\mathrm{in}} = R_{\mathrm{N}} I_1$ and $U_{\mathrm{o}} = R_{\mathrm{L}} I_2$, the voltage gain in the exact PT-symmetric region can be deduced as

$$G_{\mathrm{U}} = \frac{U_{\mathrm{o}}}{U_{\mathrm{in}}} = \frac{R_{\mathrm{L}} I_2}{R_{\mathrm{N}} I_1} = \frac{R_{\mathrm{L}}}{R_1 + R_2 + R_{\mathrm{L}}}, \tag{11.15}$$

where U_{in} and U_{o} are the RMS values of u_{in} and u_{o}, respectively

Based on (11.14) and (11.15), the output power and transfer efficiency in the exact PT-symmetric region can be expressed as follows:

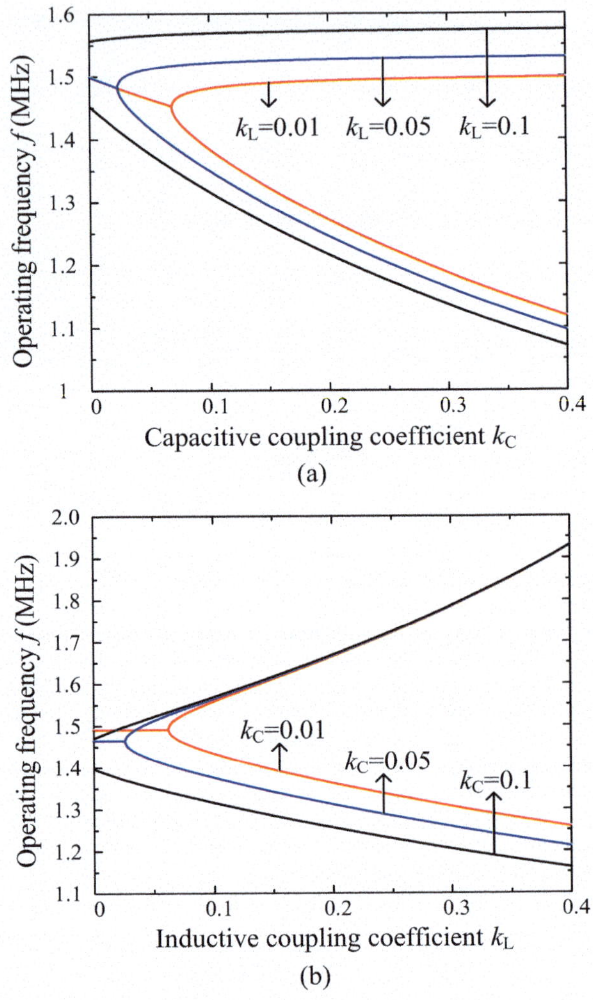

Fig. 11.4 Two-dimensional graph of the operating frequency f as a function of either coupling coefficient. **a** Operating frequency f versus capacitive coupling coefficient k_C. **b** Operating frequency f versus inductive coupling coefficient k_L

$$P_O = U_o I_2 = \frac{U_o^2}{R_L} = \frac{U_{in}^2 G_U^2}{R_L} = \frac{R_L U_{in}^2}{(R_1 + R_2 + R_L)^2}, \tag{11.16}$$

$$\eta = \frac{U_o I_2}{U_{in} I_1} = \frac{I_2^2 R_L}{I_1^2 R_N} = \frac{R_L}{R_N} G_I^2 = \frac{R_L}{R_1 + R_2 + R_L}. \tag{11.17}$$

From (11.16) and (11.17), it can be seen that the output power P_O and the transfer efficiency η depend only on the internal resistances R_1, R_2 and load resistance R_L, are not affected by the coupling coefficients k_C and k_L, and maintain at the constant values, which demonstrates that the PT-symmetric HWPT system is robust against the variation of the coupling capacitance and coupling inductance.

Similarly, when the PT-symmetric HWPT system works in the broken PT-symmetric region, i.e., $0 < k_H < k_{HR}$, submitting (11.11) into (11.6), the negative resistance can be obtained as

$$R_N = \frac{\omega_0^2 L_1 L_2 [k_L(k_C + 1) + k_C]^2}{(R_2 + R_L)(k_C + 1)(2k_C + 1)} - R_1. \quad (11.18)$$

Submitting (11.11) and (11.18) into (11.3), the voltage gain and current gain in the broken PT-symmetric region can be derived as follows:

$$G_I = \frac{I_2}{I_1} = \frac{[k_L(k_C + 1) + k_C]Q_R}{\sqrt{(k_C + 1)(2k_C + 1)}}, \quad (11.19)$$

$$G_U = \frac{U_o}{U_{in}} = \frac{R_L \omega_0 L_2 [k_L(k_C + 1) + k_C]\sqrt{(k_C + 1)(2k_C + 1)}}{\omega_0^2 L_1 L_2 [k_L(k_C + 1) + k_C]^2 + R_1(R_2 + R_L)(k_C + 1)(2k_C + 1)}. \quad (11.20)$$

Then, the output power and transfer efficiency in the broken PT-symmetric region can be deduced as follows:

$$P_O = \frac{U_{in}^2 G_U^2}{R_L} = \frac{U_{in}^2 R_L \omega_0^2 L_2^2 [k_L(k_C + 1) + k_C]^2 (k_C + 1)(2k_C + 1)}{\{\omega_0^2 L_1 L_2 [k_L(k_C + 1) + k_C]^2 + R_1(R_2 + R_L)(k_C + 1)(2k_C + 1)\}^2}, \quad (11.21)$$

$$\eta = \frac{R_L}{R_N} G_I^2 = \frac{R_L \omega_0 Q_R L_2 [k_L(k_C + 1) + k_C]^2}{\omega_0^2 L_1 L_2 [k_L(k_C + 1) + k_C]^2 + R_1(R_2 + R_L)(k_C + 1)(2k_C + 1)}. \quad (11.22)$$

From (11.21) and (11.22), it can be seen that when the hybrid coupling coefficient is less than the critical value, i.e., $k_H < k_{HR}$, the output power P_O and transfer efficiency η will change with the variation of the capacitive coupling coefficient k_C and the inductive coupling coefficient k_L. Generally, it is not desirable to operate in the broken PT-symmetric region.

Figure 11.5 shows a three-dimensional graph of the output power P_O and transfer efficiency η as a function of the capacitive coupling coefficient k_C and inductive coupling coefficient k_L. Figures 11.6 and 11.7 give two-dimensional graphs of the output power P_O and transfer efficiency η as a function of either coupling coefficient. As can be seen from Figs. 11.5, 11.6 and 11.7, the output power and transfer efficiency remain constant in the exact PT-symmetric region, independent of the capacitive and inductive coupling coefficients.

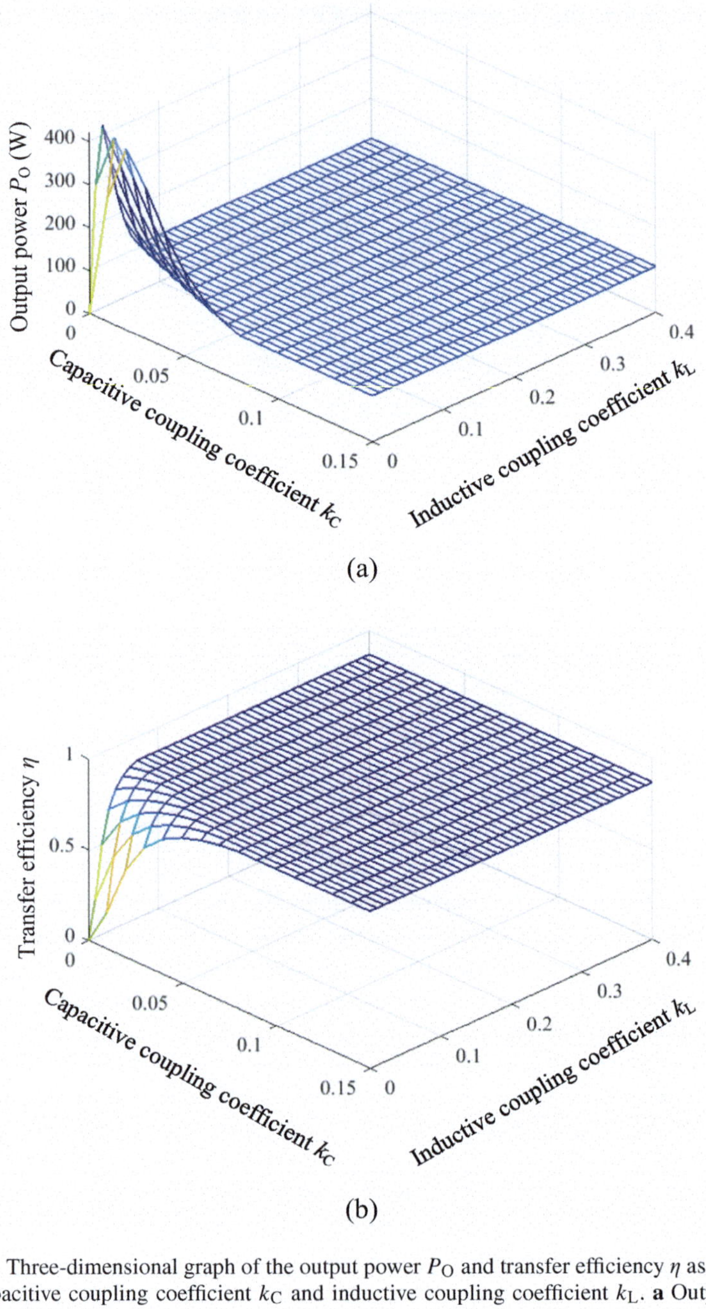

Fig. 11.5 Three-dimensional graph of the output power P_O and transfer efficiency η as a function of the capacitive coupling coefficient k_C and inductive coupling coefficient k_L. **a** Output power. **b** Transfer efficiency

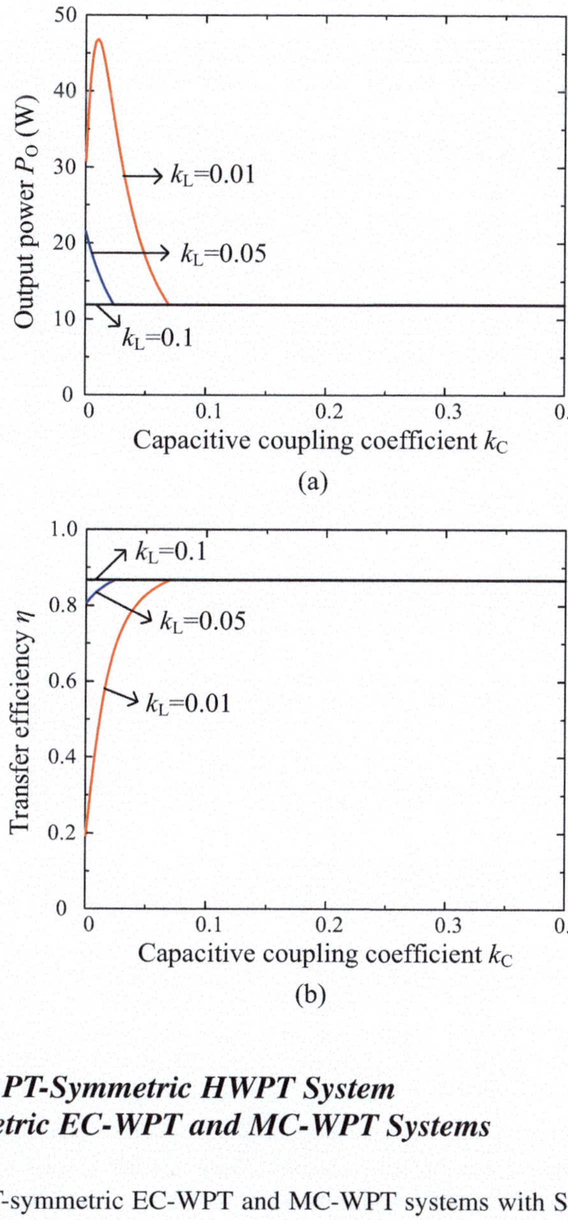

Fig. 11.6 Two-dimensional graph of the output power P_O and transfer efficiency η as a function of capacitive coupling coefficient k_C. **a** Output power. **b** Transfer efficiency

11.3.3 Comparison of PT-Symmetric HWPT System with PT-Symmetric EC-WPT and MC-WPT Systems

The equivalent circuits of PT-symmetric EC-WPT and MC-WPT systems with SS compensated topology are shown in Figs. 10.3a and 3.3a, and their system fundamental characteristics have been analyzed in Chaps. 3 and 10. The following will compare the PT-symmetric HWPT system with the PT-symmetric EC-WPT and MC-WPT systems.

Fig. 11.7 Two-dimensional graph of the output power P_O and transfer efficiency η as a function of inductive coupling coefficient k_L. **a** Output power. **b** Transfer efficiency

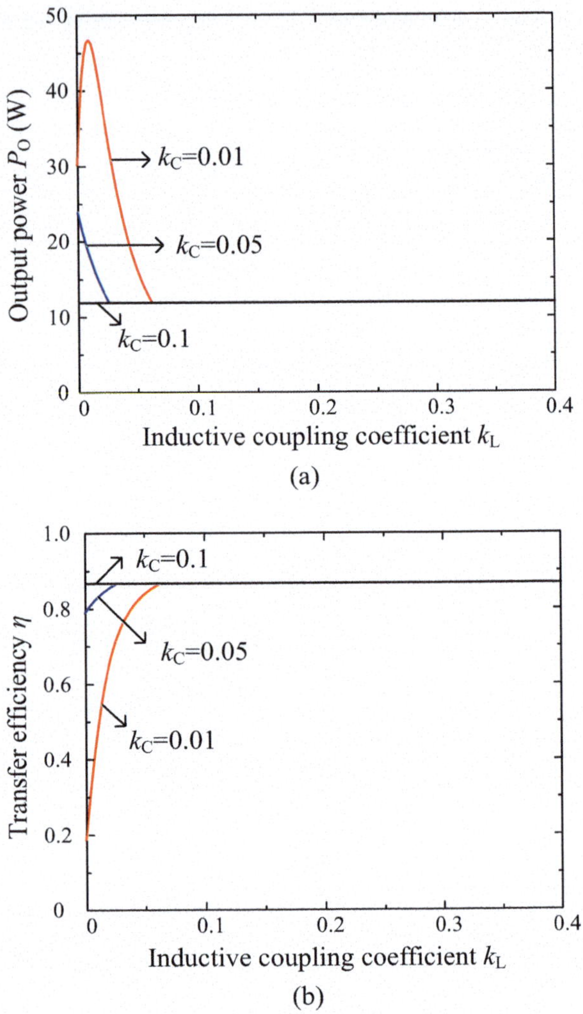

(1) Coupling range.

Since (11.9) is a binary inequality of capacitive coupling coefficient k_C and inductive coupling coefficient k_L for the PT-symmetric HWPT system, the analytical expression of the relationship between the hybrid coupling coefficient k_H and k_C and k_L cannot be directly solved. In practical applications, the numerical solution of should be solved according to the specific system parameters, so as to calculate the transfer range of the PT-symmetric HWPT system.

To solve the coupling range of the PT-symmetric HWPT system, can be rewritten as

$$z(k_C, k_L) = \frac{2k_C k_L + 2k_C + 2 - 2\sqrt{(1-k_L^2)(2k_C+1)}}{2k_C+1} \geq Q_R^{-2}. \tag{11.23}$$

Rewriting (10.22) of the critical capacitive coupling coefficient k_{CR} for the PT-symmetric EC-WPT system and (3.43) of the critical inductive coupling coefficient k_{LR} for the PT-symmetric MC-WPT system to the same forms, which are

$$z(k_C) = \frac{2k_C + 2 - 2\sqrt{2k_C+1}}{2k_C+1} \geq Q_R^{-2}, \tag{11.24}$$

$$z(k_L) = 2 - 2\sqrt{1-k_L^2} \geq Q_R^{-2}. \tag{11.25}$$

It can be seen that (11.24) and (11.25) coincide with $k_L = 0$ and $k_C = 0$ in (11.23), respectively, indicating that the coupling of the PT-symmetric HWPT system is a combination of the capacitive coupling coefficient k_C and the inductive coupling coefficient k_L. When the capacitive coupling coefficient $k_C = 0$, i.e., electric-field uncoupling between the coupling plates, the system becomes a PT-symmetric MC-WPT system. Similarly, when the inductive coupling coefficient $k_L = 0$, i.e., magnetic-field uncoupling between the coupling coils, the system becomes a PT-symmetric EC-WPT system.

By calculating the partial derivative of $Z(k_C, k_L)$ with respect to k_C and k_L respectively in (11.23), there is

$$\frac{\partial z(k_C, k_L)}{\partial k_C} = \frac{2\left[\sqrt{(1-k_L^2)(2k_C+1)} + k_L - 1\right]}{(2k_C+1)^2} > 0, \tag{11.26}$$

$$\frac{\partial z(k_C, k_L)}{\partial k_L} = \frac{2k_C}{2k_C+1} + \frac{2k_L}{\sqrt{(1-k_L^2)(2k_C+1)}} > 0, \tag{11.27}$$

where $k_C, k_L \in (0, 1)$.

From (11.26) and, it can be seen that $Z(k_C, k_L)$ is a monotonic increasing function of k_C and k_L, which illustrates that the capacitive coupling coefficient k_C and the inductive coupling coefficient k_L in the PT-symmetric HWPT system are superimposed and complementary to each other, improving the hybrid coupling coefficient k_H collectively. Therefore, compared with PT-symmetric EC-WPT and IMC-WPT systems with the same circuit parameters, the PT-symmetric HWPT system can operate in the exact PT-symmetric region at smaller k_C and k_L, which manifests that the hybrid coupling effectively expands the constant transfer range of PT-symmetric systems in weak coupling states.

The above discussion is based on the modeling and analysis of heteronymous-end connection of the loose coupled transformer in Fig. 11.2. If the capacitive plates are

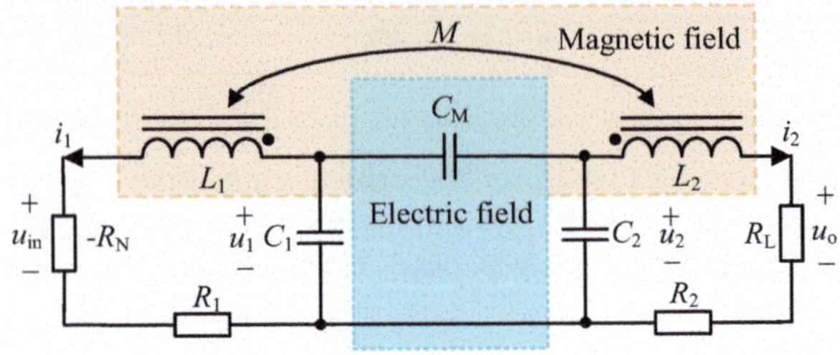

Fig. 11.8 Equivalent circuit of the PT-based HPT system with homonymous end connection of the coils

connected to the homonymous end of the coupling coils, the equivalent circuit of the PT-symmetric HWPT system is shown in Fig. 11.8, and the steady-state circuit equation becomes

$$\begin{bmatrix} jQ_T^{-1} - \frac{\omega}{\omega_0} + \frac{\omega_0(k_C+1)}{\omega(2k_C+1)} & \frac{\omega_0 k_C}{\omega(2k_C+1)} - \frac{\omega}{\omega_0}k_L \\ \frac{\omega_0 k_C}{\omega(2k_C+1)} - \frac{\omega}{\omega_0}k_L & jQ_R^{-1} - \frac{\omega}{\omega_0} + \frac{\omega_0(k_C+1)}{\omega(2k_C+1)} \end{bmatrix} \begin{bmatrix} \dot{I}_1 \\ \dot{I}_2 \end{bmatrix} = 0. \tag{11.28}$$

As can be seen from (11.28), the inductive coupling coefficient k_L term is negative, which means that the two coupling mechanisms cancel each other in the energy transfer when the coils polarity is connected in the homonymous end.

By solving (11.28), the frequency characteristics in the exact PT-symmetric region can be obtained as

$$\omega^* = \omega_0 \sqrt{\frac{B \pm \sqrt{B^2 - 4(2k_C+1)(1-k_L^2)}}{2(2k_C+1)(1-k_L^2)}}. \tag{11.29}$$

where $B = 1 - 2k_C k_L - (Q_R^{-2} - 1)(2k_C + 1)$.

The corresponding critical coupling condition can be derived as

$$z^*(k_C, k_L) = \frac{-2k_C k_L + 2k_C + 2k_L + 2 - 2\sqrt{(1-k_L^2)(2k_C+1)}}{2k_C+1} \geq Q_R^{-2}. \tag{11.30}$$

Similarly, by taking the partial derivative of (11.30) with respect to k_C and k_L respectively, we have

$$\frac{\partial z^*(k_C, k_L)}{\partial k_C} = \frac{2\left[\sqrt{(1-k_L^2)(2k_C+1)} - k_L - 1\right]}{(2k_C+1)^2}, \tag{11.31}$$

$$\frac{\partial z^*(k_C, k_L)}{\partial k_L} = -\frac{2k_C}{2k_C+1} + \frac{2k_L}{\sqrt{(1-k_L^2)(2k_C+1)}}. \tag{11.32}$$

Both (11.31) and have a region less than 0 within the coupling coefficients k_C and k_L range of (0,1), i.e., $z^*(k_C, k_L)$ is no longer monotonic, which demonstrates that there is a region where the capacitive coupling coefficient k_C and the inductive coupling coefficient k_L cancel each other out. The PT symmetry of the HWPT system tended to be broken with the increase of k_C or k_L in this region. Therefore, the PT-symmetric range of the HWPT system with the homonymous-end connection is much smaller than that of the PT-symmetric EC-WPT and MC-WPT systems with the same circuit parameters, which means that the capacitive plates of the PT-symmetric HWPT system should not be connected to the homonymous end of the coupling coils.

(2) System characteristics

By comparing the fundamental characteristics of the PT-symmetric EC-WPT and MC-WPT systems with those of the PT-symmetric HWPT system, three systems show similar frequency characteristics, that is, the frequency bifurcations occur in the exact PT-symmetric region. Meanwhile, in the broken PT-symmetric region, the operating frequency of the PT-symmetric MC-WPT system is unchanged, but the PT-symmetric EC-WPT and HWPT systems decrease with k_C, which indicates that in the PT-symmetric MC-WPT system, the mutual inductance M only participates in power transfer and does not affect the resonance of the circuit. However, in the PT-symmetric EC-WPT and HWPT systems, the mutual capacitance C_M is directly involved in the resonance of the compensated networks, and the variation of k_C also changes the actual resonant frequency of the systems.

Furthermore, in the exact PT-symmetric region, the output power and transfer efficiency of PT-symmetric EC-WPT, MC-WPT, and HWPT systems remain stable, and are independent of the coupling coefficient. Meanwhile, in the broken PT-symmetric region, the output power and transfer efficiency of the three systems are no longer constant, but change with their coupling coefficient. The above analysis shows that the PT-symmetric HWPT system has the same transfer characteristics as the two PT-symmetric single coupling systems, and all three systems can achieve constant output power and stable transfer efficiency under PT-symmetric conditions.

11.4 System Design and Verification

11.4.1 Negative Resistor

Traditional single-tube topologies, such as class-E inverters, have the advantages of simple structure, no straight-through risk, and high reliability, and are one of the preferred inverters for MHz-level WPT systems [4]. Therefore, in addition to the method of constructing a negative resistor based on the full-bridge inverter shown in Fig. 10.6, a Class E inverter can also be used to construct a negative resistor [5].

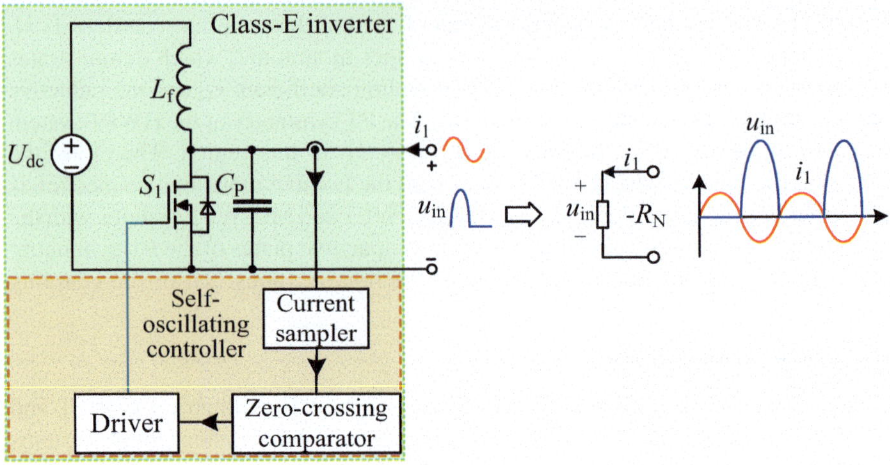

Fig. 11.9 The circuit structure diagram of negative resistor based on the class-E inverter

Figure 11.9 shows the circuit structure diagram of the negative resistor based on the class-E inverter, which consists of a choke inductor L_f, switch S_1, and parallel capacitor C_f. In addition to realizing the phase difference of 180° between voltage u_{in} and current i_1, the class-E inverter can also achieve zero voltage switching (ZVS) by paralleling resonant capacitor C_f with the switch S_1. The current can be reversed in advance before the turn-on signal of the switch S_1 arrives, which greatly reduces the switching loss. In particular, the class-E inverter is a current-source inverter topology, so the voltage stress of the switch S_1 under the action of resonance is 3.56 times that of the full-bridge inverter circuit, and the current stress is 2.86 times that of the full-bridge inverter circuit, resulting in a low output power of the class-E inverter, usually at the level of hundreds of watts [6].

In Fig. 11.9, the input DC voltage is 48 V, the maximum input current is 3 A, and the maximum operating frequency is 2 MHz, so the switch S_1 must be able to withstand the voltage stress of more than 170 V and the current stress of 8.6 A. The switch S_1 selects the GaN power IC (NV6117), its maximum input voltage is 650 V, and it has the characteristic of zero reverse recovery time, which is suitable for an MHZ high-frequency switching circuit. Besides, the GaN power IC (NV6117) integrates a high-speed gate driver, and the switching speed and drive voltage can be flexibly configured externally. Meanwhile, GaN devices do not have body diodes, the loss is large when the current is reversed, so it is connected in parallel with a Tokmas-SiC diode (CI04S65E3) to reduce the loss of the reverse current during the ZVS process. Considering the skin effect of high-frequency current, a flat wire winding inductor (SER2009) is adopted for the choke inductor L_f. Figure 11.10 gives the schematic diagram of the class-E inverter.

The self-oscillating controller is similar to Sect. 10.4.1, as shown in Figs. 10.8 and 10.9. The Tx current i_1 is sampled by a current transformer (CU8965) and a

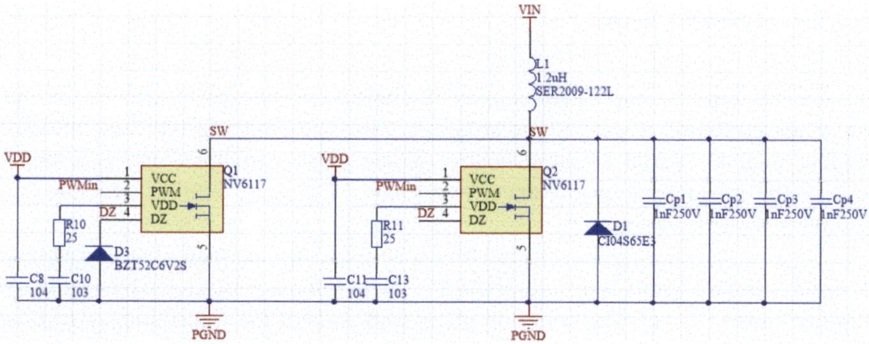

Fig. 11.10 The schematic diagram of the class-E inverter

differential amplifier (OPA2690), and then passes through a compensator and a zero-crossing comparator (TL3016) to counteract the effect of circuit delay and generate the driving pulse for switch S_1, which ensures that the output voltage u_{in} and current i_1 of the class-E inverter maintain a phase difference of 180° [2]. It can be found that the self-oscillating controlled inverter is only controlled on the transmitting side without any communication.

The key to designing a class-E inverter is the calculation of resonant network parameters. Due to the wide frequency variation range of the PT-symmetric HWPT system, the following equations are used in order to maintain ZVS throughout the frequency range of the class-E inverter [7, 8].

$$C_1 = \frac{1}{\omega_0\left[\omega_0 L_1 - \frac{\pi(\pi^2-4)R_{eq}}{16}\right]}, \tag{11.33}$$

$$\sqrt{\frac{L_f}{C_f}} = k_f R_{eq} = Z_{eq}, \tag{11.34}$$

$$C_f = \frac{1}{\omega_0 Z_{eq}}, \tag{11.35}$$

$$R_{eq} = R_1 + \frac{L_1}{L_2}(R_2 + R_L), \tag{11.36}$$

where $0.2 < k_f < 1.5$. In the case of ensuring the ZVS range, the larger k_f, the smaller loss of the inverter.

11.4.2 Hybrid Coupler

The common types of capacitive couplers include a two-pair-plate structure, a four-plate stack structure, a sleeve-type structure, etc. The common types of inductive couplers include spatial spiral structure, planar spiral structure, etc. In order to improve the misalignment tolerance, two kinds of planar coupling structures are chosen for the hybrid coupler.

Two-pair-plate structure is selected for capacitive coupling plates [9], and a planar DD coils structure with strong lateral displacement tolerance is selected for inductive coupling coils [10, 11]. The coupling plates and coils of the transmitter or receiver are integrated on a plane, respectively, and the coils are sandwiched between the two plates. In order to avoid additional self-capacitance caused by increasing electric-field coupling through coil wires on the same side of the plates, and to facilitate assembly, the coil spacing should be slightly larger than the plate spacing. Meanwhile, in order to avoid eddy currents in the metal plates caused by the magnetic field, it is necessary to prevent overlap between the coils and the plates in the longitudinal projection. Figure 11.11 shows the structure diagram of the designed hybrid coupler, where the coupling plates adopt $300 \times 175 \times 1\ \text{mm}^3$ PCB production, 8-turn $200 \times (50 + 50)\text{mm}^2$ DD structure coupling coils are wound by Litz wire ($\phi = 0.05$ mm $\times$ 400 strands) considering the skin effect of high-frequency current, and the coils and plates are fixed with nylon screws to avoid introducing metal debris. The prototype of the hybrid coupler is shown in Fig. 11.12.

ANSYS Maxwell software is used to simulate the electrostatic field and magnetostatic field of the coupling plates and coupling coils. When the coupler is aligned, the coupling capacitances are $C_{13} = C_{24} = 77.61$ pF, $C_{12} = C_{34} = 1.83$ pF, and $C_{14} = C_{23} = 1.32$ pF, the self-capacitance and mutual-capacitance can be calculated by Eq. (10.1). The self-inductances are $L_1 = L_2 = 31.71$ µH and mutual-inductance is $M = 12.71$ µH. Figure. 11.13 shows the static electric field and static magnetic field distribution. By simulating the plates and coils separately, the obtained coupling parameters remain unchanged, which demonstrates that the electric field and magnetic field do not affect each other in the hybrid coupler. By using the

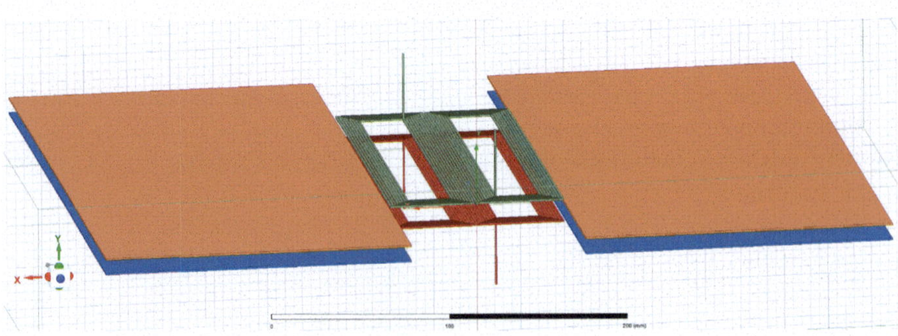

Fig. 11.11 The structure of a hybrid coupler

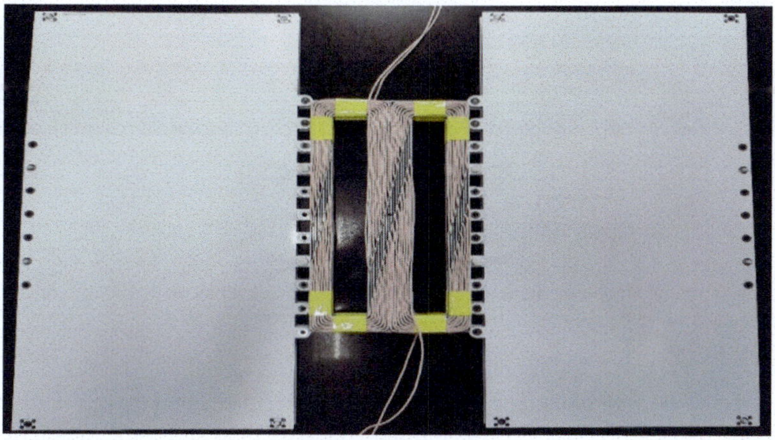

Fig. 11.12 The prototype of the hybrid coupler

impedance analyzer (Wayne Kerr 6500B) at 1.5 MHz, the measured parameters are $C_{13} = C_{24} = 79.35$ pF, $L_1 = L_2 = 32.30$ μH and $M = 13.67$ μH, which indicates the design feasibility of the hybrid coupler.

When a 100 mm lateral shift of the coupler occurs, the coupling capacitances between plates are $C_{13} = C_{24} = 58.83$ pF, $C_{12} = C_{34} = 1.40$ pF and $C_{14} = C_{23} = 0.92$ pF, respectively. The self-capacitance and mutual-capacitance can be calculated

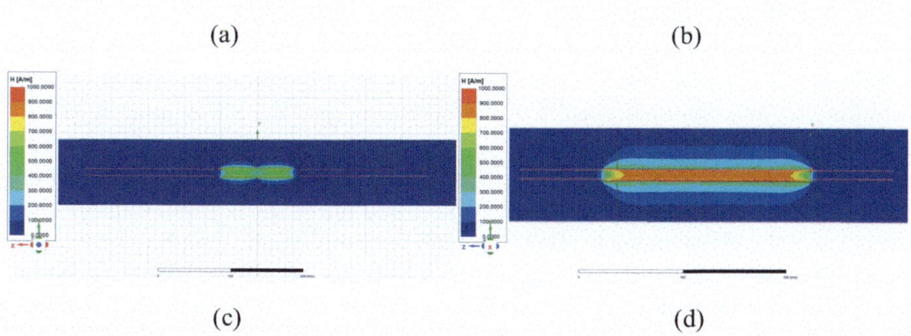

Fig. 11.13 The static electric field and static magnetic field distribution of the hybrid coupler. **a** Front view of static electric field simulation. **b** Side view of static electric field simulation. **c** Front view of static magnetic field simulation. **d** Side view of static magnetic field simulation

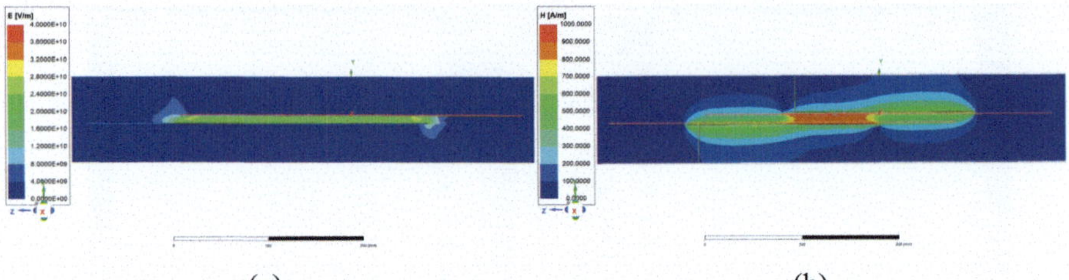

Fig. 11.14 The electromagnetic field distribution of the hybrid coupler with 100 mm lateral shift. **a** The static electric field simulation. **b** The static magnetic field simulation

by Eq. (10.1). The self-inductances are $L_1 = L_2 = 31.71$ μH and mutual-inductance is $M = 6.05$ μH. Figure 11.14 shows the static electric field and static magnetic field distribution. By measuring the prototype of the coupler, there are $C_{13} = C_{24} = 59.63$ pF, $L_1 = L_2 = 32.30$ μH and $M = 5.83$ μH. At this time, the coupler has been laterally displaced to half of the coil's length, and still can maintain a large mutual capacitance and mutual inductance, indicating that the hybrid coupler has strong lateral anti-shift ability, which is beneficial to broaden the range of the HWPT system working in the exact PT-symmetric region.

11.4.3 Experimental Verification

To demonstrate the validity of the theoretical analysis, an experimental prototype is constructed, as shown in Fig. 11. 15, where the resonant capacitances C_f, C_1, and C_2 are made of NP0 high-voltage multilayer ceramic capacitors in series and parallel. The actual circuit parameters of the prototype are measured by the impedance analyzer (Wayne Kerr 6500B), and the measured parameters are listed in Table 11.1. It is noted that the input DC voltage of the inverter is 48 V, and the RMS value of the output voltage of the inverter is $U_{in} \approx 1.152 U_{DC}$ (Fig. 11.15).

In the experiment, the lateral offset distance of the hybrid coupler is adjusted to evaluate the constant output characteristics of the PT-symmetric HWPT system. According to the circuit parameters of the hybrid coupler, the PT-symmetric HWPT system has a maximum capacitive coupling coefficient $k_{C,\mathrm{MAX}} = 0.116$ and a maximum inductive coupling coefficient $k_{L,\mathrm{MAX}} = 0.401$. Firstly, the coupling coils and plates of the hybrid coupler are decoupled, and the frequency and transfer characteristics of the PT-symmetric HWPT system under the single coupling state of the electric field and magnetic field are tested. Then, the operating frequency, output power, and transfer efficiency of the PT-symmetric HWPT system in the hybrid coupling state are tested, and the performance comparison of the hybrid coupling system and the single coupling system is analyzed.

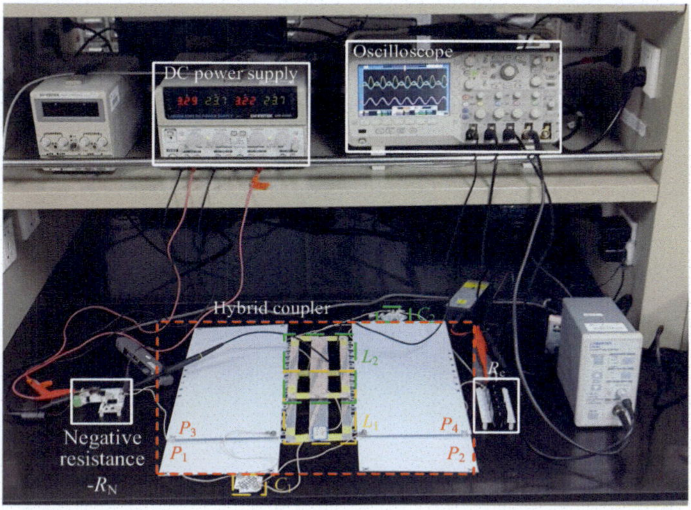

Fig. 11.15 Experimental prototype of a PT-symmetric HWPT system

Table 11.1 The circuit parameters of the experimental prototype

Description	Symbol	Values
Compensation inductance of transmitter	L_1	33.218 μH
Compensation inductance of receiver	L_2	33.839 μH
Compensation capacitance of transmitter	C_1	348.291 pF
Compensation capacitance of receiver	C_2	332.633 pF
Internal resistance of transmitter	R_1	1.609 Ω
Internal resistance of receiver	R_2	1.539 Ω
Load resistance	R_L	20.5 Ω
Natural resonant frequency of transmitter	f_1	1.497 MHz
Natural resonant frequency of receiver	f_2	1.5001 MHz
Choke inductance	L_f	1.216 μH
Parallel capacitance	C_f	4.120 nF

Figure 11.16 shows the relationship between the operating frequency f, the capacitance coupling coefficient k_C, and the inductive coupling coefficient k_L, where the theoretical curve is obtained by substituting the circuit parameters shown in Table 11.1 into (11.12), and the simulation results are obtained by Simulink. From Fig. 11.16, it can be seen that the operating frequency of the PT-symmetric HWPT system changes with the coupling state of the hybrid coupler. When the system works in the exact PT-symmetric region, the simulation results are all distributed in the high-frequency branch of the theoretical curve, which increases with the coupling coefficient. However, when k_C is larger than k_L, the operating frequency decreases

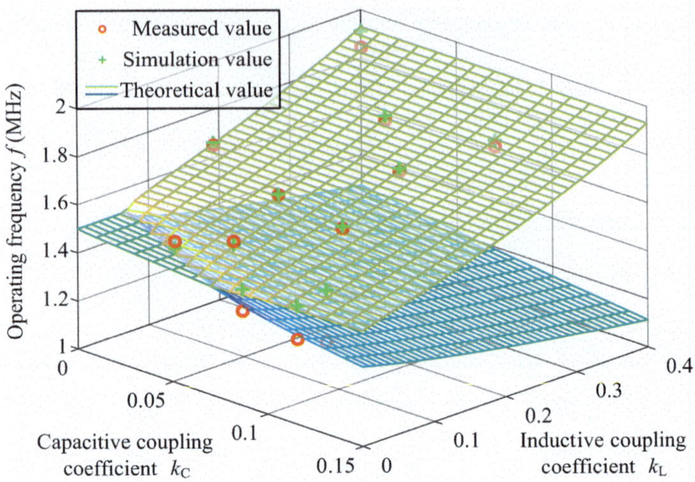

Fig. 11.16 The experimental results of operating frequency versus coupling coefficient

along the low-frequency branch of frequency bifurcation, while k_C is smaller than k_L, the system works in the high-frequency branch. When k_C and k_L are very small, the system works in the broken PT-symmetric region, where the operating frequency is no longer bifurcated and approaches the natural resonant frequency about 1.5 MHz. The variation trend of the measured frequency is basically consistent with the theoretical and simulation results.

Figures 11.17 and 11.18 show the output power and transfer efficiency under different coupling coefficients, respectively. The theoretical curve of output power is calculated by (11.16) and (11.21), and the theoretical curve of transfer efficiency is calculated by (11.17) and (11.22). The simulation and measured results of the output power and transfer efficiency are obtained by $P_O=I_2^2 R_L$ and $\eta=I_2^2 R_L/[I_1^2 R_1 + I_2^2(R_2 + R_L)]$, respectively. As can be observed from Figs. 11.17 and 11.18, in the single coupling state, both output power and transfer efficiency keep constant in the region of $k_C > 0.06$ or $k_L > 0.06$, where the output power is stable near 106 W and the transfer efficiency is basically constant at 87%. In the hybrid coupling state, because of the complementary effect of power transfer between the two coupling mechanisms, there is a certain region of $k_C < 0.06$ and $k_L < 0.06$, where the system still works in the exact PT-symmetric region, indicating that the PT-symmetric HWPT system is very suitable for weakly coupling application scenarios. Besides, when the system works in the broken PT-symmetric region, the output power and transfer efficiency are no longer constant. With the decrease of k_C and k_L, the output power increases first and then decreases, and the transfer efficiency decreases sharply, which is consistent with the theoretical analysis described in Sect. 11.3.3.

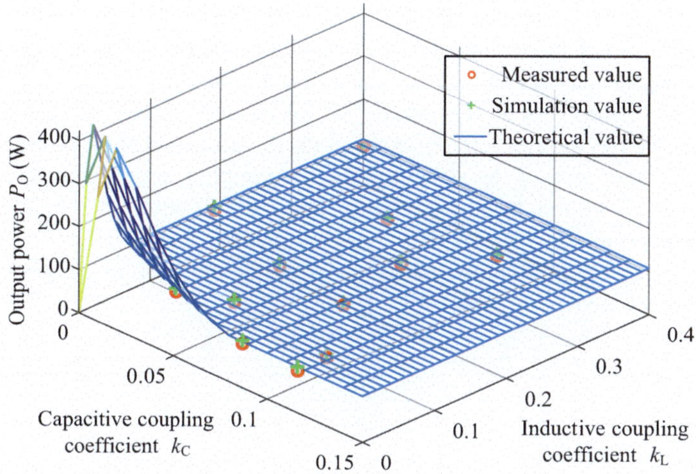

Fig. 11.17 The experimental results of output power versus coupling coefficient

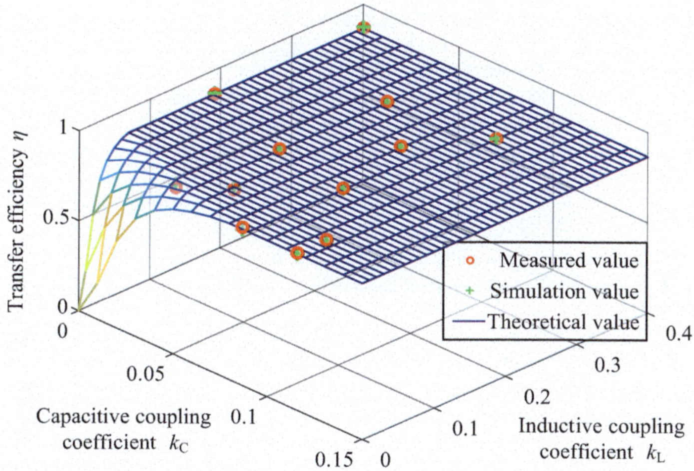

Fig. 11.18 The experimental results of transfer efficiency versus coupling coefficient

11.5 Summary

This chapter first introduces the system structure of the PT-symmetric HWPT system, and establishes the circuit model and coupled-mode model of the PT-symmetric HWPT system. Then, the basic characteristics including operating frequency, output power, and transfer efficiency are analyzed, and the comparison of PT-symmetric HWPT, EC-WPT, and MC-WPT is discussed. Finally, the method of constructing negative resistor by using a class-E inverter is introduced, and the experimental prototype of the PT-symmetric HWPT system is built. The experimental results show that the output power and transfer efficiency of the PT-symmetric HWPT system can remain constant under different coupling states in the exact PT-symmetric region, and compared with PT-symmetric MC-WPT and EC-WPT systems, the PT-symmetric HWPT system effectively expands the range of stable output and improves the transfer performance of the system in weakly coupling state. Therefore, the PT-symmetric HWPT system has the advantages of constant output power and transfer efficiency, wide transfer distance, only transmitter-side control, and has broad application prospects in high dynamic range wireless transfer such as electric vehicles, robots, and other areas.

References

1. Qiu D et al (2024) A robust parity-time-symmetric hybrid wireless power transfer system with extended coupling range. Int J Circuit Theory Appl, early view
2. Zhou J, Zhang B, Xiao W, Qiu D, Chen Y (2019) Nonlinear parity-time-symmetric model for constant efficiency wireless power transfer: application to a drone-in-flight wireless charging platform. IEEE Trans Industr Electron 66(5):4097–4107
3. Assawaworrarit S, Yu X, Fan S (2017) Robust wireless power transfer using a nonlinear parity-time-symmetric circuit. Nature 546(7658):387–390
4. Issi F, Kaplan O (2022) Design and application of wireless power transfer using Class-E inverter based on adaptive impedance-matching network. ISA Trans 126:415–427
5. Assawaworrarit S, Fan S (2020) Robust and efficient wireless power transfer using a switch-mode implementation of a nonlinear parity-time symmetric circuit. Nature Electron 3(5):273–279
6. Wang Y, Lucía Ó, Zhang Z, Guan Y, Xu DJIPE (2020) Review of very high frequency power converters and related technologies. IET Power Electron 13:1711–1721
7. Raab F (1977) Idealized operation of the class E tuned power amplifier. IEEE Trans Circuits Syst 24(12):725–735
8. Roslaniec L, Jurkov AS, Bastami AA, Perreault DJ (2015) Design of single-switch inverters for variable resistance/load modulation operation. IEEE Trans Power Electron 30(6):3200–3214
9. Wu X, Su Y, Hu AP, Qing X, Hou X (2021) Multiobjective parameter optimization of a four-plate capacitive power transfer system. IEEE J Emerg Sel Topics Power Electron 9(2):2328–2342
10. Pearce MGS, Covic GA, Boys JT (2019) Robust ferrite-less double D topology for roadway IPT applications. IEEE Trans Power Electron 34(7):6062–6075
11. Song K et al (2020) Design of DD Coil with high misalignment tolerance and low EMF emissions for wireless electric vehicle charging systems. IEEE Trans Power Electron 35(9):9034–9045

The manufacturer's authorised representative in the EU is Springer Nature Customer Service Centre GmbH, Europaplatz 3, 69115 Heidelberg, Germany. If you have any concerns regarding our products, please contact ProductSafety@springernature.com

Printed and bound by CPI Group (UK) Ltd, Croydon, CR0 4YY
30/09/2026
02225302-0011